Intelligent Control and Learning Systems

Volume 21

The Springer book series Intelligent Control and Learning Systems addresses the emerging advances in intelligent control and learning systems from both mathematical theory and engineering application perspectives. It is a series of monographs and contributed volumes focusing on the in-depth exploration of learning theory in control such as iterative learning, machine learning, deep learning, and others sharing the learning concept, and their corresponding intelligent system frameworks in engineering applications. This series is featured by the comprehensive understanding and practical application of learning mechanisms. This book series involves applications in industrial engineering, control engineering, and material engineering, etc.

The Intelligent Control and Learning System book series promotes the exchange of emerging theory and technology of intelligent control and learning systems between academia and industry. It aims to provide a timely reflection of the advances in intelligent control and learning systems. This book series is distinguished by the combination of the system theory and emerging topics such as machine learning, artificial intelligence, and big data. As a collection, this book series provides valuable resources to a wide audience in academia, the engineering research community, industry and anyone else looking to expand their knowledge in intelligent control and learning systems.

Mei Liu · Jingkun Yan · Renpeng Huang

Dynamic Neural Networks for Motion Control of Redundant Manipulators

Springer

Mei Liu
Lanzhou University
Lanzhou, China

Jingkun Yan
Lanzhou University
Lanzhou, China

Renpeng Huang
Lanzhou University
Lanzhou, China

ISSN 2662-5458 ISSN 2662-5466 (electronic)
Intelligent Control and Learning Systems
ISBN 978-981-96-9143-2 ISBN 978-981-96-9144-9 (eBook)
https://doi.org/10.1007/978-981-96-9144-9

This Springer imprint is published by the registered company Springer Nature Singapore Pte Ltd.
The registered company address is: 152 Beach Road, #21-01/04 Gateway East, Singapore 189721, Singapore

To our ancestors and parents, as always

Preface

Dynamic neural networks (DNNs) with excellent computational power, improvability, and hardware realizability have gained widespread adoption across multiple domains, particularly in robotics. Redundant manipulators possess additional degrees of freedom (DOFs) beyond the minimum required for task execution, thereby enabling enhanced flexibility. In addition to control tasks such as trajectory tracking and cyclic motion, system constraints and performance optimization of redundant manipulators usually need to be considered. Therefore, most motion control schemes are constructed in an optimized form. Beyond controlling the end-effector position of redundant manipulators, its orientation needs to be considered in some specific tasks (e.g., force control), either for orientation maintenance or orientation control. Furthermore, the motion control with different end-effectors should be taken into account, which extends application scenarios of redundant manipulators. Finally, noises and interferences are inevitable in the control process of redundant manipulators, which are likely to have negative effects.

Focusing on these unresolved issues, the book develops multiple DNN variants to optimize the trajectory tracking of manipulators. First, this book focuses on the cyclic motion and trajectory tracking problems of redundant manipulators, designs motion control schemes that include multiple performance optimization indexes, and develops three DNNs for solving these schemes, including a noise-rejection DNN. Second, this book focuses on the position and orientation control problem of redundant manipulators, designs schemes that can achieve orientation maintenance or orientation control, and develops three DNNs for solving these schemes, including a fuzzy-parameter DNN. Finally, this book focuses on the motion control of redundant manipulators with different end-effectors, designs the grasping force control scheme and the remote center of motion control scheme, and develops two DNNs for solving these schemes to achieve synchronized learning and control of redundant manipulators.

The idea for this book on building DNNs for motion control of redundant manipulators was conceived during laboratory research discussions and international scientific conferences. All materials in this book come from papers published by the authors in journals, including Neural Networks, IEEE Transactions on Cognitive and Developmental Systems, IEEE Transactions on Industrial Electronics, IEEE Transactions on Cybernetics, and so on. This rest part of book is classified into the following nine chapters.

Chapter 2—This chapter develops a DNN model to implement the double-index (DI) scheme, targeting cyclic motion generation of redundant manipulators. The DI scheme formulates two optimization objectives: minimum kinetic energy (MKE) and minimum joint-angle offset (MJAO), while incorporating the manipulator dynamics and joint physical limits as constraints. Subsequently, the DI scheme is transformed into a quadratic programming (QP) framework to enable numerical solutions. Further, a parameter-iterative DNN is designed to resolve the QP problem, generating real-time joint trajectories (i.e., angles, velocities, accelerations) for manipulator actuation. Comparative simulations between the DNN-driven DI scheme and a minimum acceleration norm (MAN) baseline validate the presented method's efficacy in redundant manipulator control.

Chapter 3—In this chapter, an acceleration-level cyclic-motion (ALCM) scheme is presented for eliminating joint-angle drift with noise rejection. Specifically, the presented ALCM scheme sets the joint-angle drift as the optimization goal and simultaneously takes into account the error feedback. Furthermore, a corresponding DNN model with noise rejection is explored to process the ALCM scheme online. As for the inevitable noise, the presented DNN model incorporates the inherent noise rejection capability, specifically designed to adapt to time-varying polynomial-type disturbances. Theoretical analysis demonstrates its convergence properties under both ideal (noise-free) and realistic (noise-corrupted) operating conditions. In addition, the simulation on a PUMA 560 redundant manipulator is carried out to validate the joint-angle drift correction capability of the presented scheme. Comparative studies across varying noise levels conclusively demonstrate the enhanced performance of the noise-resistant DNN model in implementing the control framework.

Chapter 4—This chapter presents a trajectory planning framework for redundant manipulators based on model predictive control (MPC) methodology. The manipulator's nonlinear dynamics are linearized via forward kinematics to generate predictive motion outputs. Subsequently, an MPC scheme is constructed, which minimizes tracking error, velocity norm, and acceleration norm and directly considers joint limits at three levels as well as the terminal equality constraint. This scheme is then simplified into a convex quadratic programming (QP) problem. {A zeroing-type DNN (ZDNN) model is designed for the presented MPC scheme, employing inequality-to-equality constraint transformation techniques. Extensive comparative evaluations against benchmark methods are conducted through simulations and experiments, testing both nominal and disturbance-rejection scenarios. Results demonstrate that the presented MPC scheme solved by the ZDNN model achieves exceptional trajectory tracking performance, outperforming existing approaches in computational efficiency, response speed, and robustness metrics.

Chapter 5—This chapter introduces a cerebellum-inspired model based on the echo state network (ESN), which substitutes the conventional kinematic modeling of redundant manipulators. Building upon the presented model, we develop an MPC algorithm and a neural dynamics (ND) approach. Distinct from existing studies, the presented methodology concurrently addresses performance enhancement and system constraints through an adaptive cerebellar learning mechanism, enabling real-time precision tracking without off-line pre-training. The chapter further incorporates an ND-based correction algorithm and a discrete deep neural network (DDNN) solver for the MPC optimization problem. Theoretical analyses establish global convergence for both the ND-based correction algorithm and the DDNN model. Both simulations and experiments validate the effectiveness of the presented cerebellum-inspired MPC (CIMPC) approach, showing superior trajectory tracking performance compared to conventional methods.

Chapter 6—This chapter presents an acceleration-level motion/force control (AMFC) scheme, which contains a motion/force-based cyclic motion planning (MFCMP) strategy and an improved fuzzy dynamic neural network (FDNN) model. Specifically, the MFCMP strategy incorporates four essential metrics: minimum joint error, orientation maintenance, motion/force control, and joint physical constraints. In addition, the improved FDNN model is constructed with automatic adjustment on the convergent parameter, which is noise-tolerant with the aid of dynamic variables. Furthermore, theoretical analysis is conducted to prove the convergence and robustness of the presented scheme. Finally, comparative simulations and experiments under noise-free and noisy cases are performed to demonstrate the effectiveness, robustness, and superiority of the AMFC scheme.

Chapter 7—This chapter introduces a multi-criteria control scheme aided with a DNN model, which concurrently addresses orientation tracking requirements and physical system constraints. The presented DNN model demonstrates significantly lower computational overhead compared to conventional solutions for similar tasks. Theoretical analyses confirm that the presented scheme based on the DNN model globally and exponentially converges to the theoretical solution of the robotic motion generation. Besides, illustrative simulations and experiments using a Franka Emika Panda manipulator confirm the practical effectiveness of the presented scheme.

Chapter 8—This chapter develops a model predictive tracking control of position and orientation (MPTCPO) scheme for redundant manipulators, enabling simultaneous control of position and orientation. The presented MPTCPO scheme integrates multi-objective optimization, minimizing trajectory tracking errors, joint-velocity norm, and joint-acceleration norm while explicitly enforcing three-level joint kinematic constraints, without compromising the solution feasibility or requiring constraint conversion. To efficiently solve this optimization problem, a finite-time DNN (FTDNN) model is designed. Both simulations and experiments validate the precision and real-time feasibility of the MPTCPO scheme solved by the FTDNN model.

Chapter 9—In this chapter, a noise-tolerant grasping force optimization (NTGFO) scheme is presented with a DNN for multi-fingered dexterous hands that can be placed on a redundant robot to address the challenge of grasping diverse targets under noisy conditions. The scheme includes an optimal grip force model and a noise-tolerant DNN (NTDNN). The optimal grip force model is developed to efficiently solve the grasping force optimization problem, ensuring that the target is neither damaged nor slips during manipulation. Additionally, the presented NTDNN effectively handles noises that occur during the task execution, achieving optimal solutions. Furthermore, simulations and experiments are conducted to assess the feasibility of the NTGFO scheme.

Chapter 10—This chapter presents a remote center of cyclic motion (RC^2M) scheme with a data-driven technology for manipulators with uncertain parameters. The developed solution employs a DNN model to resolve the control scheme's optimization problem. Furthermore, simulative and physical experiments on a Franka Emika Panda manipulator equipped with a rod-shaped end effector conclusively demonstrate the effectiveness of the presented RC^2M scheme.

This book is intended as a reference for researchers working in the areas of redundant manipulators, motion control, neural dynamics, and simulation and experimentation. The book summarizes the latest research achievements, development potentials, and future trends in these areas. We expect that this book will stimulate readers' desire for exploration and academic interest, help them deepen their research in related fields, and provide new perspectives and challenges for researchers seeking innovative theoretical tools and application practices. We welcome your valuable comments and suggestions. The authors can be contacted via e-mail: liumei7@foxmail.com.

Lanzhou, China
April 2025

Mei Liu
Jingkun Yan
Renpeng Huang

Acknowledgements During the work on this book, we have had the pleasure of discussing its various aspects and results with many cooperators and students, including Profs. Long Jin, Kun Liu, Li He, Fan Zhang, Guoqian Zhang, Xiufang Chen, Jiexing Li, Jiawang Tan, Jiazheng Zhang, Ying Liufu, Jiayi Wang, Liuyi Wen, Yutong Li, Jinfu Tang, Chuan Qin, Duojicairang Ma, and Qihai Jiang. We highly appreciate their contributions, which particularly allowed us to improve our book manuscript.

The continuous support of our research by the National Natural Science Foundation of China under Grant 62306130, Grant 62476236, and Grant 62506327, in part by the Natural Science Foundation of Chongqing (China) under Grant CSTB2024NSCQ-LZX0036, in part by the Fundamental Research Funds for the Central Universities under Grant lzujbky-2023-ct05, in part by the Youth Doctor Support Project of Gansu Provincial Department of Education under Grant 2024QB-003, and in part by the Supercomputing Center of Lanzhou University are gratefully acknowledged here either.

Besides, we would like to thank the editors sincerely for their very important and constructive comments and suggestions provided, in addition to their time and efforts spent in handling this book.

We are always very grateful to the nice people (especially the staff in Springer) for their strong support and encouragement during the preparation and publishing of this book.

Competing Interests The authors have no competing interests to declare that are relevant to the content of this manuscript.

Contents

Chapter 1
Introduction

Being a key area in contemporary robotics, redundant manipulators have attracted considerable interest owing to their excess degrees of freedom (DOFs) beyond the minimum necessary for accomplishing specific tasks. These extra DOFs provide a significant advantage, enabling redundant manipulators to perform primary tasks such as precise end-effector trajectory tracking or force control while simultaneously addressing secondary objectives like obstacle avoidance, joint constraints, energy optimization, and task-specific requirements such as posture control. For example, in tasks that involve delicate interactions with objects, such as assembly or medical surgery, redundant manipulators can adjust their posture dynamically, ensuring optimal tool orientation and minimizing the risk of collisions. Furthermore, the ability to implement position-force control is a key strength, allowing for effective interaction with the environment where both position and force need to be tightly controlled, such as in robotic-assisted surgery or handling fragile materials. Additionally, in scenarios like minimally invasive surgery, the concept of the remote center of motion (RCM) becomes particularly relevant, where the manipulator's motion is constrained around a fixed point, known as the RCM, offering high precision without disturbing the target area. Despite the flexibility and adaptability that these extra freedoms provide, they also significantly increase the complexity of system modeling, motion planning, and control. The inclusion of posture and force control requirements further complicates the task, especially when high precision and real-time performance are demanded.

Inverse kinematics continues to be a fundamental issue in the motion planning of redundant manipulators. A key difficulty lies in efficiently and precisely determining the joint angles that yield a specified position and orientation of the end-effector. For redundant manipulators, inverse kinematics solutions are typically non-unique, as multiple joint configurations can achieve the same end-effector position and orientation. The selection of the "optimal" solution among these feasible configurations requires the introduction of performance indices during the design and planning stages, enabling a comprehensive evaluation of the solutions. Common performance

M. Liu et al., *Dynamic Neural Networks for Motion Control of Redundant Manipulators*, Intelligent Control and Learning Systems 21,
https://doi.org/10.1007/978-981-96-9144-9_1

indices include positional error, joint drift error, and the minimum norm of joint velocities. These indices are crucial in measuring the precision, stability, energy consumption, and repeatability of the manipulator during task execution. By formulating these performance indices into an optimal control problem, one can not only ensure that the end-effector's primary task requirements, such as trajectory tracking or force control, are met but also optimize secondary constraints and goals, such as energy efficiency, stability, and joint wear reduction. Optimization-based approaches, with their ability to handle multi-objective and multi-constraint scenarios, offer promising avenues for addressing these complex control problems, ensuring that redundant manipulators can operate with high efficiency and robustness in dynamic environments.

Quadratic programming (QP) offers an efficient method for handling convex objectives under different constraints, making it suitable for real-time applications on moderate-scale problems. In more complex, multi-constraint scenarios, model predictive control (MPC) extends QP's capabilities by incorporating past, current, and predicted future system states into an online optimization problem (often involving numerous equality and inequality constraints) at every sampling instant. This receding-horizon approach naturally enforces constraints, preventing performance degradation or instability that could result if they were neglected while also adapting dynamically to changing conditions. Nevertheless, the computational burden can become substantial when dealing with high-dimensional or highly redundant manipulators. To address this challenge, dynamic neural networks (DNNs) have emerged as a powerful tool, not only offering a dynamic perspective on the evolution of neural networks but also reformulating complex optimization problems into a set of ordinary differential equations (ODEs) or their discrete variants. When solving these time-dependent issues, DNN constructs ODE systems that correspond to the dynamic evolution process, with the system's equilibrium points representing the theoretical solutions to the original problem. The ODE system can efficiently converge to the system's equilibrium point from any initial state, offering rapid convergence and optimal solutions. As a result, DNN plays a pivotal role in solving complex dynamic problems in various fields, overcoming challenges like lag, noise resilience, and stability.

DNN also extends to model predictive control (MPC) by allowing it to handle nonlinear and time-varying disturbances that are common in practical systems. Data-driven models, such as echo state networks (ESNs), are well suited for capturing nonlinear system dynamics when accurate system models are unavailable or uncertain, thereby improving predictions and enhancing overall control decisions. In this context, DNN's ability to transform dynamical issues into efficient models further contributes to enhancing the performance of both QP and MPC. However, the challenge remains in addressing these time-varying problems in noisy environments, improving the learning capacity and efficiency of neural networks, and applying DNN to critical areas without incurring excessive computational costs. Achieving high stability in the model while maintaining computational efficiency has become a significant scientific challenge in both academic research and practical applications. This book presents a systematic study on DNN applied to the motion control of redundant manipulators, emphasizing advanced planning and control techniques across three major research directions: performance-index-oriented robot control with DNN, posture-considered robot control with DNN, and different end-effectors aided robot control with DNN. The study emphasizes the integration of DNN to enhance computational efficiency and robustness, particularly in complex dynamic environments where multiple objectives and constraints must be managed simultaneously. The organization of this book is structured around these core themes, as depicted in Fig. 1.1, offering a comprehensive overview of the interconnected control strategies and technologies discussed.

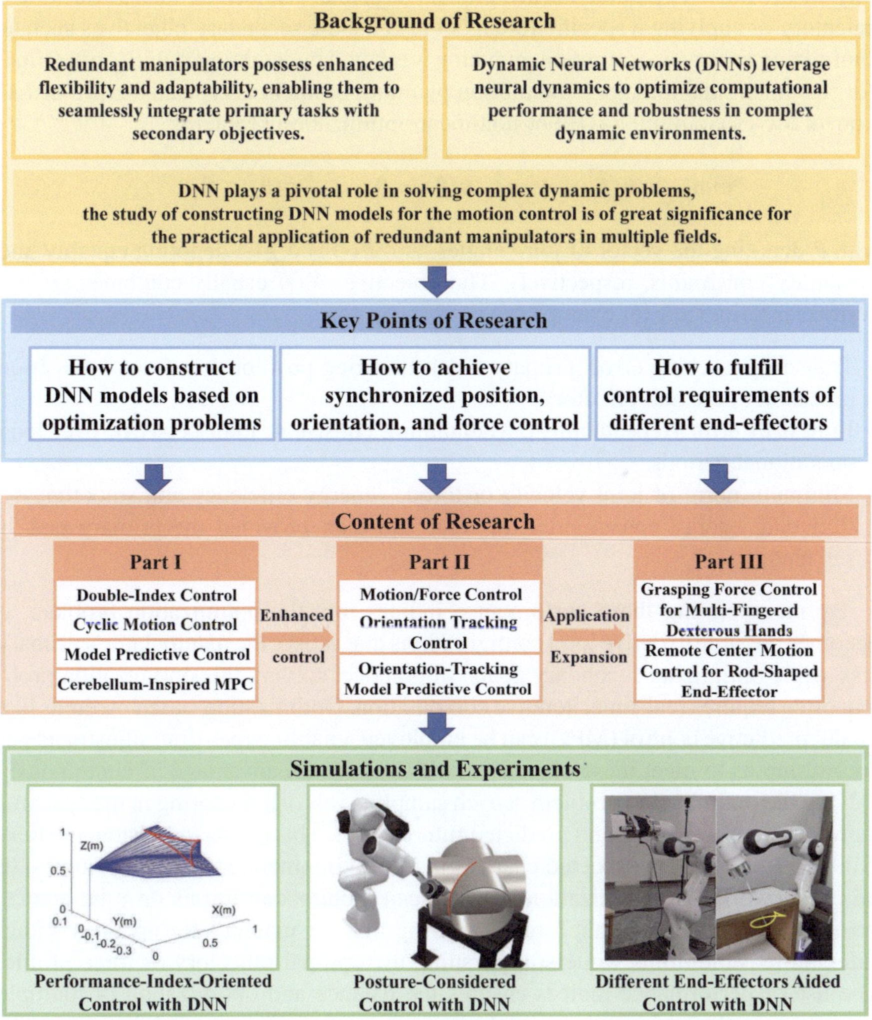

Fig. 1.1 Organization of this book

1.1 Performance-Index-Oriented Robot Control

When planning the motion of redundant manipulators, performance indices facilitate the balancing of multiple objectives simultaneously. When joint angles, velocities, end-effector accuracy, and other factors all come into play, a well-structured modeling of these performance indices, coupled with efficient real-time optimization, becomes essential to derive optimal joint configurations. In robotic systems, redundant manipulators possess extra degrees of freedom (DOFs) beyond the minimum required for task execution. While this provides operational flexibility, it simultaneously complicates the motion planning process. A well-designed optimization problem must not only ensure the manipulator can accomplish its primary objective, such as tracking a trajectory or applying a specific force, but also handle secondary objectives such as minimizing energy consumption, reducing wear, and maintaining stability over time. For redundant manipulators, the motion planning problem is typically expressed as a quadratic programming (QP) or nonlinear optimization problem:

$$\min \quad \mathscr{P}(\boldsymbol{\theta}, \dot{\boldsymbol{\theta}}, \ldots) \quad \text{s.t.} \quad \boldsymbol{h}(\boldsymbol{\theta}) = \mathbf{0}, \ldots, \quad \boldsymbol{g}(\boldsymbol{\theta}) \leq \mathbf{0}, \ldots,$$

with $\boldsymbol{\theta}$ denoting the vector of joint angles, and $\boldsymbol{h}(\cdot)$ and $\boldsymbol{g}(\cdot)$ denoting equality and inequality constraints, respectively. The objective $\mathscr{P}(\cdot)$ usually combines several performance indices, for example:

- Trajectory-tracking error (primary task): Ensuring positional or force-interaction precision of the end-effector.
- Joint drift error (repeatability): Maintaining consistent joint behavior over long operational periods.
- Minimum norm of joint velocity or torque (energy efficiency and smoothness): Reducing energy consumption or wear and tear, provided the primary task is satisfied.

By incorporating these performance indices into the optimization problem, it becomes possible to derive joint configurations that satisfy the manipulator's primary task while optimizing secondary goals. In practice, solving this optimization problem often requires real-time, iterative computation. Optimization-based methods like model predictive control (MPC) can be employed, enabling real-time adjustments of control inputs to meet these objectives. MPC offers the advantage of continuously solving the optimization problem at each sampling interval, factoring in the system's current state, past inputs, and predicted future states. This grants the system the flexibility to respond to unexpected changes in the environment or system dynamics. In this framework, the optimization problem can include constraints on joint angles, velocities, and even external forces, ensuring that the manipulator operates within safe and feasible limits while still pursuing the optimal trajectory or force profile. The use of performance indices ensures the balance and optimization of multiple objectives, such as accuracy, repeatability, energy efficiency, and stability. In practical applications, this means that the manipulator can operate efficiently in dynamic environments, adjusting to changes in real time while maintaining high performance. Over long operational periods, this capability is crucial for ensuring reliable and consistent performance, especially in fields like industrial automation, medical robotics, and autonomous operations in uncertain environments. Thus, integrating these performance indices into real-time optimization frameworks like MPC allows systems to perform complex tasks while ensuring robustness, longevity, and adaptability.

1.2 Posture-Considered Robot Control

In tasks such as precision assembly, grasping, polishing, and force interaction, redundant manipulators must not only achieve accurate positioning of the end-effector but also maintain precise orientation and regulate interaction forces with the external environment. This multifaceted requirement for simultaneous position, orientation, and force control unlocks high accuracy and adaptability in complex or uncertain environments, yet it also significantly increases system dimensionality and coupling, posing substantial challenges for motion planning and control. While position and orientation are tightly interdependent, incorporating force control introduces dynamic uncertainties from external contacts, payload changes, and factors like friction. Although rotation matrices can fully characterize the end-effector's orientation, directly controlling all nine interrelated elements is inefficient in real-time applications. Consequently, lower-dimensional orientation representations, such as quaternions, Euler angles, or axis-angle parameters, are commonly employed to simplify the control process. Nonetheless, each representation entails unique issues regarding singularities, decoupling, and numerical stability. Force control likewise demands sophisticated sensing and rapid response; without an accurate model or high-quality sensor data, real-time performance and operational safety can suffer.

Given the complexity of modeling and controlling these dynamic systems, echo state networks (ESNs) offer a promising solution. ESNs excel at capturing the nonlinear, time-varying behaviors of redundant manipulators by mapping inputs into high-dimensional spaces. Their ability to model complex, dynamic systems makes them well suited for controlling tasks that involve position, orientation, and force. In contrast to traditional neural networks, the training process for ESNs is computationally efficient, as it only involves solving a linear least-squares problem, making them ideal for real-time control in fast-paced applications. ESNs are capable of forecasting future system states using past inputs, enabling the control system to constantly adjust to environmental changes. This adaptability enhances real-time decision-making, ensuring that manipulators can maintain precision, adjust to varying forces, and handle dynamic environments. By using models like ESNs, redundant manipulators can effectively manage the complexities of motion control without relying solely on detailed physical models, improving both performance and robustness in tasks like assembly, force interaction, and grasping. This makes them especially valuable in applications ranging from industrial automation to medical robotics.

1.3 Different End-Effectors Aided Robot Control

In modern robotics, the use of multi-fingered dexterous gripping and remote center of motion (RCM) technologies is crucial for achieving precision, flexibility, and adaptability in complex tasks across a variety of fields. Compared to traditional parallel grippers, multi-fingered dexterous hands offer enhanced flexibility due to their increased DOF. This expanded DOF enables the robot to adopt more sophisticated grasping strategies, including fine force control, and provides the capability to manipulate a wider range of objects. Such hands are particularly beneficial in scenarios where delicate interaction with objects is required, such as in medical surgeries, assembly, and other tasks involving fragile or irregularly shaped targets. A key challenge in multi-fingered gripping is gripping force optimization, where the objective is to apply the least amount of force to an object while ensuring a secure grasp. This balance is crucial when handling fragile materials, where excessive force may cause damage or slippage. Achieving this requires precise control of the forces exerted by each finger, ensuring the object remains secure without causing harm. As such, multi-fingered dexterous hands are indispensable in environments where traditional grippers fall short, such as in medical applications or situations requiring high-precision manipulation.

Remote center of motion (RCM) is a key technology in tasks where the manipulator needs to pivot around a fixed point. This is particularly relevant in surgical robots or other applications that demand high precision in constrained spaces. RCM ensures that a specific point or axis remains stationary while the manipulator moves around it. This capability is critical in applications like minimally invasive surgery, where the robot needs to operate in confined spaces and maintain a fixed tool position without disturbing the surrounding area. By keeping the tool fixed at a designated point, RCM allows for precise and stable manipulation even in environments with limited space. In redundant manipulators, extra degrees of freedom (DOFs) are leveraged to perform these RCM tasks effectively, providing the flexibility to meet the precise kinematic coupling and physical constraints required for stable operation in such restricted workspaces. This ensures that the manipulator can perform complex tasks while maintaining the fixed-point motion necessary for success.

Both multi-fingered dexterous gripping and RCM contribute to the flexibility and precision required in demanding robotic tasks, but they focus on different aspects of robotic manipulation. While multi-fingered dexterous hands enable complex grasping and fine force control, RCM technology ensures that tools can remain fixed during operation, facilitating high-precision tasks in confined or dynamic environments. These technologies enhance the capabilities of robotic systems to perform a wider range of tasks, from delicate surgeries to complex industrial manipulations, where both flexibility and precision are essential.

Part I
Performance-Index-Oriented Robot Control with DNN

Chapter 2
Double-Index Control with DNN

2.1 Introduction

Currently, the extensive application of robots has sparked significant interest in the in-depth research on manipulator control (Jin et al. 2017; He et al. 2020; Ollero et al. 2022; Liu et al. 2021c). A redundant manipulator, known for its high flexibility (Yang et al. 2018; Kim et al. 2021; Zhang et al. 2022), possesses more degrees of freedom than strictly required to accomplish complex tasks (Albu-Schäffer and Sachtler 2023; Zhang et al. 2020; Yang et al. 2019), enabling it to perform complex and intricate operations with high dexterity (Khan et al. 2019; Kazemipour et al. 2022; Yan et al. 2024a). However, given the desired trajectory of a redundant manipulator, acquiring real-time joint angles remains challenging due to the complexity of inverse kinematic equations (Almarkhi et al. 2020; Jin et al. 2025; Kim and Perez 2021). The equations often involve nonlinearity and computational intensity, posing challenges to achieving the high-speed and high-precision control of the redundant manipulator (Vergnano et al. 2012; Liu et al. 2021a). As a result, numerous velocity-level and acceleration-level schemes continuously emerge to address redundancy resolution (Arteaga-Peréz et al. 2019; Khurana and Billard 2024; Liu et al. 2021b). Most velocity-level schemes, such as the minimum kinetic energy (MKE) scheme, focus on joint velocity constraints, and have limited control over joint acceleration and joint angles (Guo and Zhang 2014a; Xie et al. 2024). Acceleration-level schemes incorporating physical joint constraints, e.g., the minimum acceleration norm (MAN) scheme, are developed (Jin et al. 2021). Additionally, the scheme presented by Guo et al., which combines MAN and MKE, is reformulated at the acceleration level (Guo and Zhang 2014b). Rather than relying on a single optimization index, it integrates both the velocity and acceleration levels. In summary, the principles and methods behind the above schemes provide valuable insights for the control of redundant manipulators and are worth referencing.

Robot kinematic control is commonly simplified as the real-time solution of equations (Baek et al. 2022; Wang et al. 2022; Xiao and Zhang 2013). Traditional methods

M. Liu et al., *Dynamic Neural Networks for Motion Control of Redundant Manipulators*, Intelligent Control and Learning Systems 21,
https://doi.org/10.1007/978-981-96-9144-9_2

struggle to meet the requirements of real-time applications, particularly in dynamic and unpredictable environments (Xia et al. 2005; Reister et al. 2022; Shi et al. 2021). As a powerful numerical method with both efficiency and precision, neural networks are capable of fulfilling real-time computation demands (Jin et al. 2023; Zhang and Zhang 2013; Cursi et al. 2024), thereby being used to the solving of joint data in driving redundant manipulators (Huang et al. 2015; Koptev et al. 2023; Zhang et al. 2008). The deep research and development across various neural networks significantly enhance the precision and effectiveness of redundant manipulator control (Jin et al. 2024a; Zhang and Zhang 2012; Yan et al. 2024b). Among them, the recurrent neural network (RNN) emerges as one of the most frequently adopted models due to its capacity for dynamic information processing and suitability for trajectory-based tasks (Park et al. 2024; Yang et al. 2023; Centurelli et al. 2022; Jin et al. 2024b; Kong et al. 2020; Liang and Jin 2024). To address the quadratic programming (QP) problem encountered in redundant manipulator control, an exponential convergent RNN is explored in Xia et al. (2004). Furthermore, the RNN presented in Wang et al. (2016) enables cooperative control across multiple manipulators, leading to notable progress in the coordinated operation. In addition, Li et al. investigate an RNN capable of adapting to time-dependent noises, which markedly improves the practicality of using RNNs in the control of redundant manipulators (Li et al. 2018).

Joint-angle drift is a phenomenon observed in the cyclic motion generation of redundant manipulators, where the joint angle fails to return to its initial state after completing a cyclic task (Xie et al. 2019; Deaton et al. 2023; Xiao and Zhang 2016). This issue can lead to uncontrolled motions and, in the worst case, cause harm to people nearby (Lu et al. 2019). Therefore, it is crucial to develop methods to address this drift. In practical applications, minimizing the kinetic energy consumption of redundant manipulators is also a key objective. To mitigate both kinetic energy consumption and joint-angle offset, a double-index (DI) scheme is introduced in this chapter, incorporating two performance metrics: the minimum kinetic energy (MKE) and the minimum joint-angle offset (MJAO). Subsequently, a dynamic neural network (DNN) model is formulated to solve the quadratic programming (QP) problem based on the DI scheme, which effectively reduces both kinetic energy consumption and joint-angle offset.

The chapter is structured as follows: Section 2.2 introduces the DI scheme for redundant manipulators, reformulating it into a QP problem. Section 2.3 details the derivation of a DNN model to solve the QP problem. Section 2.4 presents comparative simulations to assess the performance of the DI scheme solved by the DNN model. The final section concludes the chapter. The main contributions are summarized as follows:

(1) This chapter investigates a DI scheme that reduces both kinetic energy and joint-angle offset for redundant manipulators. The DI scheme features an objective function that dynamically weighs MKE and MJAO, with constraints defined by joint physical limits. The MKE performance indicator is derived, and the DI scheme is expressed as a QP problem.

(2) A DNN model is developed to solve the QP problem and provides joint acceleration information for controlling the manipulator's motion. This model allows the control of the manipulator's motion based on the joint data when the target trajectory is known.

2.2 Scheme and Reformulation

This section presents a DI scheme for redundant manipulators based on the kinematic equations, with a brief explanation of the design details.

2.2.1 *DI Scheme*

Before introducing the DI scheme, the foundational concepts related to redundant manipulators need to be outlined. Let k denote the number of joints, with the joint angles represented as $\boldsymbol{\theta} = [\boldsymbol{\theta}_1, \boldsymbol{\theta}_2, \cdots, \boldsymbol{\theta}_k]^{\mathrm{T}} \in \mathbb{R}^k$ where the superscript ^{T} represents the transpose operator. Accordingly, the position of the end-effector can be expressed as

$$\boldsymbol{r}_{\mathrm{p}} = \Upsilon(\boldsymbol{\theta}) \in \mathbb{R}^n \ (n < k), \tag{2.1}$$

where $\Upsilon(\cdot)$ denotes the mapping from $\boldsymbol{\theta}$ to $\boldsymbol{r}_{\mathrm{p}}$. Based on Eq. (2.1), we further have

$$\ddot{\boldsymbol{r}}_{\mathrm{p}} = J\ddot{\boldsymbol{\theta}} + \dot{J}\dot{\boldsymbol{\theta}}, \tag{2.2}$$

where $J = \partial\Upsilon(\boldsymbol{\theta})/\partial\boldsymbol{\theta} \in \mathbb{R}^{n\times k}$ denotes the Jacobian matrix of $\Upsilon(\cdot)$, and $\dot{J}$ is its time derivative. The vector $\ddot{\boldsymbol{r}}_{\mathrm{p}}$ represents the acceleration of the end-effector, while $\dot{\boldsymbol{\theta}}$ and $\ddot{\boldsymbol{\theta}}$ correspond to joint velocity and acceleration, respectively. With above definitions, the DI scheme can be formulated as

$$\begin{aligned}
\min\ & \frac{1}{2}w_{\mathrm{e}}\dot{\boldsymbol{\theta}}^{\mathrm{T}}K\dot{\boldsymbol{\theta}} + \frac{1}{2}w_{\mathrm{d}}\|\ddot{\boldsymbol{\theta}} + \delta\|^2, \\
\text{s.t. } & J\ddot{\boldsymbol{\theta}} = \ddot{\boldsymbol{r}}_{\mathrm{e}} - \dot{J}\dot{\boldsymbol{\theta}} + \zeta_0 + \zeta_1, \\
& \boldsymbol{\theta}^- \le \boldsymbol{\theta} \le \boldsymbol{\theta}^+, \\
& \dot{\boldsymbol{\theta}}^- \le \dot{\boldsymbol{\theta}} \le \dot{\boldsymbol{\theta}}^+, \\
& \ddot{\boldsymbol{\theta}}^- \le \ddot{\boldsymbol{\theta}} \le \ddot{\boldsymbol{\theta}}^+, \\
\text{with } & \delta = (c_1 + c_2)\boldsymbol{\theta} + c_1c_2(\boldsymbol{\theta} - \boldsymbol{\theta}(0)),
\end{aligned} \tag{2.3}$$

where $K \in \mathbb{R}^{k\times k}$ is a positive definite symmetric matrix, and $c_1, c_2 \in \mathbb{R}$ are positive scalar coefficients; $|\cdot|$ denotes the vector norm; $\dot{\boldsymbol{\theta}}^{\mathrm{T}}K\dot{\boldsymbol{\theta}}/2$ corresponds to the kinetic energy, while $|\ddot{\boldsymbol{\theta}} + \delta|^2/2$ serves as the performance index for MJAO, weighed by w_{e}

and $w_{\rm d}$, respectively; $\boldsymbol{\theta}(0)$ denotes the initial joint configuration, and $\boldsymbol{\theta}^{\mp}$, $\dot{\boldsymbol{\theta}}^{\mp}$, and $\ddot{\boldsymbol{\theta}}^{\mp}$ represent the lower and upper bounds of joint angles, velocities, and accelerations, respectively. The desired end-effector position, velocity, and acceleration are given by $\boldsymbol{r}_{\rm s}$, $\dot{\boldsymbol{r}}_{\rm s}$, and $\ddot{\boldsymbol{r}}_{\rm s}$. To ensure trajectory tracking, position and velocity feedback terms are introduced as $\zeta_0 = c_3(\boldsymbol{r}_{\rm s} - \boldsymbol{r}_{\rm p})$ and $\zeta_1 = c_4(\dot{\boldsymbol{r}}_{\rm s} - J\dot{\boldsymbol{\theta}})$, where $c_3, c_4 > 0 \in \mathbb{R}$ are feedback gains. These feedback terms enable the manipulator to follow the desired trajectory.

In the DI scheme (2.3), minimizing $|\ddot{\boldsymbol{\theta}} + \delta|^2/2$ serves to suppress joint-angle drift, as demonstrated in Jin and Zhang (2015). Unlike constant values, the weight coefficients $w_{\rm e}$ and $w_{\rm d}$ are time-varying. Specifically, $w_{\rm e}$ is required to be greater than $w_{\rm d}$ at the beginning and end of the motion, while during the motion $w_{\rm d}$ is greater than $w_{\rm e}$. This indicates that MJAO is prioritized at the start and conclusion of the task, whereas MKE is emphasized throughout the motion process.

2.2.2 *Reformulation*

In this section, the DI scheme (2.3) is reformulated into a QP problem by modifying the MKE term $\dot{\boldsymbol{\theta}}^{\rm T} K\dot{\boldsymbol{\theta}}/2$. The MKE index can be equivalently expressed as

$$\frac{1}{2}\dot{\boldsymbol{\theta}}^{\rm T} K\dot{\boldsymbol{\theta}} = \frac{1}{2}\|M\dot{\boldsymbol{\theta}}\|, \tag{2.4}$$

where $K = M^{\rm T}M$ with $M \in \mathbb{R}^{k\times k}$ is a symmetric matrix. Based on the lemma presented in Guo and Zhang (2014b), the MKE term can be rewritten in the following form:

$$\frac{1}{2}\ddot{\boldsymbol{\theta}}^{\rm T} K\ddot{\boldsymbol{\theta}} + \dot{\boldsymbol{\theta}}^{\rm T}(\dot{M}^{\rm T}M + c_5K)\ddot{\boldsymbol{\theta}}, \tag{2.5}$$

where $\dot{M}$ denotes the time derivative of M and c_5 is a positive scalar. In addition, the physical constraints on joints should be reformulated in terms of joint acceleration. Let $\Psi^{\mp}$ represent the bounds of joint acceleration, where the upper and lower limits for the jth joint are defined as

$$\begin{aligned}\Psi_j^+ &= \min\{\ddot{\boldsymbol{\theta}}_j^+, \gamma_{\rm p}(\boldsymbol{\theta}_j^+ - \phi - \boldsymbol{\theta}_j), \gamma_{\rm v}(\dot{\boldsymbol{\theta}}_j^+ - \dot{\boldsymbol{\theta}}_j)\},\\ \Psi_j^- &= \max\{\ddot{\boldsymbol{\theta}}_j^-, \gamma_{\rm p}(\boldsymbol{\theta}_j^- + \phi - \boldsymbol{\theta}_j), \gamma_{\rm v}(\dot{\boldsymbol{\theta}}_j^- - \dot{\boldsymbol{\theta}}_j)\}.\end{aligned} \tag{2.6}$$

The parameter $\phi > 0$ defines the critical regions $(\boldsymbol{\theta}^+ - \phi, \boldsymbol{\theta}^+)$ and $(\boldsymbol{\theta}^-, \boldsymbol{\theta}^- + \phi)$. With the help of the coefficient $\gamma_{\rm p} > 0 \in \mathbb{R}$, the joint angle is adjusted to the safe region $(\boldsymbol{\theta}^-, \boldsymbol{\theta}^+)$ when it enters the critical areas. When the joint velocity exceeds its limit, $\gamma_{\rm v} > 0 \in \mathbb{R}$ helps to make the necessary adjustments. Both $\gamma_{\rm p}$ and $\gamma_{\rm v}$ are used to control the adjustment magnitude. Thus, joint limits are transformed into $\ddot{\boldsymbol{\theta}} \in \Psi$ $(\Psi^- \le \ddot{\boldsymbol{\theta}} \le \Psi^+)$, where Ψ represents the set of allowable values for $\ddot{\boldsymbol{\theta}}$.

With the previous preparations, the DI scheme (2.3) is reformulated as

$$\begin{aligned} &\min\ w_{\rm e}\{\frac{1}{2}\ddot{\boldsymbol{\theta}}^{\rm T}K\ddot{\boldsymbol{\theta}}+\dot{\boldsymbol{\theta}}^{\rm T}(\dot{M}^{\rm T}M+c_5K)\ddot{\boldsymbol{\theta}}\}+w_{\rm d}\{\frac{1}{2}\ddot{\boldsymbol{\theta}}^{\rm T}\ddot{\boldsymbol{\theta}}+\delta^{\rm T}\ddot{\boldsymbol{\theta}}\}\\ &\text{s.t. } J\ddot{\boldsymbol{\theta}}=\ddot{\boldsymbol{r}}_{\rm s}-\dot{J}\dot{\boldsymbol{\theta}}+\zeta_0+\zeta_1,\\ &\quad \Psi^-\le\ddot{\boldsymbol{\theta}}\le\Psi^+, \end{aligned} \tag{2.7}$$

which is equivalent to the following QP problem:

$$\begin{aligned} &\min\ \frac{1}{2}\boldsymbol{y}^{\rm T}W\boldsymbol{y}+\rho^{\rm T}\boldsymbol{y},\\ &\text{s.t. } A\boldsymbol{y}=\boldsymbol{b},\\ &\quad \Psi^-\le\boldsymbol{y}\le\Psi^+, \end{aligned} \tag{2.8}$$

where $\boldsymbol{y}=\ddot{\boldsymbol{\theta}}$, $W=w_{\rm e}K+w_{\rm d}I\in\mathbb{R}^{k\times k}$, $A=J$, $\rho=w_{\rm e}(\dot{M}^{\rm T}M+c_5K)\dot{\boldsymbol{\theta}}+w_{\rm d}\delta\in\mathbb{R}^k$, $\boldsymbol{b}=\ddot{\boldsymbol{r}}_{\rm s}-\dot{J}\dot{\boldsymbol{\theta}}+\zeta_0+\zeta_1\in\mathbb{R}^n$. The letter I denotes a unit matrix in this chapter. The DI scheme, incorporating joint velocity and acceleration levels, is converted into a generalized quadratic optimization problem, providing the basis for the subsequent solution.

2.3 DNN Model

This section explores the DNN model to solve the QP problem (2.8), derived as follows. First, a Lagrangian function is introduced as

$$F(\boldsymbol{y},\boldsymbol{\kappa})=\frac{1}{2}\boldsymbol{y}^{\rm T}W\boldsymbol{y}+\rho^{\rm T}\boldsymbol{y}+\boldsymbol{\kappa}^{\rm T}(A\boldsymbol{y}-\boldsymbol{b}), \tag{2.9}$$

where $\boldsymbol{\kappa}\in\mathbb{R}^n$ represents the Lagrange multiplier. Using the Karush–Kuhn–Tucker condition, the optimal solution to the QP problem (2.8) can be determined by

$$N_\Psi(\boldsymbol{y})+\frac{\partial F}{\partial\boldsymbol{y}}\ni\boldsymbol{0}, \tag{2.10}$$

$$\frac{\partial F}{\partial\boldsymbol{\kappa}}=\boldsymbol{0}, \tag{2.11}$$

where $N_\Psi(\boldsymbol{y})$ is the normal cone of set Ψ at $\boldsymbol{y}$ and satisfy

$$(\boldsymbol{\eta}-\boldsymbol{y})^{\rm T}(\frac{\partial F}{\partial\boldsymbol{y}})\ge 0,\ \ \forall\boldsymbol{\eta}\in\Psi. \tag{2.12}$$

Based on (2.12), one gets

$$\|\boldsymbol{\eta} - (\boldsymbol{y} - \frac{\partial F}{\partial \boldsymbol{y}})\|^2 - \|\boldsymbol{y} - (\boldsymbol{y} - \frac{\partial F}{\partial \boldsymbol{y}})\|^2 = \|\boldsymbol{y} - \boldsymbol{\eta}\|^2 + 2(\boldsymbol{\eta} - \boldsymbol{y})^{\mathrm{T}}(\frac{\partial F}{\partial \boldsymbol{y}}) \geq 0. \tag{2.13}$$

To facilitate the subsequent derivation, the projection function $\Gamma_{\Psi}(\cdot)$ needs to be defined.

$$\Gamma_{\Psi}(\boldsymbol{a}) = \boldsymbol{d}, \ \text{with} \|\boldsymbol{h} - \boldsymbol{a}\|^2 - \|\boldsymbol{d} - \boldsymbol{a}\|^2 \geq 0, \ \boldsymbol{d} \in \Psi, \ \forall \boldsymbol{h} \in \Psi. \tag{2.14}$$

It can also be written as

$$\Gamma_{\Psi}(x_i) = \begin{cases} \Psi_i^+, \ x_i \geq \Psi_i^+ \\ x_i, \quad \Psi_i^- < x_i < \Psi_i^+ \\ \Psi_i^-, \ x_i \leq \Psi_i^-, \end{cases} \tag{2.15}$$

where x_i is the ith subsystem of the vector $\boldsymbol{x} \in \mathbb{R}^k$. Then, combining (2.13), (2.14), and (2.10) is transformed into

$$\boldsymbol{y} = \Gamma_{\Psi}(\boldsymbol{y} - \frac{\partial F}{\partial \boldsymbol{y}}) = \Gamma_{\Psi}(\boldsymbol{y} - W\boldsymbol{y} - \rho - A^{\mathrm{T}}\boldsymbol{\kappa}). \tag{2.16}$$

Therefore, (2.10) and (2.11) can be rewritten as

$$\begin{aligned} \boldsymbol{0} &= -\boldsymbol{y} + \Gamma_{\Psi}(\boldsymbol{y} - W\boldsymbol{y} - \rho - A^{\mathrm{T}}\boldsymbol{\kappa}), \\ \boldsymbol{0} &= A\boldsymbol{y} - \boldsymbol{b}. \end{aligned} \tag{2.17}$$

Addressing the nonlinear equations (2.17) presents a complex and time-consuming challenge, and thus we provide the following DNN model:

$$\begin{aligned} \epsilon\dot{\boldsymbol{y}} &= -\boldsymbol{y} + \Gamma_{\Psi}(\boldsymbol{y} - W\boldsymbol{y} - \rho - A^{\mathrm{T}}\boldsymbol{\kappa}), \\ \epsilon\dot{\boldsymbol{\kappa}} &= A\boldsymbol{y} - \boldsymbol{b}, \end{aligned} \tag{2.18}$$

where $\epsilon > 0$ is the coefficient to determine the convergence rate of the DNN model (2.18). Although the DNN model (2.18) might not generate the exact solution to the system, it is capable of yielding a highly accurate approximation. It should be noted that the DNN model (2.18) can converge to the optimal solution of the DI scheme (2.3), as demonstrated by the theorem in Li et al. (2018). The DNN model (2.18) is illustrated in Fig. 2.1. In DNN model (2.18), with $\boldsymbol{r}_{\mathrm{s}}$, $\dot{\boldsymbol{r}}_{\mathrm{s}}$, and $\ddot{\boldsymbol{r}}_{\mathrm{s}}$ as inputs, through iterations on the vectors ($\boldsymbol{\theta}$, $\dot{\boldsymbol{\theta}}$, $\ddot{\boldsymbol{\theta}}$, and $\boldsymbol{\kappa}$), the output is $\ddot{\boldsymbol{\theta}}$.

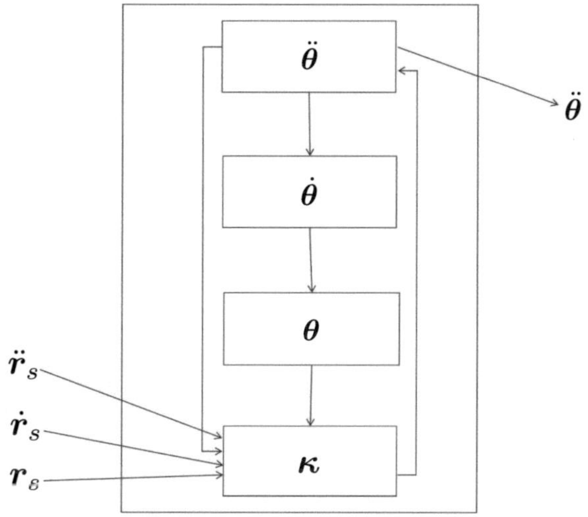

Fig. 2.1 Architecture of DNN model

2.4 Simulation

To verify the effectiveness and reliability of the DI scheme (2.3) solved by the DNN model (2.18), contrastive simulations are conducted. The comparison results further demonstrate that the DI scheme (2.3) solved by the DNN model (2.18) can reduce the kinetic energy and address the joint drift problem.

2.4.1 *Simulation Setting Up*

The simulations are conducted using the redundant manipulator, namely, the PUMA 560, with joint limits detailed in Saravanan et al. (2008). Assume that the trajectory of the cyclic motion is a circle, with the task duration being $T_m = 50$ s. The weight coefficients of the two minimization indicators are $w_e = \sin(\pi t/T_m)$ and $w_e = 1 - \sin(\pi t/T_m)$, respectively. Other parameters are set as follows: $\boldsymbol{\theta}(0) = [0, 0, 0, 0, 0, 0]^T$ rad, $K = I$, $c_1 = c_2 = 4$, $c_3 = c_4 = 8$, $c_5 = 40$, $\gamma_v = \gamma_p = 20$, $\epsilon = 0.001$, and $\phi = 0.05$ rad.

2.4.2 *Results Analysis*

The simulation results of the DI scheme (2.3) solved by the DNN model (2.18) are presented in Figs. 2.2 and 2.3. Specifically, the redundant manipulator is able to move along the desired trajectory as shown in Fig. 2.2a, and the position error is illustrated

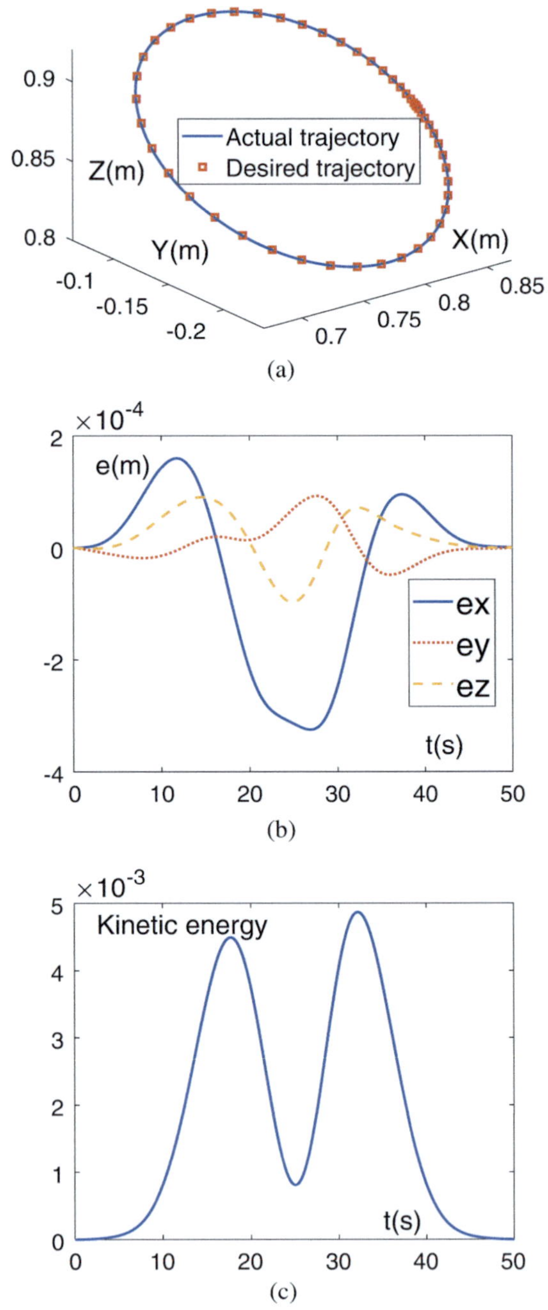

Fig. 2.2 The results of DI scheme (2.3) solved by DNN model (2.18) in tracking a circular trajectory. **a** Actual trajectory and desired trajectory. **b** Position error. **c** Kinetic energy

Fig. 2.3 The joint information of DI scheme (2.3) solved by DNN model (2.18) in tracking a circular trajectory. **a** Joint angle. **b** Joint velocity. **c** Joint acceleration

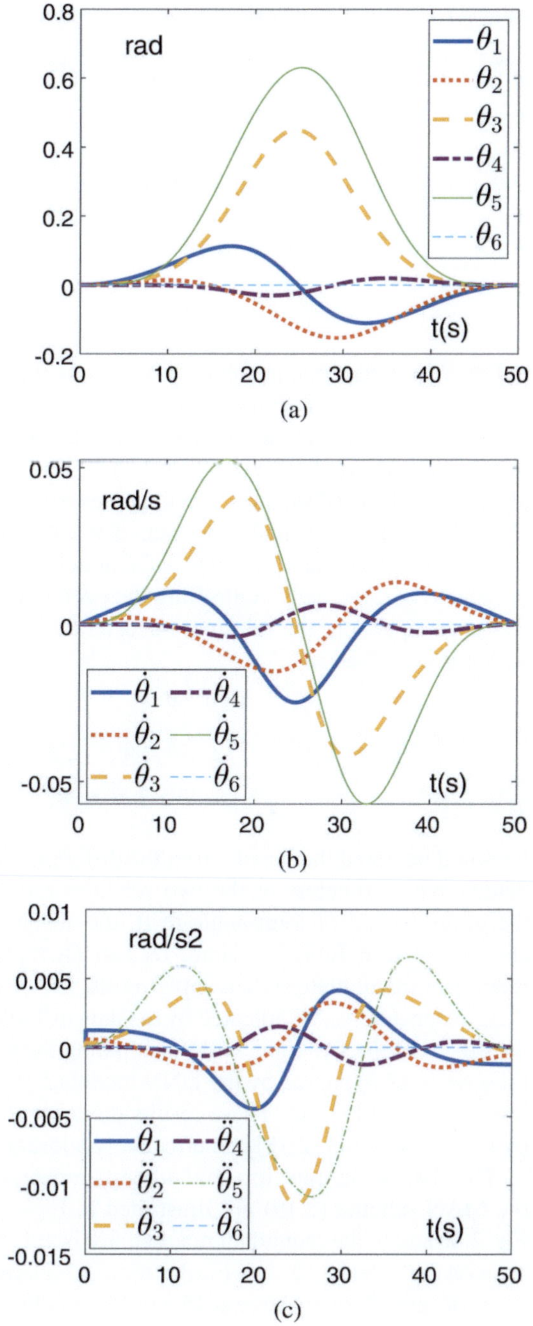

Table 2.1 The offset of joint-angle

	$\lvert D_1 \rvert$ (rad)	$\lvert D_2 \rvert$ (rad)
$\boldsymbol{\theta}_1$	0.010250	4.579725×10^{-7}
$\boldsymbol{\theta}_2$	0.339213	1.950322×10^{-5}
$\boldsymbol{\theta}_3$	0.666742	3.972688×10^{-5}
$\boldsymbol{\theta}_4$	0.501614	3.601455×10^{-5}
$\boldsymbol{\theta}_5$	0.093417	1.943157×10^{-5}
$\boldsymbol{\theta}_6$	0	0

in Fig. 2.2b, where the absolute value is less than 2×10^{-4} m, with e_X, e_Y, and e_Z denoting the position errors in three different directions, respectively. Additionally, during the movement, the profiles of the related parameters ($\boldsymbol{\theta}$, $\dot{\boldsymbol{\theta}}$, and $\ddot{\boldsymbol{\theta}}$) of each joint over time are depicted in Fig. 2.3a–c. The states of the joint angles at 50 s are identical to those of the initial joint angles $\boldsymbol{\theta}(0)$. Furthermore, the kinetic energy in Fig. 2.2c is relatively small (less than 5×10^{-3}) and down to zero at the end of the task. The results indicate that the DNN model (2.18) designed for the DI scheme (2.3) is capable of effectively controlling the cyclic motion of redundant manipulators.

For comparison, the minimum-acceleration-norm (MAN) scheme is simulated, which is described as

$$\begin{aligned} &\min \frac{1}{2}\ddot{\boldsymbol{\theta}}^{\mathrm{T}}\ddot{\boldsymbol{\theta}}, \\ &\text{s.t. } J\ddot{\boldsymbol{\theta}} = \ddot{\boldsymbol{r}}_{\mathrm{s}} - \dot{J}\dot{\boldsymbol{\theta}} + \zeta_0 + \zeta_1, \\ &\quad \Psi^- \leq \ddot{\boldsymbol{\theta}} \leq \Psi^+. \end{aligned} \tag{2.19}$$

It should be noted that, aside from the difference in the objective function, the other details and parameters of the two schemes remain identical. In order to compare the improvement of joint-angle drift, the absolute values of the joint-angle offset are presented in Table 2.1. Here, D_1 and D_2 represent the joint-angle offset of the redundant manipulator when applying the MAN scheme (2.19) and the DI scheme (2.3), respectively. As indicated by the data in Table 2.1, for the MAN scheme (2.19), the offsets of the five joint angles, excluding the sixth joint, exceed 0.01 rad. For the DI scheme (2.3) solved by the DNN model (2.18), the offsets of all joint angles are less than 4×10^{-5} rad. These results demonstrate that the DI scheme (2.3) solved by the DNN model (2.18) can effectively address the joint-angle drift problem.

The changes in trajectory and other parameters during the motion when applying the MAN scheme (2.19) are illustrated in Figs. 2.4 and 2.5. As can be seen from Fig. 2.4a and b, the redundant manipulator is able to complete the movement with an error on the order of 10^{-4} m. Furthermore, accelerations at individual joint angles like $\ddot{\boldsymbol{\theta}}_2$ and $\ddot{\boldsymbol{\theta}}_5$ are relatively large at 45.5 to 46.5 s in Fig. 2.5c, and the joint velocity and the

Fig. 2.4 The results of MAN scheme (2.19) in tracking a circular trajectory. **a** Actual trajectory and desired trajectory. **b** Position error. **c** Kinetic energy

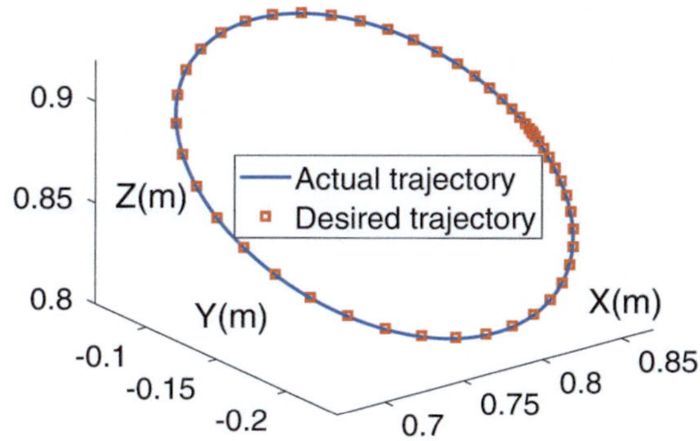

(a)

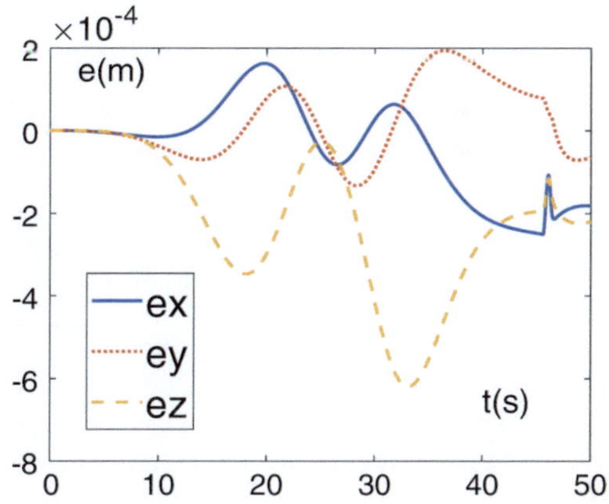

(b)

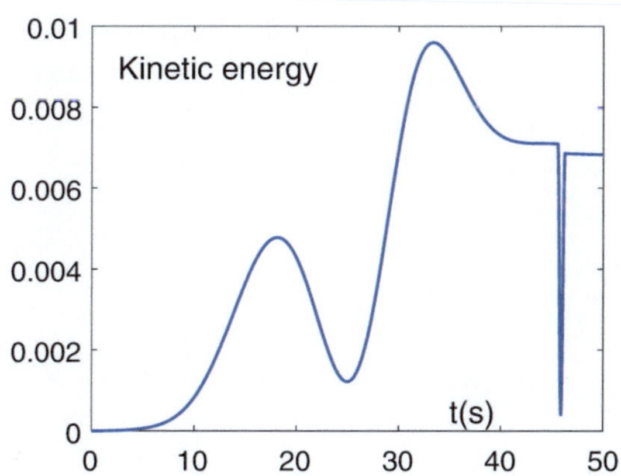

(c)

Fig. 2.5 The joint information of MAN scheme (2.19) in tracking a circular trajectory. **a** Joint angle. **b** Joint velocity. **c** Joint acceleration

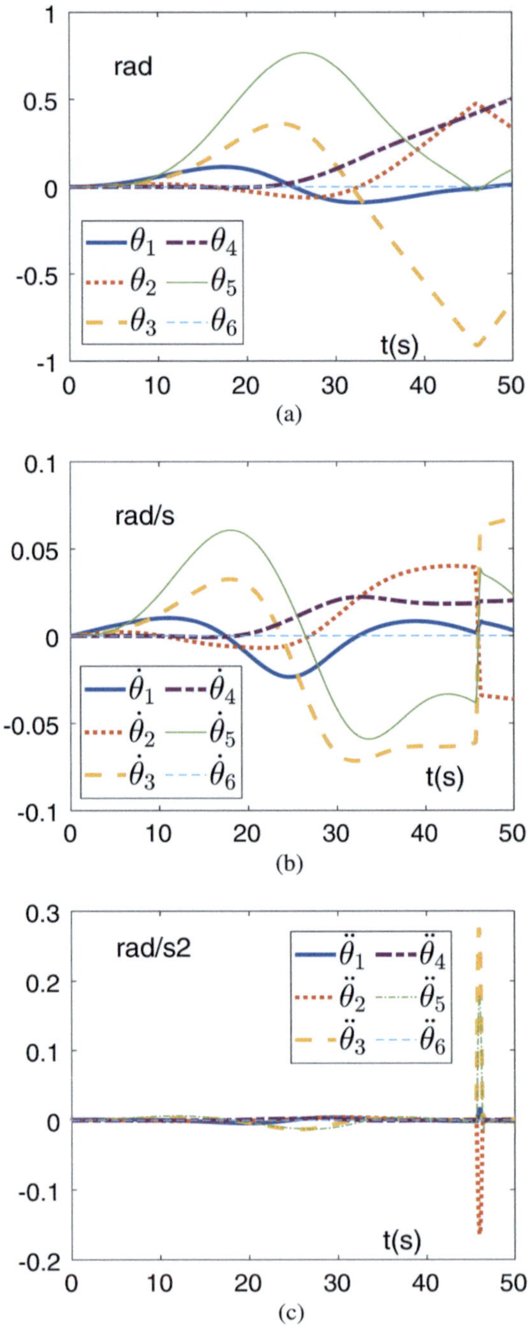

kinetic energy changes accordingly in Figs. 2.5b and 2.4c. The reason is that the MAN scheme (2.19) aims to minimize $\|\ddot{\boldsymbol{\theta}}\|$ rather than $\ddot{\boldsymbol{\theta}}$, which can lead to local increases in the acceleration of individual joints. The increase in joint acceleration leads to a significant change in joint velocity. Furthermore, the kinetic energy displayed in Fig. 2.4c is larger than that in Fig. 2.2c and the kinetic energy keeps about 0.007 even though the motion is over, which is impractical. These comparative results highlight that the DI scheme (2.3), when solved by the DNN model (2.18), effectively reduces joint-angle drift and minimizes kinetic energy consumption.

2.5 Conclusion

In this chapter, a DI scheme for the MKE and MJAO has been presented to control the cyclic motion generation of redundant manipulators, and a DNN model has been derived to solve the DI scheme. Specifically, in the objective function of the DI scheme, the two indices of MKE and MJAO are weighed using two dynamic coefficients, and two time-dependent weights are employed to adjust the weights during the motion. Moreover, the constraints of the DI scheme encompass the kinematic equation constraint and joint limit constraints. Furthermore, the DI scheme has been transformed into a QP problem, which has been addressed by the derived DNN model. Finally, comparative experiments have been carried out to demonstrate the effectiveness of the DI scheme, solved by the DNN model, in mitigating joint-angle drift and reducing kinetic energy.

References

Albu-Schäffer A, Sachtler A (2023) Redundancy resolution at position level. IEEE Trans Rob 39(6):4240–4261

Almarkhi AA, Maciejewski AA, Chong EKP (2020) An algorithm to design redundant manipulators of optimally fault-tolerant kinematic structure. IEEE Robot Autom Lett 5(3):4727–4734

Arteaga-Peréz MA, Pliego-Jiménez J, Romero JG (2019) Experimental results on the robust and adaptive control of robot manipulators without velocity measurements. IEEE Trans Control Syst Technol 28(6):2770–2773

Baek S, Lee H, Han S (2022) Communication-efficient event-triggered time-delay control and its application to robot manipulators. IEEE Trans Industr Electron 69(9):9288–9297

Centurelli A, Arleo L, Rizzo A, Tolu S, Laschi C, Falotico E (2022) Closed-loop dynamic control of a soft manipulator using deep reinforcement learning. IEEE Robot Autom Lett 7(2):4741–4748

Cursi F, Bai W, Yeatman EM, Kormushev P (2024) Task accuracy enhancement for a surgical macro-micro manipulator with probabilistic neural networks and uncertainty minimization. IEEE Trans Autom Sci Eng 21(1):241–256

Deaton NJ, Sheft M, Desai JP (2023) Towards FBG-based shape sensing and sensor drift for a steerable needle. IEEE/ASME Trans Mechatron 28(6):3041–3052

Guo D, Zhang Y (2014a) Acceleration-level inequality-based MAN scheme for obstacle avoidance of redundant robot manipulators. IEEE Trans Industr Electron 61(12):6903–6914

Guo D, Zhang Y (2014b) Simulation and experimental verification of weighted velocity and acceleration minimization for robotic redundancy resolution. IEEE Trans Autom Sci Eng 11(4):1203–1217

He L, Liu M, Yi C, Shi Y, Sun Z, Chen L (2020) Recurrent neural network for kinematic control of redundant robot manipulators. In: 2020 39th Chinese control conference, pp 3845–3850

Huang Y, Na J, Gao G, Wu X, Guo Y (2015) Robust adaptive control for vehicle active suspension systems with uncertain dynamics. In: 2015 34th Chinese control conference, pp 8033–8038

Jin L, Li S, Xiao L, Liao R, Liao B (2017) Cooperative motion generation in a distributed network of redundant robot manipulators with noises. IEEE Trans Syst Man Cybern Syst 48(10):1715–1724

Jin L, Xie Z, Liu M, Chen K, Li C, Yang C (2021) Novel joint-drift-free scheme at acceleration level for robotic redundancy resolution with tracking error theoretically eliminated. IEEE/ASME Trans Mechatron 26(1):90–101

Jin L, Wei L, Li S (2023) Gradient-based differential neural-solution to time-dependent nonlinear optimization. IEEE Trans Autom Control 68(1):62–627

Jin L, Li Y, Zhang X, Luo X (2024a) Fuzzy k-winner-take-all network for competitive coordination in multirobot systems. IEEE Trans Fuzzy Syst 32(4):2005–2016

Jin L, Su Z, Fu D, Xiao X (2024b) Coevolutionary neural solution for nonconvex optimization with noise tolerance. IEEE Trans Neural Netw Learn Syst 35(12):17571–17581

Jin L, Zhang Y (2015) G2-type SRMPC scheme for synchronous manipulation of two redundant robot arms. IEEE Trans Cybern 45(2):153–164

Jin L, Zhao J, Chen L, Li S (2025) Collective neural dynamics for sparse motion planning of redundant manipulators without Hessian matrix inversion. IEEE Trans Neural Netw Learn Syst. 36(3):4326–4335

Kazemipour A, Khatib M, Al Khudir K, Gaz C, De Luca A (2022) Kinematic control of redundant robots with online handling of variable generalized hard constraints. IEEE Robot Autom Lett 7(4):9279–9286

Khan AH, Li S, Luo X (2019) Obstacle avoidance and tracking control of redundant robotic manipulator: an RNN-based metaheuristic approach. IEEE Trans Industr Inf 16(7):4670–4680

Khurana H, Billard A (2024) Motion planning and inertia-based control for impact aware manipulation. IEEE Trans Rob 40:2201–2216

Kim J, Jie W, Kim H, Lee MC (2021) Modified configuration control with potential field for inverse kinematic solution of redundant manipulator. IEEE/ASME Trans Mechatron 26(4):1782–1790

Kim S, Perez J (2021) Learning reachable manifold and inverse mapping for a redundant robot manipulator. In: 2021 IEEE international conference on robotics and automation, pp 4731–4737

Kong Y, Jiang Y, Xia X (2020) Terminal recurrent neural networks for time-varying reciprocal solving with application to trajectory planning of redundant manipulators. IEEE Trans Syst Man Cybern Syst 52(1):387–399

Koptev M, Figueroa N, Billard A (2023) Neural joint space implicit signed distance functions for reactive robot manipulator control. IEEE Robot Autom Lett 8(2):480–487

Li S, Wang H, Rafique MU (2018) A novel recurrent neural network for manipulator control with improved noise tolerance. IEEE Transactions on Neural Networks and Learning Systems 29(5):1908–1918

Liang S, Jin L (2024) Distributed k-winner-take-all network under weight-unbalanced topology. IEEE Trans Circuits Syst II Express Briefs 71(2):717–721

Liu T, Yang T, Xu W, Mylonas G, Liang B (2021a) Efficient inverse kinematics and planning of a hybrid active and passive cable-driven segmented manipulator. IEEE Trans Syst Man Cybern Syst 52(7):4233–4246

Liu M, Chen L, Du X, Jin L, Shang M (2021b) Activated gradients for deep neural networks. IEEE Trans Neural Netw Learn Syst 34(4):2156–2168

Liu M, He L, Shang M (2021c) Dynamic neural network for bicriteria weighted control of robot manipulators. IEEE Trans Neural Netw Learn Syst 34(8):4570–4583

Lu H, Jin L, Zhang J, Sun Z, Li S, Zhang Z (2019) New joint-drift-free scheme aided with projected ZNN for motion generation of redundant robot manipulators perturbed by disturbances. IEEE Trans Syst Man Cybern Syst 51(9):5639–5651

Ollero A, Tognon M, Suarez A, Lee D, Franchi A (2022) Past, present, and future of aerial robotic manipulators. IEEE Trans Rob 38(1):626–645

Park J, Jang S, Park H, Bae S, Hwang M (2024) Hysteresis compensation of flexible continuum manipulator using RGBD sensing and temporal convolutional network. IEEE Robot Autom Lett 9(7):6091–6098

Reister F, Grotz M, Asfour T (2022) Combining navigation and manipulation costs for time-efficient robot placement in mobile manipulation tasks. IEEE Robot Autom Lett 7(4):9913–9920

Saravanan R, Ramabalan S, Balamurugan C (2008) Evolutionary collision-free optimal trajectory planning for intelligent robots. Int J Adv Manuf Technol 36(11–12):1234–1251

Shi Y, Zhao W, Li S, Li B, Sun X (2021) Novel discrete-time recurrent neural network for robot manipulator: a direct discretization technical route. IEEE Trans Neural Netw Learn Syst 34(6):2781–2790

Vergnano A, Thorstensson C, Lennartson B, Falkman P, Pellicciari M, Leali F, Biller S (2012) Modeling and optimization of energy consumption in cooperative multi-robot systems. IEEE Trans Autom Sci Eng 9(2):423–428

Wang Y, Cheng L, Hou Z, Yu J, Tan M (2016) Optimal formation of multirobot systems based on a recurrent neural network. IEEE Trans Neural Netw Learn Syst 27(2):322–333

Wang Y, Leibold M, Lee J, Ye W, Xie J, Buss M (2022) Incremental model predictive control exploiting time-delay estimation for a robot manipulator. IEEE Trans Control Syst Technol 30(6):2285–2300

Xia Y, Feng G, Wang J (2004) A recurrent neural network with exponential convergence for solving convex quadratic program and related linear piecewise equations. Neural Netw 17:1003–1015

Xia Y, Feng G, Wang J (2005) A primal-dual neural network for online resolving constrained kinematic redundancy in robot motion control. IEEE Trans Syst Man Cybern Syst 35(1):54–64

Xiao L, Zhang Y (2013) Acceleration-level repetitive motion planning and its experimental verification on a six-link planar robot manipulator. IEEE Trans Control Syst Technol 21(3):906–914

Xiao L, Zhang Y (2016) Dynamic design, numerical solution and effective verification of acceleration-level obstacle-avoidance scheme for robot manipulators. Int J Syst Sci 47(4):932–945

Xie Z, Jin L, Du X, Xiao X, Li H (2019) On generalized RMP scheme for redundant robot manipulators aided with dynamic neural networks and nonconvex bound constraints. IEEE Trans Industr Inf 15(9):5172–5181

Xie Z, Jin L, Luo X, Zhou M, Zheng Y (2024) A biobjective scheme for kinematic control of mobile robotic arms with manipulability optimization. IEEE/ASME Trans Mechatron 29(2):1534–1545

Yan J, Jin L, Luo X, Li S (2024a) Modified RNN for solving comprehensive Sylvester equation with TDOA application. IEEE Trans Neural Netw Learn Syst 35(9):12553–12563

Yan J, Jin L, Hu B (2024b) Data-driven model predictive control for redundant manipulators with unknown model. IEEE Trans Cybern 54(10):5901–5911

Yang C, Jiang Y, He W, Na J, Li Z, Xu B (2018) Adaptive parameter estimation and control design for robot manipulators with finite-time convergence. IEEE Trans Industr Electron 65(10):8112–8123

Yang C, Zeng C, Cong Y, Wang N, Wang M (2019) A learning framework of adaptive manipulative skills from human to robot. IEEE Trans Industr Inf 15(2):1153–1161

Yang X, Zhao Z, Ma B, Xu Z, Zhao J, Liu H (2023) Kinematic and dynamic manipulability optimizations of redundant manipulators based on RNN model. IEEE Trans Industr Inf 20(4):5763–5773

Zhang Z, Zhang Y (2012) Acceleration-level cyclic-motion generation of constrained redundant robots tracking different paths. IEEE Trans Syst Man Cybern Syst 42(4):1257–1269

Zhang Z, Zhang Y (2013) Design and experimentation of acceleration-level drift-free scheme aided by two recurrent neural networks. IET Control Theory & Appl 7(1):25–42

Zhang H, Jin H, Liu Z, Liu Y, Zhu Y, Zhao J (2020) Real-time kinematic control for redundant manipulators in a time-varying environment: multiple-dynamic obstacle avoidance and fast tracking of a moving object. IEEE Trans Industr Inf 16(1):28–41
Zhang J, Jin L, Yang C (2022) Distributed cooperative kinematic control of multiple robotic manipulators with an improved communication efficiency. IEEE/ASME Trans Mechatron 27(1):149–158
Zhang Y, Tan Z, Yang Z, Lv X (2008) A dual neural network applied to drift-free resolution of five-link planar robot arm. In: International conference on information and automation, pp 1274–1279

Chapter 3
Cyclic-Motion Control with Noise-Rejection DNN

3.1 Introduction

Due to their higher DOF than task requirements (Huang et al. 2020; Bodie et al. 2021; Liu et al. 2021), redundant manipulators offer enhanced operational flexibility, making them widely used in multiple industrial applications (Qin et al. 2015; Qi et al. 2020). With expanding applications across diverse fields, the motion control in redundant manipulators has emerged as a prominent research focus (Xie and Jin 2024; Yang et al. 2018; Stanimirovic et al. 2015). The core of redundant manipulators in finishing a given task is to obtain joint parameters (i.e., angular displacement, angular velocity, and angular acceleration of joints) and to drive the end-effector to move along the predefined trajectory (Haviland and Corke 2024; Yan et al. 2024; Haviland et al. 2022). Considering the redundancy of manipulators, these parameters of redundant manipulators are not unique (Liu and Qin 2019). On this basis, plenty of schemes have been presented to get the corresponding variables for controlling manipulators (Yilmaz et al. 2021; Cheng et al. 2018; Xiao et al. 2019). In You et al. (2021), a scheme with the manipulability as the optimization index investigated offers a method to get the relevant joint velocity. On the premise of effectiveness and reliability, the scheme makes great progress in maximizing the manipulability. Control schemes for redundant manipulators typically operate at either the velocity or acceleration level. Velocity-level schemes are computationally inexpensive and efficient, but cannot take into account limitations of joint accelerations. In addition, velocity-level control schemes may exhibit discontinuous velocity profiles that are physically unrealizable. In contrast, the acceleration-level scheme effectively overcomes these limitations.

In industrial manufacturing applications, redundant manipulators frequently need to perform repetitive operational cycles, making the development of control schemes enabling cyclic-motion patterns a fundamental research challenge (Lu et al. 2021). A prominent challenge in cyclic-motion control involves the joint-angle drift, resulting in non-convergence between initial and final joint configurations. Xie et al. present

M. Liu et al., *Dynamic Neural Networks for Motion Control of Redundant Manipulators*, Intelligent Control and Learning Systems 21,
https://doi.org/10.1007/978-981-96-9144-9_3

a redundancy resolution with the ability to eliminate the joint-angle drift and handle joint physical limits (Jin et al. 2020). Notably, the majority of existing control schemes lack explicit noise-rejection mechanisms. Given the unavoidable presence of the noise, developing robust noise-resistant algorithms for cyclic-motion generation in redundant manipulators represents a critical requirement for practical applications. Furthermore, acceleration-level control schemes offer enhanced operational flexibility, as they can directly execute acceleration commands or generate velocity profiles through the integration, demonstrating the superior feasibility compared to velocity-level schemes. To this end, the acceleration-level scheme with the noise rejection is presented in this chapter.

Neural networks are appealing methods to a large number of complicated issues at present (Liang and Jin 2024; Li et al. 2018; Ha et al. 2022). The ability to model nonlinear relationships and adapt to dynamic environments makes them indispensable in fields such as robotics, where the precision and adaptability are critical (Xiao et al. 2018; Shi et al. 2022; Reiter et al. 2018). Moreover, these methods play an increasingly vital role in resolving the kinematics redundancy for redundant manipulators (Zhang et al. 2020; Safeea et al. 2021; Jarzebowska 2008). RNNs, distinguished by their exceptional computational capabilities among various neural architectures, possess an inherent memory mechanism for retaining processed information (Ma et al. 2022; Jin et al. 2023; Guo et al. 2019). This feature enables RNN to be widely used as a solver for optimization problems (Zhang et al. 2019a, b; Ohara et al. 2021; Jin et al. 2024a). RNN performs well in solving optimization problems (Wei and Jin 2024). In Liu et al. (2023), an RNN is used to assist in achieving control over the joint acceleration of redundant manipulators. Accordingly, this chapter presents an acceleration-level cyclic-motion (ALCM) scheme to realize precise cyclic manipulation through the joint acceleration regulation.

Notably, numerous existing neural network architectures lack inherent disturbance rejection capabilities (Xiao et al. 2018; Yang et al. 2023; Yoo et al. 2021). However, practical implementations inevitably encounter various noise types including quantization errors, stochastic disturbances, and additive noise components (Jin et al. 2024b; Wang et al. 2023; Li et al. 2019). Time-varying disturbances particularly affect the system performance due to their cumulative nature over prolonged operation periods (Yan et al. 2025; Nainer and Giusti 2022; Cui et al. 2024). When the noise interference surpasses the system's handling capacity, the system would be unable to operate. In consideration of this, a DNN (a kind of RNN) model resisting noise interference, particularly for the time-varying polynomial noise, is presented in this chapter. The fundamental principle of the DNN involves incorporating dynamic variables within hidden layers to compensate for noise-induced errors. Aided with these dynamic variables, the DNN model can eliminate the influence of the time-varying polynomial noise.

The rest content of the chapter is organized as follows: Section 3.2 presents the fundamental kinematic formulations for redundant manipulators. Section 3.3 introduces the ALCM scheme for joint-angle drift compensation. The DNN-based solution framework for the presented control scheme is subsequently developed, with the rigorous convergence analysis. Section 3.4 presents the NRDNN model and the proof of its convergence. To demonstrate the robustness of the presented ALCM scheme implemented through the NRDNN framework, simulations are carried out in Sect. 3.5. The last section concludes this chapter.

3.2 Preparatory Work

This section examines fundamental manipulator kinematics formulated at the acceleration level. For a k-DOF manipulator, the end-effector position $\boldsymbol{r}_{\mathrm{e}}(t) \in \mathbb{R}^m$ $(m < k)$ relates to joint angles $\boldsymbol{\theta}(t) = [\boldsymbol{\theta}_1(t); \boldsymbol{\theta}_2(t); \ldots; \boldsymbol{\theta}_k(t)] \in \mathbb{R}^k$ through the nonlinear mapping $\boldsymbol{r}_{\mathrm{e}}(t) = \Gamma(\boldsymbol{\theta}(t))$, where $\Gamma(\cdot)$ is defined in (2.1). The time derivative yields

$$\dot{\boldsymbol{r}}_{\mathrm{e}}(t) = J\dot{\boldsymbol{\theta}}(t), \tag{3.1}$$

where $J = \partial\Gamma(\boldsymbol{\theta}(t))/\boldsymbol{\theta}(t) \in \mathbb{R}^{m\times k}$ denotes the Jacobian matrix of $\Gamma(\cdot)$; $\dot{\boldsymbol{r}}_{\mathrm{e}}(t)$ represents the end-effector velocity; and $\dot{\boldsymbol{\theta}}(t)$ indicates the joint velocity. Differentiating with respect to time t, we obtain the following equation, which is similar to (2.2):

$$\ddot{\boldsymbol{r}}_{\mathrm{e}}(t) = J\ddot{\boldsymbol{\theta}}(t) + \dot{J}\dot{\boldsymbol{\theta}}(t), \tag{3.2}$$

where $\dot{J}$ represents the time derivative of the Jacobian matrix J; $\ddot{\boldsymbol{\theta}}(t)$ is the joint acceleration vector; and $\ddot{\boldsymbol{r}}_{\mathrm{e}}(t)$ denotes the end-effector acceleration. Redundant manipulators inherently exhibit constrained joint acceleration capabilities due to physical limitations. Consequently, the kinematic equation should obey

$$J\ddot{\boldsymbol{\theta}}(t) = \ddot{\boldsymbol{r}}_{\mathrm{s}}(t) - \dot{J}\dot{\boldsymbol{\theta}}(t), \tag{3.3}$$

where $\ddot{\boldsymbol{r}}_{\mathrm{s}}(t) \in \mathbb{R}^m$ defining the desired end-effector acceleration corresponding to reference trajectory $\boldsymbol{r}_{\mathrm{s}}(t)$. Practical implementations face inevitable quantization errors and noise disturbances that compromise the control precision, leading to non-compliance with the joint-angle constraint in (3.3). Thus, for eliminating the influence of noises, $\rho_0(t) = d_0(\boldsymbol{r}_{\mathrm{s}}(t) - \boldsymbol{r}_{\mathrm{e}}(t))$ and $\rho_1(t) = d_1(\dot{\boldsymbol{r}}_{\mathrm{s}}(t) - J\dot{\boldsymbol{\theta}}(t))$ function as feedback control terms for the position correction, where positive constants $d_0 > 0$ and $d_1 > 0$ denote feedback gains, and $\dot{\boldsymbol{r}}_{\mathrm{s}}(t)$ specifies the reference end-effector velocity. Then, (3.3) is revised as

$$J\ddot{\boldsymbol{\theta}}(t) = \ddot{\boldsymbol{r}}_{\mathrm{s}}(t) - \dot{J}\dot{\boldsymbol{\theta}}(t) + \rho_0(t) + \rho_1(t). \tag{3.4}$$

Continuous end-effector position adjustment drives the actual trajectory toward the ideal reference motion. The feedback gains d_0 and d_1 can theoretically approach arbitrarily large values to enhance the control precision.

3.3 ALCM Scheme and DNN Model Construction

Based on the kinematic model (3.4) in Sect. 3.2, the presented ALCM scheme enables redundant manipulators to achieve precise cyclic trajectories through the joint acceleration control. This is complemented by a dedicated DNN-based solver for the ALCM scheme, accompanied by rigorous convergence guarantees.

3.3.1 ALCM Scheme

For achieving the cyclic-motion control of redundant manipulators, the ALCM scheme is presented. The design of the ALCM scheme is as follows. The joint-angle displacement vector is formally defined as

$$\boldsymbol{\delta}(t) = \boldsymbol{\theta}(t) - \boldsymbol{\theta}(0) \in \mathbb{R}^k. \tag{3.5}$$

The manipulator should maintain the periodicity through $\boldsymbol{\theta}(T_{\boldsymbol{\theta}}) = \boldsymbol{\theta}(0)$, with $T_{\boldsymbol{\theta}}$ representing the task's motion period and $\boldsymbol{\theta}(0)$ denoting the initial joint configuration. Differentiating (3.5) with respect to time, we obtain

$$\dot{\boldsymbol{\delta}}(t) = \dot{\boldsymbol{\theta}}(t). \tag{3.6}$$

Then, according to Jin et al. (2016), defining $\boldsymbol{\delta}(t)$ as the error function $\boldsymbol{\epsilon}(t)$ and substituting it into the formula $\dot{\boldsymbol{\epsilon}}(t) = -c_1\boldsymbol{\epsilon}(t)$ lead to

$$\dot{\boldsymbol{\delta}}(t) = -c_1\boldsymbol{\delta}(t) = -c_1(\boldsymbol{\theta}(t) - \boldsymbol{\theta}(0)), \tag{3.7}$$

where $c_1 > 0$ is a coefficient to control the convergence speed of $\boldsymbol{\delta}(t)$. This control formulation guarantees the asymptotic convergence of joint angles to their initial configurations. Combining (3.6) and (3.7) gives

$$\dot{\boldsymbol{\theta}}(t) + c_1(\boldsymbol{\theta}(t) - \boldsymbol{\theta}(0)) = \mathbf{0}. \tag{3.8}$$

Thirdly, defining $\boldsymbol{d}_2(t) = \dot{\boldsymbol{\theta}}(t) + c_1(\boldsymbol{\theta}(t) - \boldsymbol{\theta}(0))$ and repeating the above operations lead to

$$\ddot{\boldsymbol{\theta}}(t) + (c_1 + c_2)\dot{\boldsymbol{\theta}}(t) + c_1c_2(\boldsymbol{\theta}(t) - \boldsymbol{\theta}(0)) = \mathbf{0}, \tag{3.9}$$

where the positive gain coefficient $c_2 > 0$ regulates the convergence rate of $\boldsymbol{d}_2(t)$. As proven in Jin and Zhang (2015), the system guarantees that $\boldsymbol{\theta}(t)$ converges to $\boldsymbol{\theta}(0)$ and $\dot{\boldsymbol{\theta}}(t)$ converges to zero upon task completion. Thus, the optimization index in the ALCM scheme is selected as $\|\ddot{\boldsymbol{\theta}}(t) + \alpha\dot{\boldsymbol{\theta}}(t) + \beta(\boldsymbol{\theta}(t) - \boldsymbol{\theta}(0))\|$, where α and β are used to replace $(c_1 + c_2)$ and c_1c_2, separately.

Integrating the kinematic equation from (3.4), the acceleration-level ALCM scheme is formally expressed as

$$\begin{aligned}
&\min \frac{1}{2}\ddot{\boldsymbol{\theta}}^{\mathrm{T}}(t)\ddot{\boldsymbol{\theta}}(t) + (\alpha\dot{\boldsymbol{\theta}}(t) + \beta(\boldsymbol{\theta}(t) - \boldsymbol{\theta}(0)))^{\mathrm{T}}\ddot{\boldsymbol{\theta}}(t),\\
&\text{s.t. } J\ddot{\boldsymbol{\theta}}(t) = \ddot{\boldsymbol{r}}_{\mathrm{s}}(t) - \dot{\boldsymbol{J}}\dot{\boldsymbol{\theta}}(t) + \rho_0(t) + \rho_1(t),\\
&\quad\boldsymbol{\theta}^- \le \boldsymbol{\theta}(t) \le \boldsymbol{\theta}^+,\\
&\quad\dot{\boldsymbol{\theta}}^- \le \dot{\boldsymbol{\theta}}(t) \le \dot{\boldsymbol{\theta}}^+,\\
&\quad\ddot{\boldsymbol{\theta}}^- \le \ddot{\boldsymbol{\theta}}(t) \le \ddot{\boldsymbol{\theta}}^+,
\end{aligned} \tag{3.10}$$

where ^{T} denotes matrix transposition, with $\boldsymbol{\theta}^{\pm}$, $\dot{\boldsymbol{\theta}}^{\pm}$, and $\ddot{\boldsymbol{\theta}}^{\pm}$, respectively, representing the physical bounds for joint angles, velocities, and accelerations. Through the definitions $\ddot{\boldsymbol{\theta}}(t) = \psi(t)$ and $\boldsymbol{\varpi}(t) = \alpha\dot{\boldsymbol{\theta}}(t) + \beta(\boldsymbol{\theta}(t) - \boldsymbol{\theta}(0))$, along with time-argument simplifications ($\boldsymbol{\theta} = \boldsymbol{\theta}(t)$, $\ddot{\boldsymbol{r}}_{\mathrm{s}} = \ddot{\boldsymbol{r}}_{\mathrm{s}}(t)$), (3.10) transforms to

$$\begin{aligned}
&\min_{\psi} \frac{1}{2}\psi^{\mathrm{T}}\psi + \boldsymbol{\varpi}^{\mathrm{T}}\psi,\\
&\text{s.t. } J\psi = \ddot{\boldsymbol{r}}_{\mathrm{s}} - \dot{\boldsymbol{J}}\dot{\boldsymbol{\theta}} + \rho_0 + \rho_1,\\
&\quad\Theta^- \le \psi \le \Theta^+,
\end{aligned} \tag{3.11}$$

where Θ^+ and Θ^- are the upper and lower limits of joint accelerations and modified from the constraints in (3.10). The jth joint acceleration constraints are bounded by

$$\begin{aligned}
\Theta_j^+ &= \min\{\ddot{\boldsymbol{\theta}}_j^+, \gamma_{\mathrm{p}}(\boldsymbol{\theta}_j^+ - \phi - \boldsymbol{\theta}_j), \gamma_{\mathrm{v}}(\dot{\boldsymbol{\theta}}_j^+ - \dot{\boldsymbol{\theta}}_j)\},\\
\Theta_j^- &= \max\{\ddot{\boldsymbol{\theta}}_j^-, \gamma_{\mathrm{p}}(\boldsymbol{\theta}_j^- + \phi - \boldsymbol{\theta}_j), \gamma_{\mathrm{v}}(\dot{\boldsymbol{\theta}}_j^- - \dot{\boldsymbol{\theta}}_j)\},
\end{aligned} \tag{3.12}$$

where $\phi > 0$ helps to form the critical regions $(\boldsymbol{\theta}^+ - \phi, \boldsymbol{\theta}^+)$ and $(\boldsymbol{\theta}^-, \boldsymbol{\theta}^- + \phi)$; γ_{p} is exploited to adjust the joint angle when it approaches limits; and γ_{v} can adjust joint velocity when it enters critical regions.

3.3.2 DNN Model Construction

To deal with the ALCM scheme (3.10), a DNN model is presented. Prior to developing the DNN architecture, we define a projection operator to enforce joint constraints:

$$P_{\Theta}(\boldsymbol{\zeta}) = \arg\min_{\boldsymbol{b}\in\Theta} \parallel \boldsymbol{b} - \boldsymbol{\zeta} \parallel^2, \tag{3.13}$$

where Θ denotes the feasible set of ψ. This projection operator is inherently incorporated into the DNN framework to enforce joint state constraints. During the real-time control, it operates through the following nonlinear mapping:

$$P_\Theta(\zeta_i) = \begin{cases} \Theta_i^+, \ \zeta_i \geq \Theta_i^+ \\ \zeta_i, \quad \Theta_i^- < \zeta_i < \Theta_i^+ \\ \Theta_i^-, \ \zeta_i \leq \Theta_i^-, \end{cases}$$

where ζ_i, Θ_i^+, and Θ_i^- represent the ith subsystem of $\boldsymbol{\zeta}$, Θ^+, and Θ^-, respectively.

Through the Lagrangian duality, the ALCM formulation in (3.10) admits the equivalent representation:

$$\min_{\psi\in\Theta}\max_{\iota} Y(\psi, \iota) = \frac{1}{2}\psi^{\mathrm{T}}\psi + \boldsymbol{\varpi}^{\mathrm{T}}\psi + \iota^{\mathrm{T}}(J\psi - \ddot{\boldsymbol{r}}_{\mathrm{s}} + \dot{J}\dot{\boldsymbol{\theta}} - \rho_0 - \rho_1), \quad (3.14)$$

where $Y(\psi, \iota)$ denotes the Lagrangian derived from (3.10), with $\boldsymbol{\iota} \in \mathbb{R}^m$ being the Lagrange multiplier vector. The Karush–Kuhn–Tucker (KKT) conditions yield the saddle point solution for (3.14) through

$$\begin{aligned} &N_\Theta(\psi) + \frac{\partial Y}{\partial \psi} \ni 0, \\ &\frac{\partial Y}{\partial \iota} = 0, \end{aligned} \quad (3.15)$$

where $N_\Theta(\psi)$ stands for the normal cone of set Θ at ψ. Further, we get

$$(\tau - \psi)\left(\frac{\partial Y}{\partial \psi}\right)^{\mathrm{T}} \geq 0, \ \forall \tau \in \Theta. \quad (3.16)$$

Thus, we can obtain the following result:

$$\begin{aligned} &\| \tau - \left(\psi - \frac{\partial Y}{\partial \psi}\right) \|^2 - \| \psi - \left(\psi - \frac{\partial Y}{\partial \psi}\right) \|^2 \\ =& \| \psi - \tau \|^2 + 2\left(\frac{\partial Y}{\partial \psi}\right)^{\mathrm{T}} (\tau - \psi) \\ \geq& 0, \ \forall \tau \in \Theta. \end{aligned} \quad (3.17)$$

By synthesizing (3.17) and (3.13), ψ admits the reformulated expression:

$$\psi = P_\Theta\left(\psi - \frac{\partial Y}{\partial \psi}\right) = P_\Theta(\psi - (\psi + \boldsymbol{\varpi} + J^{\mathrm{T}}\iota)) = P_\Theta(-\boldsymbol{\varpi} - J^{\mathrm{T}}\iota). \quad (3.18)$$

Further, (3.15) is expressed as

$$\begin{aligned} \mathbf{0} &= -\psi + P_\Theta(-\boldsymbol{\varpi} - J^{\mathrm{T}}\iota), \\ \mathbf{0} &= J\psi - \ddot{\boldsymbol{r}}_{\mathrm{s}} + \dot{J}\dot{\boldsymbol{\theta}} - \rho_0 - \rho_1. \end{aligned} \quad (3.19)$$

To solve this coupled system, we develop a DNN architecture formulated as

$$\begin{aligned} \xi\dot{\psi} &= -\psi + P_{\Theta}(-\boldsymbol{\varpi} - J^{\mathrm{T}}\boldsymbol{\iota}), \\ \xi\dot{\boldsymbol{\iota}} &= J\psi - \ddot{\boldsymbol{r}}_{\mathrm{s}} + \dot{J}\dot{\boldsymbol{\theta}} - \rho_0 - \rho_1, \end{aligned} \tag{3.20}$$

with $\xi > 0$ serving as a scaling parameter. Note that the above DNN model does not take into account the effect of noises.

3.3.3 Analyses on DNN Model Without Noise

The following analysis establishes convergence properties of the DNN solver (3.20).

Theorem 3.1 *Under noise-free conditions, the DNN solver's output ψ in (3.20) exhibits the global convergence to the ALCM scheme's optimal solution specified in (3.10).*

Proof The neural network requires the compact reformulation for subsequent analysis, beginning with three foundational definitions:

$$\boldsymbol{h} = \begin{bmatrix} \psi \\ \boldsymbol{\iota} \end{bmatrix} \in \mathbb{R}^{k+m}, \tag{3.21}$$

$$\widehat{\psi} = \{\boldsymbol{y} = [\boldsymbol{y}_1^{\mathrm{T}}, \boldsymbol{y}_2^{\mathrm{T}}]^{\mathrm{T}}, \boldsymbol{y}_1 \in \Theta, \boldsymbol{y}_2 \in \mathbb{R}^m\}, \tag{3.22}$$

$$\Upsilon(\boldsymbol{h}) = \begin{bmatrix} \psi + \boldsymbol{\varpi} + J^{\mathrm{T}}\boldsymbol{\iota} \\ -J\psi + \ddot{\boldsymbol{r}}_{\mathrm{s}} - \dot{J}\dot{\boldsymbol{\theta}} + \rho_0 + \rho_1 \end{bmatrix} \in \mathbb{R}^{k+m}. \tag{3.23}$$

These new definitions enable the DNN model (3.20) to be written as an equation:

$$\xi\dot{\boldsymbol{h}} = -\boldsymbol{h} + P_{\Theta}(\boldsymbol{h} - \Upsilon(\boldsymbol{h})). \tag{3.24}$$

The gradient operator $\Upsilon(\cdot)$ takes the form as

$$\triangledown\Upsilon(\boldsymbol{h}) = \begin{bmatrix} E & J^{\mathrm{T}} \\ -J & 0 \end{bmatrix}, \tag{3.25}$$

where $E \in \mathbb{R}^{k\times k}$ denotes the identity matrix. Subsequently, based on (3.25), we obtain

$$\triangledown\Upsilon(\boldsymbol{h}) + \triangledown^{\mathrm{T}}\Upsilon(\boldsymbol{h}) = \begin{bmatrix} 2E & 0 \\ 0 & 0 \end{bmatrix}. \tag{3.26}$$

The inequality $\triangledown\Upsilon(\boldsymbol{h}) + \triangledown^{\mathrm{T}}\Upsilon(\boldsymbol{h}) \succeq 0$ clearly indicates a positive semi-definite matrix. To establish the monotonicity of $\Upsilon(\cdot)$, we derive

$$
\begin{aligned}
(\boldsymbol{h}_1-\boldsymbol{h}_2)^{\mathrm{T}}(\Upsilon(\boldsymbol{h}_1)-\Upsilon(\boldsymbol{h}_2)) &= (\boldsymbol{h}_1-\boldsymbol{h}_2)^{\mathrm{T}}\nabla\Upsilon(\hbar)(\boldsymbol{h}_1-\boldsymbol{h}_2)\\
&= \frac{1}{2}(\boldsymbol{h}_1-\boldsymbol{h}_2)^{\mathrm{T}}(\nabla\Upsilon(\hbar)+\nabla^{\mathrm{T}}\Upsilon(\hbar))(\boldsymbol{h}_1-\boldsymbol{h}_2)\geq 0,
\end{aligned}
\tag{3.27}
$$

where $\boldsymbol{h}_1$, $\boldsymbol{h}_2$, and $\hbar$ denote arbitrary vectors in $\widehat{\psi}$. Through (3.27), it can be concluded that $\Upsilon(\cdot)$ is monotonic. Following Gao (2003), the system (3.24) exhibits Lyapunov stability with global convergence of $\boldsymbol{h}$ to $\widetilde{\boldsymbol{h}}=[\widetilde{\psi};\widetilde{\iota}]$, where $\widetilde{\boldsymbol{h}}$ is a solution to a variational inequality

$$
(\boldsymbol{h}-\widetilde{\boldsymbol{h}})^{\mathrm{T}}\Upsilon(\widetilde{\boldsymbol{h}})\geq 0,\quad \forall\boldsymbol{h}\in\widehat{\psi}. \tag{3.28}
$$

By the variational inequality, the equivalent formulation becomes

$$
\dot{\widetilde{\boldsymbol{h}}}=-\widetilde{\boldsymbol{h}}+P_{\Theta}(\widetilde{\boldsymbol{h}}-\Upsilon(\widetilde{\boldsymbol{h}}))=\mathbf{0}, \tag{3.29}
$$

where $\dot{\widetilde{\boldsymbol{h}}}$ denotes the time derivative of $\widetilde{\boldsymbol{h}}$. Direct comparison between (3.24) and (3.29) reveals $\widetilde{\boldsymbol{h}}$ as an equilibrium point of (3.24). Applying these definitions yields

$$
\begin{aligned}
\mathbf{0} &= -\widetilde{\psi}+P_{\Theta}(-\boldsymbol{\varpi}-J^{\mathrm{T}}\widetilde{\iota}),\\
\mathbf{0} &= J\widetilde{\psi}-\ddot{\boldsymbol{r}}_{\mathrm{s}}+\dot{J}\dot{\boldsymbol{\theta}}-\rho_0-\rho_1.
\end{aligned}
\tag{3.30}
$$

The variables $\widetilde{\psi}$ and $\widetilde{\iota}$ in (3.30) constitute optimal solutions to the ALCM scheme (3.10), as they satisfy the KKT conditions specified in (3.19). The proof is completed. □

3.4 NRDNN Model and Theoretical Analysis

Practical implementations inherently encounter diverse noise scenarios, with time-varying disturbances exerting particularly significant impacts. To address this, we develop an NRDNN model capable of handling the time-varying polynomial noise, while rigorously establishing its convergence properties.

3.4.1 NRDNN Model

To enhance the robustness, we modify the baseline DNN architecture (3.20) to obtain the NRDNN model as

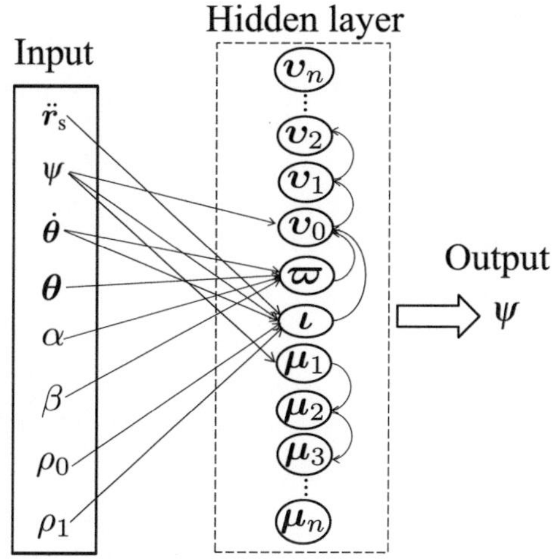

Fig. 3.1 The structure of the NRDNN model (3.31)

$$
\begin{aligned}
\xi\dot{\psi} &= P_{\Theta}(-\boldsymbol{\varpi} - J^{\mathrm{T}}\boldsymbol{\iota}) - (x_0\xi + 1)\psi - x_0\boldsymbol{v}_0 - \sum_{j=1}^{n}(\xi x_j\boldsymbol{\mu}_j + x_j\boldsymbol{v}_j) + \sum_{j=0}^{n} z_j,\\
\xi\dot{\boldsymbol{\iota}} &= J\psi - \ddot{\boldsymbol{r}}_{\mathrm{s}} + \dot{J}\dot{\boldsymbol{\theta}} - \rho_0 - \rho_1,\\
\dot{\boldsymbol{v}}_0 &= \psi - P_{\Theta}(-\boldsymbol{\varpi} - J^{\mathrm{T}}\boldsymbol{\iota}),\\
\dot{\boldsymbol{v}}_j &= \boldsymbol{v}_{j-1},\ j \in Z^+, j \le n,\\
\dot{\boldsymbol{\mu}}_1 &= \psi,\\
\dot{\boldsymbol{\mu}}_j &= \boldsymbol{\mu}_{j-1},\ j \in Z^+, 2 \le j \le n,
\end{aligned}
\tag{3.31}
$$

where x_j represents a design parameter. The noise model adopts an n-order polynomial structure $z = z_0 + z_1 + \cdots + z_n$, which is expanded by Taylor expansion as $z = a_0t^0 + a_1t^1 + \cdots + a_nt^n$. Two sets of dynamic variables $\boldsymbol{\mu}_1,\ \boldsymbol{\mu}_2,\ \ldots,\ \boldsymbol{\mu}_n$ and $\boldsymbol{v}_0,\ \boldsymbol{v}_1,\ \ldots,\ \boldsymbol{v}_n$ are incorporated to actively suppress noise disturbances within the system. These variables exhibit a derivative relationship. The structure of the NRDNN model (3.31) is drawn in Fig. 3.1. The network architecture follows standard computational stratification: the input layer buffers raw data, the hidden layers maintain dynamic state variables through nonlinear transformations, and the output layer generates processed solutions.

3.4.2 Convergence Analyses

The following theorems establish the exponential convergence of the NRDNN model (3.31) under time-varying polynomial-noise disturbances.

Theorem 3.2 *When parameters* $x_0,\ x_1,\ \ldots,\ x_n$ *satisfy that all the roots of* $x_n + sx_{n-1} + s^2x_{n-2} + \cdots + s^n x_0 + s^{n+1} = 0$ *are in the left S-plane, the NRDNN's output* ψ *in (3.31) achieves noise-immunized convergence to the ALCM scheme's optimal solution (3.10), replicating the noise-free performance.*

Proof We first introduce an auxiliary variable $\boldsymbol{u} = \psi - P_{\Theta}(-\boldsymbol{\varpi} - J^{\mathrm{T}}\boldsymbol{\iota})$, and dynamic state vectors $\boldsymbol{l}_0 = \xi\psi + \int_0^t \boldsymbol{u}\mathrm{d}t, \boldsymbol{l}_j = \xi\boldsymbol{\mu}_j + \boldsymbol{v}_j$. These enable the reformulation of (3.31) as

$$\begin{aligned}
\xi\dot{\psi} &= -\boldsymbol{u} - \sum_{j=0}^{n} x_j \boldsymbol{l}_j + \sum_{j=0}^{n} z_j,\\
\xi\dot{\boldsymbol{\iota}} &= J\psi - \ddot{\boldsymbol{r}}_{\mathrm{s}} + \dot{J}\dot{\boldsymbol{\theta}} - \rho_0 - \rho_1,\\
\dot{\boldsymbol{l}}_0 &= -\sum_{j=0}^{n} x_j \boldsymbol{l}_j + \sum_{j=0}^{n} z_j,\\
\dot{\boldsymbol{l}}_j &= \boldsymbol{l}_{j-1},\ j \in Z^+, j \le n.
\end{aligned} \tag{3.32}$$

The final two equations in (3.32) can be represented as an output equation and state equation of a communication system with z_j and $\boldsymbol{l}_j$ corresponding to input and state variables, respectively. Their Laplace transforms can be obtained as

$$\begin{aligned}
sL_0(s) - \boldsymbol{l}_0(0) &= -\sum_{j=0}^{n} x_j L_j(s) + \sum_{j=0}^{n} Z_j(s),\\
sL_j(s) - \boldsymbol{l}_j(0) &= L_{j-1}(s),\ j \in Z^+, j \le n,
\end{aligned} \tag{3.33}$$

where $L_j(s)$ is the Laplace transformation of $\boldsymbol{l}_j$ with $j = 0,\ 1,\ 2,\ \ldots,\ n$. The Laplace transformation of noise $z_j = a_j t^j$ with $j = 0,\ 1,\ 2,\ \ldots,\ n$ is $Z_j(s) = j!/s^{j+1}$. Iteration of (3.33) leads to

$$\begin{aligned}
L_j(s) &= \frac{\boldsymbol{l}_j(0)}{s} + \frac{L_{j-1}(s)}{s}\\
&= \frac{\boldsymbol{l}_j(0)}{s} + \frac{1}{s}\left(\frac{\boldsymbol{l}_{j-1}(0)}{s} + \frac{L_{j-2}(s)}{s}\right)\\
&\ \ \vdots\\
&= \frac{\boldsymbol{l}_j(0)}{s} + \frac{\boldsymbol{l}_{j-1}(0)}{s^2} + \cdots + \frac{\boldsymbol{l}_1(0)}{s^j} + \frac{L_0(s)}{s^j}.
\end{aligned} \tag{3.34}$$

Then, combining (3.33) and (3.34), we obtain

$$\begin{aligned}
sL_0(s) = &\ -L_0(s)(x_0 + \frac{x_1}{s} + \frac{x_2}{s^2} + \cdots + \frac{x_n}{s^n}) + \boldsymbol{l}_0(0)\\
&- \frac{1}{s}\sum_{j=1}^{n} x_j \boldsymbol{l}_j(0) - \frac{1}{s^2}\sum_{j=1}^{n-1} x_{j+1}\boldsymbol{l}_j(0) - \cdots - \frac{1}{s^n}x_n\boldsymbol{l}_1(0) + \sum_{j=0}^{n} Z_j(s).
\end{aligned} \tag{3.35}$$

Further derivation of (3.35) leads to

$$L_0(s) = \frac{s^n}{s^{n+1} + \sum_{j=0}^{n} s^j x_{n-j}} \times \left(\boldsymbol{l}_0(0) - \frac{1}{s}\sum_{j=1}^{n} x_j \boldsymbol{l}_j(0) - \frac{1}{s^2}\sum_{j=1}^{n-1} x_{j+1}\boldsymbol{l}_j(0) - \cdots - \frac{1}{s^n} x_n \boldsymbol{l}_1(0) + \sum_{j=0}^{n} Z_j(s) \right). \tag{3.36}$$

Let $q(t) = \dot{\boldsymbol{l}}_0$, whose Laplace transform is expressed as

$$\begin{aligned} Q(s) &= sL_0(s) - \boldsymbol{l}_0(0) \\ &= \frac{s}{s^{n+1} + \sum_{j=0}^{n} s^j x_{n-j}} \times \left(s^n \boldsymbol{l}_0(0) - s^{n-1}\sum_{j=1}^{n} x_j \boldsymbol{l}_j(0) - \cdots - x_n \boldsymbol{l}_1(0) \right) \\ &\quad - \boldsymbol{l}_0(0) + \frac{s^{n+1}}{s^{n+1} + \sum_{j=0}^{n} s^j x_{n-j}} \times \sum_{j=0}^{n} Z_j(s). \end{aligned} \tag{3.37}$$

The equation characterizes the system's frequency-domain response to the noise input, comprising both the zero-input response $Q_0(s)$ and zero-state response $Q_i(s)$ as

$$\begin{aligned} Q_0(s) &= -\boldsymbol{l}_0(0)\frac{s}{s^{n+1} + \sum_{j=0}^{n} s^j x_{n-j}} \times \left(-s^{n-1}\sum_{j=1}^{n} x_j \boldsymbol{l}_j(0) \right. \\ &\quad \left. - s^{n-2}\sum_{j=1}^{n-1} x_{j+1}\boldsymbol{l}_j(0) - \cdots - x_n \boldsymbol{l}_1(0) + s^n \boldsymbol{l}_0(0) \right), \\ Q_i(s) &= \frac{s^{n+1}}{s^{n+1} + \sum_{j=0}^{n} s^j x_{n-j}} \times \sum_{j=0}^{n} Z_j(s). \end{aligned} \tag{3.38}$$

Hence, the transfer function is

$$H(s) = \frac{Q_i(s)}{\sum_{j=0}^{n} Z_j(s)} = \frac{s^{n+1}}{s^{n+1} + \sum_{j=0}^{n} s^j x_{n-j}}. \tag{3.39}$$

Since it is assumed that all the roots of $x_n + sx_{n-1} + s^2 x_{n-2} + \cdots + s^n x_0 + s^{n+1} = 0$ lie in the left S-plane, $H(s)$ is stable according to the Routh–Hurwitz criterion (Franklin et al. 2001). This demonstrates that the system's output asymptotically converges to a steady-state value, computable through the final value theorem:

$$
\begin{aligned}
\lim_{t\to\infty} q(t) &= \lim_{s\to 0} sQ(s) \\
&= \lim_{s\to 0} \frac{\boldsymbol{l}_0(0)s^{n+2}}{s^{n+1}+s^n x_0+\cdots+x_n} - \frac{s^{n+1}\sum_{j=1}^{n} x_j \boldsymbol{l}_j(0)}{s^{n+1}+s^n x_0+\cdots+x_n} \\
&\quad -\cdots- \frac{s x_n \boldsymbol{l}_1(0)}{s^{n+1}+s^n x_0+\cdots+x_n} + \frac{\sum_{j=0}^{n} s^{n-j+1} j!}{s^{n+1}+s^n x_0+\cdots+x_n} - s l_0(0) \\
&= 0.
\end{aligned}
\tag{3.40}
$$

It guarantees that $q(t)$ infinitely approaches zero. Substituting the state definition $\boldsymbol{l}_0 = \xi\psi + \int_0^t \boldsymbol{u}\mathrm{d}t$, the output dynamics is transformed as

$$q(t) = \dot{\boldsymbol{l}}_0 = \xi\dot{\psi} + \boldsymbol{u} = \xi\dot{\psi} + \psi - P_\Theta(-\boldsymbol{\varpi} - J^{\mathrm{T}}\iota) = \mathbf{0}. \tag{3.41}$$

Comparing (3.41) with (3.20), we can naturally conclude that these two equations are equivalent and have the same solution. This confirms the identical convergence behavior to the noise-free case when the stability conditions are met. Theoretical analysis thus proves that the NRDNN model (3.31) achieves effective attenuation of the time-varying polynomial noise. The proof is completed. □

Detailed parameter selections need to be explained, and thus the following remark is given.

Remark 3.1 Parameters $x_0,\ x_1,\ x_2,\ \ldots,\ x_n$ must adhere to stability conditions in Theorem 3.2, with the following design procedure:

$$(s+\gamma)^{n+1} = s^{n+1} + (n+1)\gamma s^n + \frac{n(n+1)}{2}\gamma^2 s^{n-1} + \cdots + \gamma^{n+1} = 0, \tag{3.42}$$

where $\gamma > 0$, the solution $s = -\gamma < 0$ of (3.42) satisfies Theorem 3.2's left half-plane requirement. Consequently, parameters $x_0,\ x_1,\ x_2,\ \ldots,\ x_n$ are configured via (3.42) as

$$
\begin{aligned}
x_0 &= (n+1)\gamma, \\
x_1 &= \frac{(n+1)!}{(n-1)!2!}\gamma^2, \\
&\vdots \\
x_n &= \gamma^{n+1}.
\end{aligned}
\tag{3.43}
$$

3.5 Simulations and Experiments

Numerical simulations and hardware experiments demonstrate that the ALCM scheme (3.10) coupled with the DNN solver (3.20) successfully compensates for the joint-angle drift, while the NRDNN (3.31) provides a robust to time-varying noise disturbances.

3.5.1 *Simulation Setup*

Simulations are divided into four parts. The first part focuses on verifying the effect of the presented ALCM scheme (3.10) on eliminating the joint-angle drift. In the second part, the simulation results generated by DNN model (3.20) in the noise-free condition are demonstrated. The third part concentrates on explaining that NRDNN model (3.31) can handle constant noises. In the last part, the resistance of the presented NRDNN model (3.31) to time-varying polynomial noises is examined. Moreover, in the simulations to demonstrate the noise-rejection performance of the presented NRDNN model (3.31), the comparison model selected is the solver applied in Xia et al. (2005).

Specifically, a PUMA 560 redundant manipulator is employed in simulations. The end-effector tracks a curved triangular trajectory over $T_\sigma = 50$ s, with parameters configured as $d_1 = d_0 = 8, \xi = 0.001, \gamma = 2, \alpha = \beta = 4, \gamma_p = \gamma_v v = 20, \phi = 0.05$ rad, and $\boldsymbol{\theta}(0) = [0, 0, 0, 0, 0, 0]^T$ rad. The NRDNN coefficients $x_0, x_1, \ldots, x_n$ are computed from γ and the selected model order n in (3.31), where varying n yields distinct parameter sets.

3.5.2 *Joint-Angle Drift*

Comparative simulations against the MAN method (Zhang and Zhang 2013) validate the ALCM scheme's (3.10) superior joint-angle drift compensation capability. Different from the presented ALCM scheme (3.10), the MAN scheme aims to minimize the square of the 2-norm of the joint acceleration. Under identical trajectory-tracking conditions and parameter settings, Table 3.1 compares the joint-angle offset performance between the presented ALCM scheme (3.10) and benchmark MAN scheme for the PUMA 560 manipulator control. Note that, $\boldsymbol{\delta}_1$ represents the joint-angle offset of the presented ALCM scheme (3.10), and $\boldsymbol{\delta}_2$ is the result of the MAN scheme. As shown in Table 3.1, $\boldsymbol{\delta}_1$ is smaller than 10^{-4} rad, which suggests that the joint-angle drift is improved with the assistance of the presented ALCM scheme (3.10). In contrast, the significant magnitude of $\boldsymbol{\delta}_2$ demonstrates a pronounced joint-angle drift induced by the MAN scheme. In summary, data in Table 3.1 verify that the presented ALCM scheme (3.10) can effectively remedy the joint-angle drift.

Table 3.1 Comparison of joint-angle offsets

Joint	$\boldsymbol{\theta}(0)$ (rad)	δ_2 (rad)	δ_1 (rad)
θ_1	0	0.0508	2.5594×10^{-6}
θ_2	0	0.1848	1.1330×10^{-5}
θ_3	0	0.3599	9.0841×10^{-6}
θ_4	0	1.0116	6.9102×10^{-6}
θ_5	0	0.2623	2.8805×10^{-5}
θ_6	0	0	0

3.5.3 *Noise-Free Condition*

This section experimentally validates the DNN solver (3.20) in implementing the ALCM scheme (3.10) under the noise-free scenario, demonstrating its theoretical effectiveness in ideal conditions. The end-effector position error vector $\boldsymbol{e} = \boldsymbol{r}_{\mathrm{e}} - \boldsymbol{r}_{\mathrm{s}} \in \mathbb{R}^3$ and its Cartesian components (e_{X}, e_{Y}, and e_{Z}) quantify tracking accuracy in three-dimensional space. Figures 3.2 and 3.3 present the PUMA 560's trajectory-tracking performance under the ALCM scheme, where the manipulator accurately follows a curved triangular trajectory. Figure 3.2a illustrates the manipulator's sequential configurations during the trajectory tracking, demonstrating the proper kinematic execution of the curved triangular trajectory. Figure 3.2b demonstrates the sub-millimeter trajectory-tracking accuracy, with actual and desired trajectories indistinguishable, confirming the DNN solver's (3.20) perfect task execution capability. Figure 3.3a–c displays the bounded joint state evolution, with angle, velocity, and acceleration profiles strictly adhering to physical limits throughout the trajectory execution. Figure 3.2c reveals end-effector positioning errors of 10^{-4} m, quantitatively validating theoretical predictions.

3.5.4 *Constant-Noise Conditions*

The NRDNN model's (3.31) noise-rejection capability is evaluated under constant-noise conditions. For the zero-order case ($n = 0$), the architecture reduces as

$$\begin{aligned} \xi\dot{\psi} &= P_{\Theta}(-\boldsymbol{\varpi} - J^{\mathrm{T}}\boldsymbol{\iota}) - (x_0\xi + 1)\psi - x_0\boldsymbol{v}_0 + z, \\ \xi\dot{\boldsymbol{\iota}} &= J\psi - \ddot{\boldsymbol{r}}_{\mathrm{s}} + \dot{J}\dot{\boldsymbol{\theta}} - \rho_0 - \rho_1, \\ \dot{\boldsymbol{v}}_0 &= \psi - P_{\Theta}(-\boldsymbol{\varpi} - J^{\mathrm{T}}\boldsymbol{\iota}), \end{aligned} \tag{3.44}$$

where $x_0 = 2$ is obtained via (3.43). Two methods (i.e., NRDNN model (3.31) and the method in Xie et al. 2019) are simulated with three groups of noises (i.e., $z = [0.1]^{6\times1}$, $z = [1]^{6\times1}$, and $z = [10]^{6\times1}$), and their position errors are displayed in Figs. 3.4

Fig. 3.2 Simulations of a curved sided triangle carried out by PUMA 560 in nominal situation synthesized by ALCM scheme (3.10) aided with DNN model (3.20). **a** Motion process. **b** Actual trajectory and desired trajectory. **c** Position error

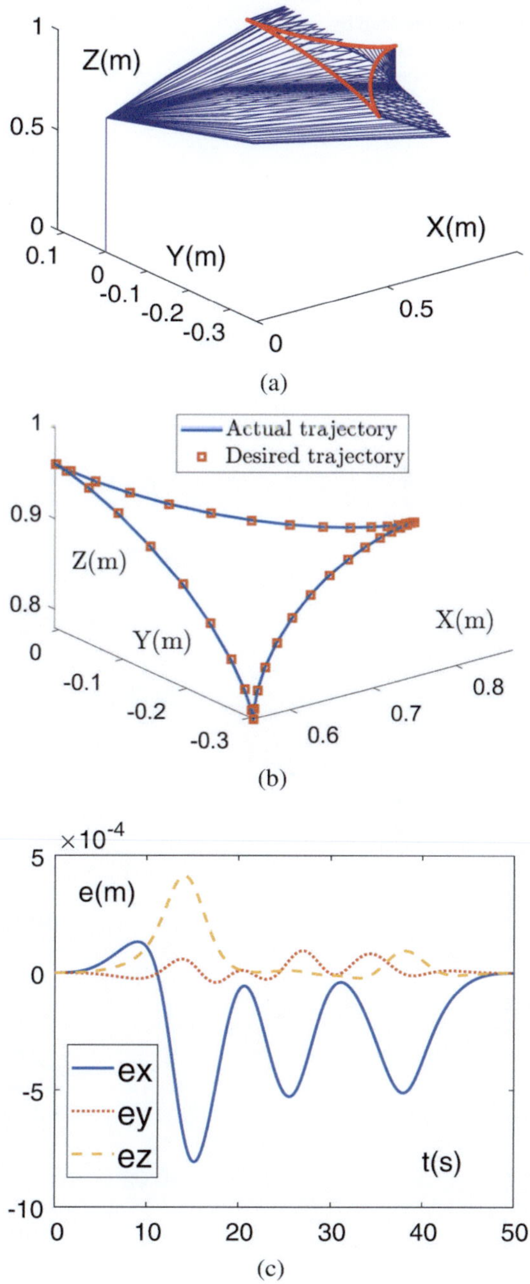

Fig. 3.3 Joint information of the PUMA 560 in nominal situation synthesized by ALCM scheme (3.10) aided with DNN model (3.20). **a** Joint angle. **b** Joint velocity. **c** Joint acceleration

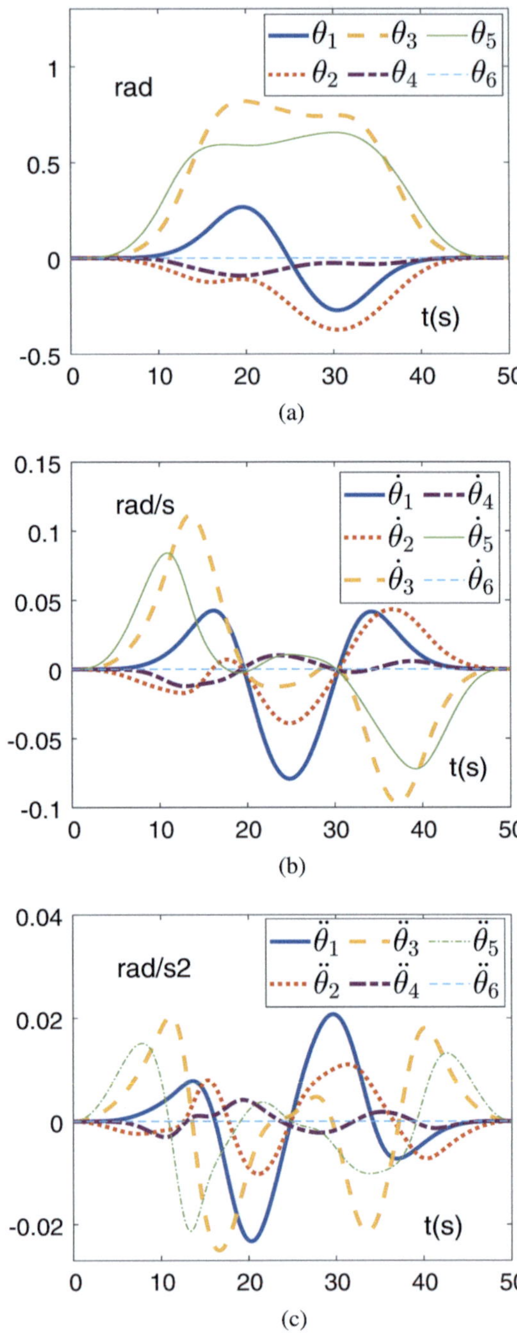

Fig. 3.4 Simulations performed by PUMA 560 with different constant noises by applying NRDNN model (3.31). **a** Position error with noise $z = [0.1]^{6\times1}$. **b** Position error with noise $z = [1]^{6\times1}$. **c** Position error with noise $z = [10]^{6\times1}$

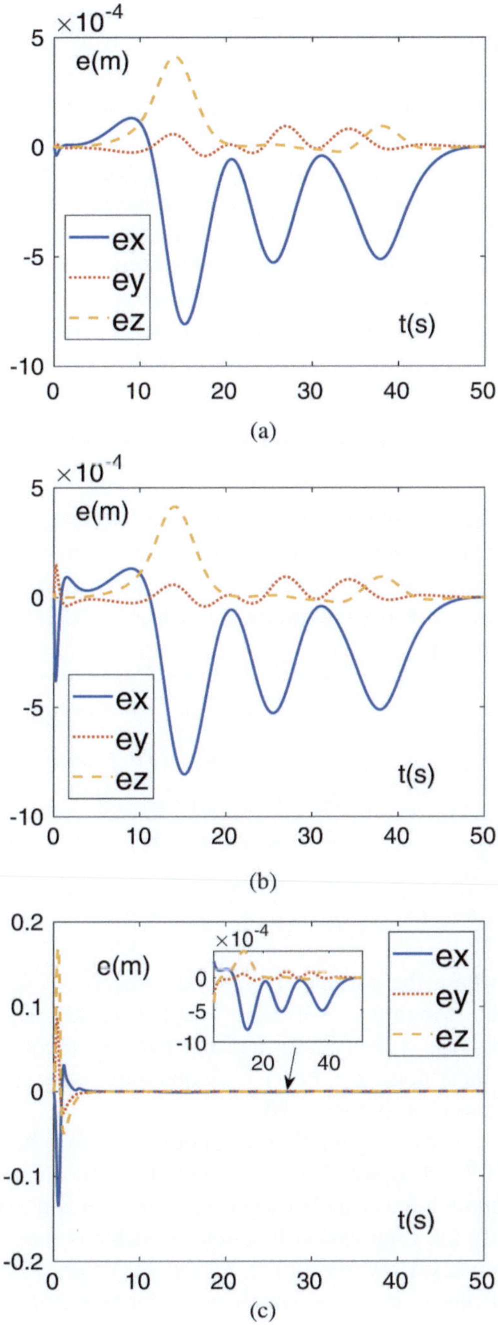

and 3.5. More specifically, the simulation results of NRDNN model (3.31) are depicted in Fig. 3.4a–c. Figure 3.4a–b demonstrates end-effector position errors with magnitudes consistently below the 10^{-3} m throughout the task execution. Figure 3.4c exhibits the transient error behavior, where initial peaks decay asymptotically to zero as the NRDNN model adapts to constant-noise disturbances. For comparison, simulation results of the method in Xie et al. (2019) are shown in Fig. 3.5a–c. The errors in Fig. 3.5a and b are at the order of 10^{-3} m and 10^{-2} m, respectively, while the errors in Fig. 3.5c are huge and change significantly. This indicates finite noise tolerance bounds, beyond which trajectory deviations occur when disturbance magnitudes surpass the model's compensation capability. These comparisons certify that the presented NRDNN model (3.31) performs well in dealing with constant noises.

3.5.5 *Time-Varying Polynomial-Noise Condition*

For illustrating that the presented NRDNN model (3.31) possesses the capability to adapt to time-varying polynomial noises, two models are adopted for simulations under the simultaneous excitation by multiple time-varying polynomial noise. In this part of simulations, a third-order NRDNN model is selected and described as follows:

$$
\begin{aligned}
\xi\dot{\psi} &= P_{\Theta}(-\boldsymbol{\varpi} - J^{\mathrm{T}}\boldsymbol{\iota}) - (x_0\boldsymbol{\upsilon}_0 + x_1\boldsymbol{\upsilon}_1 + x_2\boldsymbol{\upsilon}_2 + x_3\boldsymbol{\upsilon}_3) \\
&\quad - (x_0\xi + 1)\psi - (x_1\boldsymbol{\mu}_1 + x_2\boldsymbol{\mu}_2 + x_3\boldsymbol{\mu}_3) + z, \\
\xi\dot{\boldsymbol{\iota}} &= J\psi - \ddot{\boldsymbol{r}}_{\mathrm{s}} + \dot{J}\dot{\boldsymbol{\theta}} - \rho_0 - \rho_1, \\
\dot{\boldsymbol{\upsilon}}_0 &= \psi - P_{\Theta}(-\boldsymbol{\varpi} - J^{\mathrm{T}}\boldsymbol{\iota}), \\
\dot{\boldsymbol{\upsilon}}_j &= \boldsymbol{\upsilon}_{j-1}, \; j = 1, 2, 3, \\
\dot{\boldsymbol{\mu}}_1 &= \psi, \\
\dot{\boldsymbol{\mu}}_j &= \boldsymbol{\mu}_{j-1}, \; j = 2, 3,
\end{aligned}
\tag{3.45}
$$

where the parameters (i.e., $x_0 = 8$, $x_1 = 24$, $x_2 = 32$, and $x_3 = 16$) are acquired by bringing $n = 3$ and $\gamma = 2$ into (3.43). The compared model is the solver used in Xie et al. (2019), and the analysis incorporates three distinct noise classes: a linear noise $z = [10t]^{6\times 1}$, a quadratic noise $z = [10t + 10t^2]^{6\times 1}$, and a cubic noise $z = [10t + 10t^2 + 10t^3]^{6\times 1}$.

The corresponding simulation results of the two models are shown in Figs. 3.6 and 3.7. As depicted in Fig. 3.6a–c, the position errors of the end-effector with NRDNN model (3.31) are between -1×10^{-3} m and 5×10^{-4} m. These near-zero errors verify the suppression function of NRDNN model (3.31) on time-varying polynomial noises. However, in Fig. 3.7a–c, the errors with the model in Xie et al. (2019) range from -2 m to 2 m, which means the trajectory is completely out of control. The comparative simulations conclusively demonstrate the NRDNN model's (3.31) robustness against the interference of time-varying polynomial noises, with a consistent performance across all tested disturbance profiles.

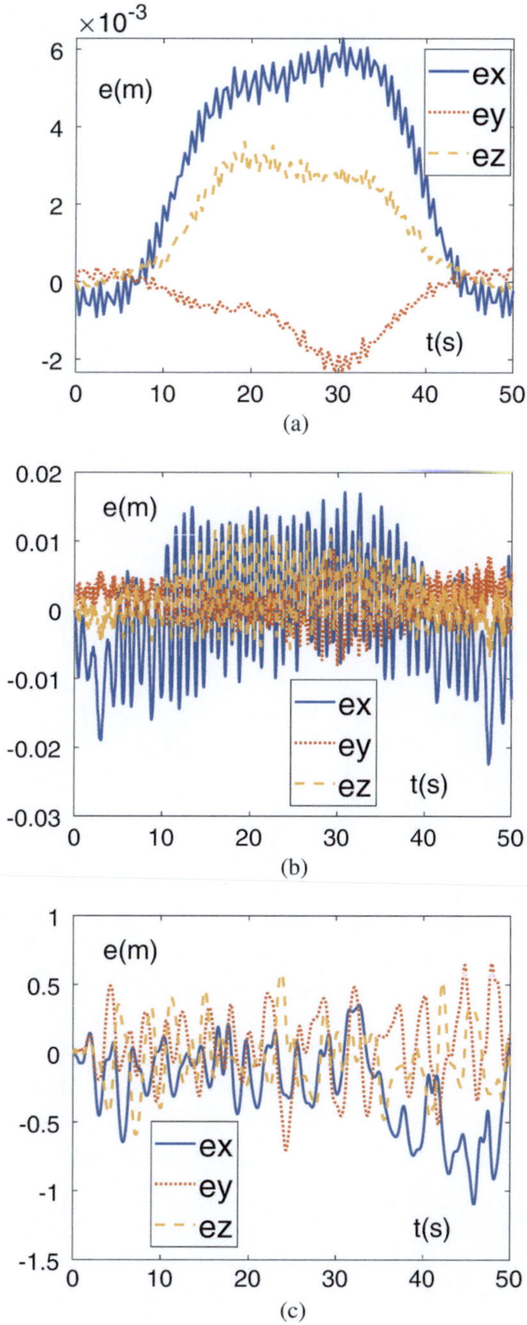

Fig. 3.5 Simulations performed by PUMA 560 with different constant noises by applying the method in Xie et al. (2019). **a** Position error with noise $z = [0.1]^{6\times 1}$. **b** Position error with noise $z = [1]^{6\times 1}$. **c** Position error with noise $z = [10]^{6\times 1}$

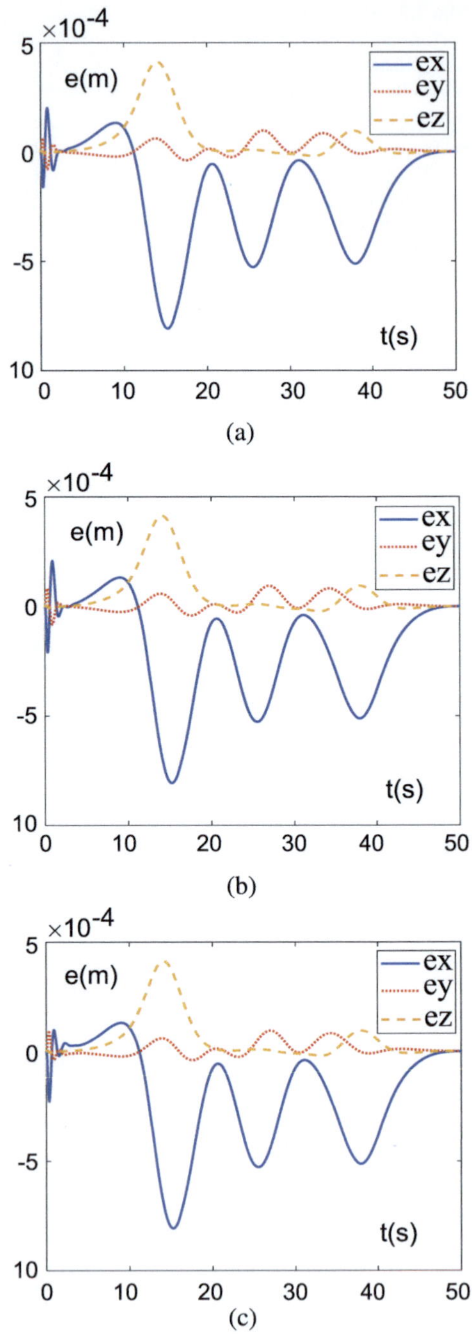

Fig. 3.6 Simulations performed by PUMA 560 with various time-varying polynomial noises by applying NRDNN model (3.31). **a** Position error with noise $z = [10t]^{6\times 1}$. **b** Position error with noise $z = [10t + 10t^2]^{6\times 1}$. **c** Position error with noise $z = [10t + 10t^2 + 10t^3]^{6\times 1}$

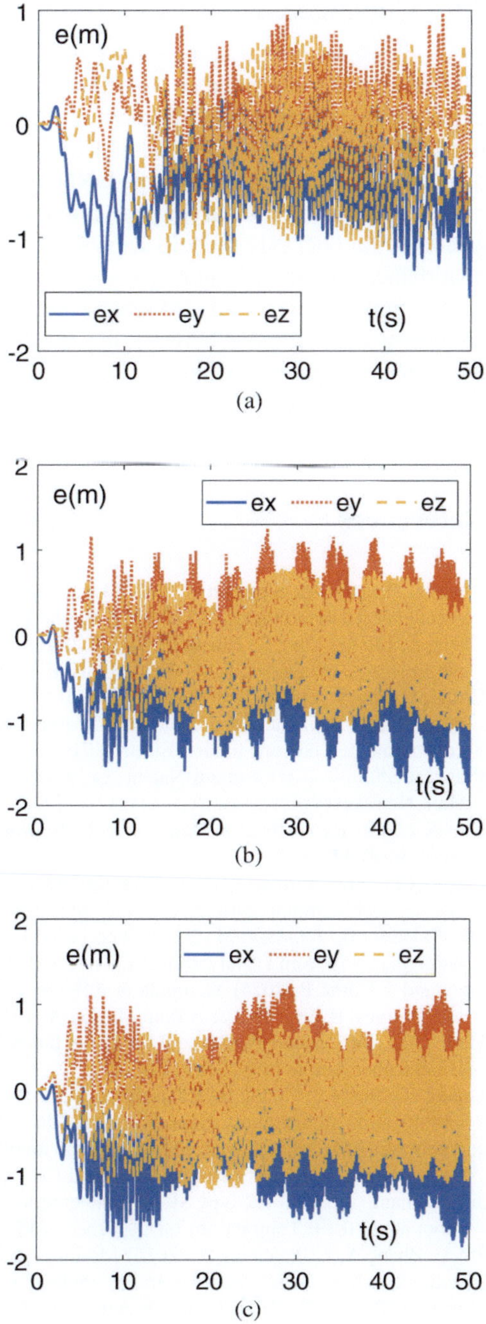

Fig. 3.7 Simulations performed by PUMA 560 with various time-varying polynomial noises by applying the method in Xie et al. (2019). **a** Position error with noise $z = [10t]^{6\times1}$. **b** Position error with noise $z = [10t + 10t^2]^{6\times1}$. **c** Position error with noise $z = [10t + 10t^2 + 10t^3]^{6\times1}$

3.6 Conclusions

This chapter has presented an ALCM control scheme to address the persistent joint-angle drift in redundant manipulators' cyclic operations. Accordingly, a DNN model has been presented to solve the ALCM scheme with the proof of its convergence displayed. Then, by modifying the DNN model, an NRDNN model has been developed to deal with time-varying polynomial noises. Theoretical analyses rigorously establish the NRDNN model's capability to reject time-varying polynomial-noise disturbances while maintaining the system's stability. Experimental results on the PUMA 560 validate the ALCM scheme's efficacy in reducing joint-angle offsets. The comparative simulations between the existing solution and the presented NRDNN model have demonstrated the superiority of the presented model under time-varying polynomial-noise conditions.

References

Bodie K, Tognon M, Siegwart R (2021) Dynamic end effector tracking with an omnidirectional parallel aerial manipulator. IEEE Robot Autom Lett 6(4):8165–8172

Cheng L, Liu W, Yang C, Huang T, Hou Z, Tan M (2018) A neural-network-based controller for piezoelectric-actuated stick-slip devices. IEEE Trans Industr Electron 65(3):2598–2607

Cui Z, Huang Y, Li W, Chiu P, Zheng Li (2024) Noise-resistant adaptive gain recurrent neural network for visual tracking of redundant flexible endoscope robot with time-varying state variable constraints. IEEE Trans Industr Electron 71(3):2694–2704

Franklin GF, Powell JD, Emami-Naeini A (2001) Feedback control of dynamic systems. Prentice Hall, New Jersey

Gao X (2003) Exponential stability of globally projected dynamic systems. IEEE Trans Neural Netw 14(2):426–431

Guo D, Li S, Stanimirović P (2019) Analysis and application of modified ZNN design with robustness against harmonic noise. IEEE Trans Industr Inf 16(7):4627–4638

Ha JS, Driess D, Toussaint M (2022) Deep visual constraints: neural implicit models for manipulation planning from visual input. IEEE Robot Autom Lett 7(4):10857–10864

Haviland J, Corke P (2024) Manipulator differential kinematics: part i: kinematics, velocity, and applications. IEEE Robot & Autom Mag 31(4):149–158

Haviland J, Sünderhauf N, Corke P (2022) A holistic approach to reactive mobile manipulation. IEEE Robot Autom Lett 7(2):3122–3129

Huang K, Zhang S, Zhang R, Husssain A (2020) Novel deep neural network based pattern field classification architectures. Neural Netw 127:82–95

Jarzebowska EM (2008) Advanced programmed motion tracking control of nonholonomic mechanical systems. IEEE Trans Rob 24(6):1315–1328

Jin L, Zhang Y (2015) G2-type SRMPC scheme for synchronous manipulation of two redundant robot arms. IEEE Transactions on Cybernetics 45(2):153–164

Jin L, Zhang Y, Li S, Zhang Y (2016) Modified ZNN for time-varying quadratic programming with inherent tolerance to noises and its application to kinematic redundancy resolution of robot manipulators. IEEE Trans Industr Electron 63(11):6978–6988

Jin L, Xie Z, Liu M, Chen K, Li C, Yang C (2020) Novel joint-drift-free scheme at acceleration level for robotic redundancy resolution with tracking error theoretically eliminated. IEEE/ASME Trans Mechatron 26(1):90–101

Jin L, Wei L, Li S (2023) Gradient-based differential neural-solution to time-dependent nonlinear optimization. IEEE Trans Autom Control 68(1):62–627
Jin L, Su Z, Fu D, Xiao X (2024a) Coevolutionary neural solution for nonconvex optimization with noise tolerance. IEEE Trans Neural Netw Learn Syst 35(12):17571–17581
Jin L, Liu L, Wang X, Shang M, Wang FY (2024b) Physical-informed neural network for MPC-based trajectory tracking of vehicles with noise considered. IEEE Trans Intell Veh 9(3):4493–4503
Li S, Wang H, Rafique MU (2018) A novel recurrent neural network for manipulator control with improved noise tolerance. IEEE Trans Neural Netw Learn Syst 29(5):1908–1918
Li Z, Li C, Li S, Cao X (2019) A fault-tolerant method for motion planning of industrial redundant manipulator. IEEE Trans Industr Inf 16(12):7469–7478
Liang S, Jin L (2024) Distributed k-winner-take-all network under weight-unbalanced topology. IEEE Trans Circuits Syst II Express Briefs 71(2):717–721
Liu N, Qin S (2019) A neurodynamic approach to nonlinear optimization problems with affine equality and convex inequality constraints. Neural Netw 109:147–158
Liu M, He L, Hu B, Li S (2021) Recurrent neural network with noise rejection for cyclic motion generation of robotic manipulators. Neural Netw 138:164–178
Liu M, Liu K, Fu D, Jin L (2023) Data-driven motion-force control for acceleration minimization of robots. In: 2023 13th international conference on information science and technology (ICIST)
Lu H, Jin L, Zhang J, Sun Z, Li S, Zhang Z (2021) New joint-drift-free scheme aided with projected ZNN for motion generation of redundant robot manipulators perturbed by disturbances. IEEE Trans Syst Man Cybern Syst 51(9):5639–5651
Ma Y, Farshidian F, Miki T, Lee J, Hutter M (2022) Combining learning-based locomotion policy with model-based manipulation for legged mobile manipulators. IEEE Robot Autom Lett 7(2):2377–2384
Nainer C, Giusti A (2022) Automatically deployable robust control of modular reconfigurable robot manipulators. IEEE Robot Autom Lett 7(2):5286–5293
Ohara S, Ogata T, Awano H (2021) Binary neural network in robotic manipulation: flexible object manipulation for humanoid robot using partially binarized auto-encoder on FPGA. In: 2021 IEEE/RSJ international conference on intelligent robots and systems, pp 6010–6015
Qi Y, Wang Y, Xiao L, Zhang J (2020) Complex-valued discrete-time neural dynamics for perturbed time-dependent complex quadratic programming with applications. IEEE Trans Neural Netw Learn Syst 31(9):3555–3569
Qin S, Fan D, Wu G, Zhao L (2015) Neural network for constrained nonsmooth optimization using Tikhonov regularization. Neural Netw 63:272–281
Reiter A, Müller A, Gattringer H (2018) On higher order inverse kinematics methods in time-optimal trajectory planning for kinematically redundant manipulators. IEEE Trans Industr Inf 14(4):1681–1690
Safeea M, Bearee R, Neto P (2021) A modified DLS scheme with controlled cyclic solution for inverse kinematics in redundant robots. IEEE Trans Industr Inf 17(12):8014–8023
Shi Y, Jin L, Li S, Li J, Qiang J, Gerontitis DK (2022) Novel discrete-time recurrent neural networks handling discrete-form time-variant multi-augmented Sylvester matrix problems and manipulator application. IEEE Trans Neural Netw Learn Syst 33(2):587–599
Stanimirovic P, Zivkoviäc I, Wei Y (2015) Recurrent neural network for computing the Drazin inverse. IEEE Trans Neural Netw Learn Syst 26(11):2830–2843
Wang Y, Han Y, Tang Z, Liu W, Wang P (2023) Reciprocal of exponential varying-parameter RNN solving repetitive tracking control problems with tolerance of random initial error compounded with noise perturbation. IEEE Trans Industr Inf 19(11):10998–11007
Wei L, Jin L (2024) Collaborative neural solution for time-varying nonconvex optimization with noise rejection. IEEE Trans Emerg Topics Comput Intell 8(4):2935–2948
Xia Y, Feng G, Wang J (2005) A primal-dual neural network for online resolving constrained kinematic redundancy in robot motion control. IEEE Trans Syst Man Cybern B Cybern 35(1):54–64

Xiao L, Liao B, Li S, Chen K (2018) Nonlinear recurrent neural networks for finite-time solution of general time-varying linear matrix equations. Neural Netw 98:102–113
Xiao L, Zhang Z, Li W, Li S (2018) Design, verification and robotic application of a novel recurrent neural network for computing dynamic Sylvester equation. Neural Netw 105:185–196
Xiao L, Zhang Y, Dai J, Chen K, Yang S, Li W, Liao B, Ding L, Li J (2019) A new noise-tolerant and predefined-time ZNN model for time-dependent matrix inversion. Neural Netw 117:124–134
Xie Z, Jin L (2024) A fuzzy neural controller for model-free control of redundant manipulators with unknown kinematic parameters. IEEE Trans Fuzzy Syst 32(3):1589–1601
Xie Z, Du X, Xiao X, Li H (2019) On generalized RMP scheme for redundant robot manipulators aided with dynamic neural networks and nonconvex bound constraints. IEEE Trans Industr Inf 15(9):5172–5181
Yan J, Jin L, Hu B (2024) Data-driven model predictive control for redundant manipulators with unknown model. IEEE Trans Cybern 54(10):5901–5911
Yan J, Su Z, Ma X, Jin L (2025) An obstacle avoidance scheme for manipulators aided by noise-tolerant neural dynamics. IEEE Trans Industr Inf 21(1):703–712
Yang C, Jiang Y, He W, Na J, Li Z, Xu B (2018) Adaptive parameter estimation and control design for robot manipulators with finite-time convergence. IEEE Trans Industr Electron 65(10):8112–8123
Yang W, Li S, Li Z, Luo X (2023) Highly accurate manipulator calibration via extended Kalman filter-incorporated residual neural network. IEEE Trans Industr Inf 19(11):10831–10841
Yilmaz BM, Tatlicioglu E, Savran A, Alci M (2021) Self-adjusting fuzzy logic based control of robot manipulators in task space. IEEE Trans Industr Electron 69(2):1620–1629
Yoo Y, Lee CY, Zhang BT (2021) Multimodal anomaly detection based on deep auto-encoder for object slip perception of mobile manipulation robots. In: 2021 IEEE international conference on robotics and automation, pp 11443–11449
You JM, Kim H, Kim J, Kwon DS (2021) Design and analysis of high-stiffness hyperredundant manipulator with sigma-shaped wire path and rolling joints. IEEE Robot Autom Lett 6(4):7357–7364
Zhang Z, Chen S, Zhu X, Yan Z (2019a) Two hybrid end-effector posture-maintaining and obstacle-limits avoidance schemes for redundant robot manipulators. IEEE Trans Industr Inf 16(2):754–763
Zhang Z, Li S, Zhou X (2019b) Recurrent-neural-network-based velocity-level redundancy resolution for manipulators subject to a joint acceleration limit. IEEE Trans Industr Electron 66(2):3573–3582
Zhang H, Jin H, Liu Z, Liu Y, Zhu Y, Zhao J (2020) Real-time kinematic control for redundant manipulators in a time-varying environment: Multiple-dynamic obstacle avoidance and fast tracking of a moving object. IEEE Trans Industr Inf 16(1):28–41
Zhang Y, Zhang Z (2013) Repetitive motion planning and control of redundant robot manipulators. Springer, New York, NY, USA

Chapter 4
Trajectory-Tracking MPC with Z-Type DNN

4.1 Introduction

In recent decades, robots have garnered significant research attention and have been extensively studied across multiple scientific and engineering domains (Wang et al. 2019; Tian et al. 2024; Yang et al. 2018). Equipped with high dexterity and efficiency (Xu et al. 2020; Cursi et al. 2022; Yan et al. 2022), redundant manipulators that possess more degrees of freedom (DOFs) than required can perform complex tasks while accomplishing additional objectives (Peng et al. 2019; Yang et al. 2019; Karimi and Ahmadi 2022), such as obstacle avoidance and orientation control (Liu et al. 2021; Li et al. 2017; Yang et al. 2020), making them suitable for extensive applications (Rubagotti et al. 2023; Ahmed et al. 2017; Howell et al. 2022). Trajectory tracking for redundant manipulators is achieved through a real-time solution of the inverse kinematics equations to obtain joint information, which serves as the manipulator's input (Van Duijkeren et al. 2018; Yan et al. 2025; Li et al. 2016). High-accuracy trajectory tracking represents a critical capability in manipulator control, especially for applications demanding complex motion patterns (Xie et al. 2024; Wang et al. 2024). Several schemes in the form of optimization problems are designed for redundant manipulators, including the minimum-velocity-norm (MVN) scheme (Zhang et al. 2022; Bhattacharya et al. 2023; Ma et al. 2023) and the minimum-acceleration-norm (MAN) scheme (Jin and Zhang 2015; Xiao and Zhang 2016; Jin et al. 2020), which are separately constructed at velocity level and acceleration level. However, most existing schemes, such as those developed in Zhang et al. (2022), Bhattacharya et al. (2023), Ma et al. (2023), Jin and Zhang (2015), Xiao and Zhang (2016), consider either MVN or MAN separately, with few investigations simultaneously incorporating both approaches. Existing schemes treat joint limits as optimization constraints, in addition to addressing performance index optimization and inverse kinematics solutions (Ma et al. 2023; Jin and Zhang 2015; Xiao and Zhang 2016). However, joint limits at different levels can only be enforced indirectly through combined formulation using conversion techniques (Jin and Zhang 2015; Xiao and Zhang 2016). This approach

M. Liu et al., *Dynamic Neural Networks for Motion Control of Redundant Manipulators*, Intelligent Control and Learning Systems 21,
https://doi.org/10.1007/978-981-96-9144-9_4

requires introducing additional parameters and may shrink the feasible region for redundant manipulators. In acceleration-level schemes, when angle-level constraints are incorporated through conversion techniques, a safety margin must be introduced. This margin design may consequently restrict the feasible range of joint angles. As an advanced control technique, model predictive control (MPC) has been extensively researched and successfully applied across multiple domains (Faulwasser et al. 2017; Garcia et al. 2015; Sleiman et al. 2021). MPC is a control technology that implements system regulation through online optimization (Jin et al. 2024). Different from most optimization control schemes considering the whole time horizon, the optimization operation involved in MPC is repeated at each time instant, based on the receding prediction horizon and the predictive output of the system model, with system constraints considered (Pan and Wang 2008). The resulting optimal solution comprises an input sequence, where the first element is applied as the system's current control input. The repeated online optimization characteristic enables MPC to effectively handle complex scenarios arising during control processes in real time. The work in Faulwasser et al. (2017) develops a model predictive path-following control framework for robots, analyzing cases with specified speed profiles and unconstrained speed, followed by physical implementation. Furthermore, in Garcia et al. (2015), a robust nonlinear MPC scheme is adopted for unmanned aircraft control utilizing adaptive neural networks for predictive model update and parameter estimation. The constructed MPC algorithm incorporates smoothness in the cost function to achieve precise vehicle steering control, demonstrating superior performance compared to both pure-pursuit controllers and conventional MPC approaches without smoothness optimization (Lima et al. 2017). MPC offers significant advantages for modern industrial applications, including effective handling of multi-input–multi-output systems, explicit constraint management, optimal performance seeking, and future state consideration, ensuring its growing adoption in industrial practice. Given MPC's inherent real-time requirements, the online solution of optimization problems becomes crucial, especially for dynamic and time-sensitive applications (Arcari et al. 2023; Gold et al. 2023; Yan et al. 2024; Li et al. 2018; Gulbudak and Gokdag 2021; Jin et al. 2023). In Arcari et al. (2023), a noise-resistant neural algorithm is presented to solve future dynamic nonlinear optimization problems. This approach builds upon the zeroing-type dynamic neural network (ZDNN), variant of recurrent neural network (RNN), and incorporates a matrix inversion estimation technique to enhance robustness. Qi et al. (2023) introduce two globally convergent discrete-time RNNs to solve time-varying complex-valued QP problems with linear constraints. For harmonic noise suppression in redundant manipulators, an enhanced primal-dual neural network is developed to solve the disturbance-affected optimization problem (Li et al. 2018). These outcomes corroborate the performance of RNN in terms of high efficiency, demonstrating high accuracy and global convergence across optimization problems. Given these computational advantages, RNN emerges as a promising online solution method for MPC applications (Pan and Wang 2008). Existing schemes for redundant manipulators typically solve full-horizon optimization problems, whereas this work implements trajectory tracking via MPC.

Due to the inherent nonlinearity of robot kinematics, a linearization procedure is required. The MPC scheme for trajectory tracking of redundant manipulators is constructed by explicitly enforcing joint limits at position, velocity, and acceleration levels while incorporating terminal equality constraints to guarantee stability and optimizing a composite objective function that simultaneously minimizes end-effector position error, joint velocity norm, and joint acceleration norm. For the MPC scheme formulated as a convex QP problem with equality and inequality constraints, a ZDNN model is developed to provide online solutions. The remainder of this chapter is organized as follows: Section 4.2 develops the MPC scheme and derives its equivalent convex QP formulation. The ZDNN model is developed for solving the constructed MPC scheme, with its convergence analyzed in Sect. 4.3. Section 4.4 presents simulation results using a 6-DOF redundant manipulator platform. The contributions of this chapter are given below:

1. This chapter develops an MPC scheme for acceleration-level trajectory tracking of redundant manipulators, formulates it as a convex QP problem with equality and inequality constraints, and presents an effective ZDNN model to compute its online solution.
2. The constructed MPC scheme explicitly enforces joint limits for redundant manipulators, eliminating the need for complex kinematic conversions or the introduction of auxiliary parameters while preserving the fully feasible workspace.
3. The constructed MPC scheme simultaneously minimizes end-effector position error, joint velocity norm, and joint acceleration norm, thereby generating physically reasonable control inputs for redundant manipulators.

4.2 Model Predictive Control

For a redundant manipulator, the end-effector position is kinematically related to its joint configuration through the forward kinematics equation, analogous to (3.1):

$$\boldsymbol{r}(t) = f(\boldsymbol{\theta}(t)), \tag{4.1}$$

with $\boldsymbol{r}(t) \in \mathbb{R}^m$ representing the position of the end-effector; $\boldsymbol{\theta}(t) \in \mathbb{R}^p$ standing for the joint angle, $f(\cdot) : \mathbb{R}^p \to \mathbb{R}^m$ symbolizing a projection operation and $p > m$. The kinematic relationship between the end-effector velocity and the joint velocity follows from (3.2) as

$$\dot{\boldsymbol{r}}(t) = \dot{f}(\boldsymbol{\theta}(t)) = J(\boldsymbol{\theta}(t))\dot{\boldsymbol{\theta}}(t), \tag{4.2}$$

of which $J(\boldsymbol{\theta}(t)) = \partial f(\boldsymbol{\theta}(t))/\partial\boldsymbol{\theta}(t) \in \mathbb{R}^{m\times p}$ is the Jacobian matrix of the redundant manipulator.

Before applying MPC to the trajectory-tracking task, the joint velocity is treated as the controller input, i.e.:

$$\boldsymbol{u}(t) = \dot{\boldsymbol{\theta}}(t). \tag{4.3}$$

The discrete form is formulated as follows:

$$\boldsymbol{\theta}(k+1) = \boldsymbol{\theta}(k) + \tau\,\boldsymbol{u}(k), \tag{4.4}$$

where τ is the sampling interval, and k represents the updating index at time instant $t = k\tau$. Besides, for input $\boldsymbol{u}(k)$, the following formula holds on:

$$\boldsymbol{u}(k) = \Delta\boldsymbol{u}(k) + \boldsymbol{u}(k-1), \tag{4.5}$$

of which $\Delta\boldsymbol{u}(k)$ signifies the input increment. Further, one can get

$$\begin{aligned}\boldsymbol{u}(k+i|k) =& \Delta\boldsymbol{u}(k+i|k) + \Delta\boldsymbol{u}(k+i-1|k) + \cdots + \Delta\boldsymbol{u}(k|k) + \boldsymbol{u}(k-1),\\ i =& 1, 2, \ldots, N_u,\end{aligned} \tag{4.6}$$

where N_u denotes the control horizon; $\boldsymbol{u}(k+i|k)$ and $\Delta\boldsymbol{u}(k+i|k)$ are the future input and future input increment corresponding to sampling instant k, respectively. Differentiating (4.3) with respect to time yields $\dot{\boldsymbol{u}}(t) = \ddot{\boldsymbol{\theta}}(t)$. Applying the backward Euler discretization then gives

$$\frac{\Delta\boldsymbol{u}(k)}{\tau} \approx \ddot{\boldsymbol{\theta}}(k).$$

Combine (4.4) and (4.5) to obtain the following expression of predicted state $\boldsymbol{\theta}(k+i|k)$ at sampling instant k:

$$\begin{aligned}\boldsymbol{\theta}(k+i|k) =& \boldsymbol{\theta}(k) + i\tau\,\Delta\boldsymbol{u}(k|k) + (i-1)\tau\,\Delta\boldsymbol{u}(k+1|k)\\ &+ \cdots + \tau\,\Delta\boldsymbol{u}(k+i-1|k) + i\tau\,\boldsymbol{u}(k-1),\\ i =& 1, 2, \ldots, N,\end{aligned} \tag{4.7}$$

where N denotes the predictive horizon and satisfies $N \geq N_u$. It is assumed that $\Delta\boldsymbol{u}(k+i|k) = 0$ for $i \geq N_u$, which means that $\boldsymbol{u}(k+i|k) = \boldsymbol{u}(k+N_u-1|k)$ for $i \geq N_u$. For predicted output $f(\boldsymbol{\theta}(k+i|k))$ at sampling instant k, the linearization operation is needed:

$$f(\boldsymbol{\theta}(k+i|k)) = f(\boldsymbol{\theta}(k)) + J(\boldsymbol{\theta}(k))[\boldsymbol{\theta}(k+i|k) - \boldsymbol{\theta}(k)],\, i = 1, 2, \ldots, N. \tag{4.8}$$

The linearization corresponds to a first-order Taylor approximation, with higher-order terms omitted.

For convenience, the following definitions are given:

$$
\begin{aligned}
\widetilde{\theta} &= [\boldsymbol{\theta}(k+1|k), \boldsymbol{\theta}(k+2|k), \ldots, \boldsymbol{\theta}(k+N|k)]^{\mathrm{T}} \in \mathbb{R}^{N\times p},\\
\widetilde{u} &= [\boldsymbol{u}(k|k), \boldsymbol{u}(k+1|k), \ldots, \boldsymbol{u}(k+N_u-1|k)]^{\mathrm{T}} \in \mathbb{R}^{N_u\times p},\\
\widetilde{\Delta u} &= [\Delta\boldsymbol{u}(k|k), \Delta\boldsymbol{u}(k+1|k), \ldots, \Delta\boldsymbol{u}(k+N_u-1|k)]^{\mathrm{T}} \in \mathbb{R}^{N_u\times p},\\
\widetilde{r_{\mathrm{d}}} &= [\boldsymbol{r}_{\mathrm{d}}(k+1), \boldsymbol{r}_{\mathrm{d}}(k+2), \ldots, \boldsymbol{r}_{\mathrm{d}}(k+N)]^{\mathrm{T}} \in \mathbb{R}^{N\times m},\\
\widetilde{f(\theta)} &= [f(\boldsymbol{\theta}(k+1|k)), f(\boldsymbol{\theta}(k+2|k)), \ldots, f(\boldsymbol{\theta}(k+N|k))]^{\mathrm{T}} \in \mathbb{R}^{N\times m},\\
\widetilde{\theta^-} &= [\boldsymbol{\theta}^-, \boldsymbol{\theta}^-, \ldots, \boldsymbol{\theta}^-]^{\mathrm{T}} \in \mathbb{R}^{N\times p}, \widetilde{\theta^+} = [\boldsymbol{\theta}^+, \boldsymbol{\theta}^+, \ldots, \boldsymbol{\theta}^+]^{\mathrm{T}} \in \mathbb{R}^{N\times p},\\
\widetilde{\dot{\theta}^-} &= [\dot{\boldsymbol{\theta}}^-, \dot{\boldsymbol{\theta}}^-, \ldots, \dot{\boldsymbol{\theta}}^-]^{\mathrm{T}} \in \mathbb{R}^{N_u\times p}, \widetilde{\dot{\theta}^+} = [\dot{\boldsymbol{\theta}}^+, \dot{\boldsymbol{\theta}}^+, \ldots, \dot{\boldsymbol{\theta}}^+]^{\mathrm{T}} \in \mathbb{R}^{N_u\times p},\\
\widetilde{\ddot{\theta}^-} &= [\ddot{\boldsymbol{\theta}}^-, \ddot{\boldsymbol{\theta}}^-, \ldots, \ddot{\boldsymbol{\theta}}^-]^{\mathrm{T}} \in \mathbb{R}^{N_u\times p}, \widetilde{\ddot{\theta}^+} = [\ddot{\boldsymbol{\theta}}^+, \ddot{\boldsymbol{\theta}}^+, \ldots, \ddot{\boldsymbol{\theta}}^+]^{\mathrm{T}} \in \mathbb{R}^{N_u\times p},
\end{aligned}
$$

where $\boldsymbol{r}_{\mathrm{d}}(k+i) \in \mathbb{R}^m$ for $i = 1, 2, \ldots, N$ denotes the reference trajectory to be tracked by the end-effector of the redundant manipulator; $\boldsymbol{\theta}^-$ and $\boldsymbol{\theta}^+$ symbolize the lower and upper bounds of joint angles of the redundant manipulator, respectively; $\dot{\boldsymbol{\theta}}^-$ and $\dot{\boldsymbol{\theta}}^+$ signify the lower and upper bounds of joint velocities of the redundant manipulator, respectively; $\ddot{\boldsymbol{\theta}}^-$ and $\ddot{\boldsymbol{\theta}}^+$ denote the lower and upper bounds of joint accelerations of the redundant manipulator, respectively; and the superscript ^{T} represents the transposition operation. On this basis, (4.6)–(4.8) can be reformulated as the following compact forms:

$$
\widetilde{u} = I_u + E\widetilde{\Delta u},\ \widetilde{\theta} = I_\theta + A\widetilde{\Delta u} + B,\ \widetilde{f(\theta)} = I_f + A\widetilde{\Delta u}J^{\mathrm{T}} + BJ^{\mathrm{T}},
$$

where

$$
I_\theta = \begin{bmatrix} \boldsymbol{\theta}^{\mathrm{T}}(k) \\ \boldsymbol{\theta}^{\mathrm{T}}(k) \\ \vdots \\ \boldsymbol{\theta}^{\mathrm{T}}(k) \end{bmatrix} \in \mathbb{R}^{N\times p}, I_f = \begin{bmatrix} f^{\mathrm{T}}(\boldsymbol{\theta}(k)) \\ f^{\mathrm{T}}(\boldsymbol{\theta}(k)) \\ \vdots \\ f^{\mathrm{T}}(\boldsymbol{\theta}(k)) \end{bmatrix} \in \mathbb{R}^{N\times m}, I_u = \begin{bmatrix} \boldsymbol{u}^{\mathrm{T}}(k-1) \\ \boldsymbol{u}^{\mathrm{T}}(k-1) \\ \vdots \\ \boldsymbol{u}^{\mathrm{T}}(k-1) \end{bmatrix} \in \mathbb{R}^{N_u\times p},
$$

$$
A = \begin{bmatrix} \tau & 0 & \cdots & 0 \\ 2\tau & \tau & \cdots & 0 \\ \vdots & \vdots & \ddots & \vdots \\ N_u\tau & (N_u-1)\tau & \cdots & \tau \\ (N_u+1)\tau & N_u\tau & \cdots & 2\tau \\ \vdots & \vdots & \ddots & \vdots \\ N\tau & (N-1)\tau & \cdots & (N-N_u+1)\tau \end{bmatrix} \in \mathbb{R}^{N\times N_u}, J = J(\boldsymbol{\theta}(k)),
$$

$$B = \begin{bmatrix} \tau \boldsymbol{u}^{\mathrm{T}}(k-1) \\ 2\tau \boldsymbol{u}^{\mathrm{T}}(k-1) \\ \vdots \\ N_u \tau \boldsymbol{u}^{\mathrm{T}}(k-1) \\ \vdots \\ N\tau \boldsymbol{u}^{\mathrm{T}}(k-1) \end{bmatrix} \in \mathbb{R}^{N\times p},\ E = \begin{bmatrix} 1 & 0 & \cdots & 0 \\ 1 & 1 & \cdots & 0 \\ \vdots & \vdots & \ddots & \vdots \\ 1 & 1 & \cdots & 1 \end{bmatrix} \in \mathbb{R}^{N_u\times N_u}.$$

Let $\boldsymbol{\epsilon} = \mathrm{vec}(\widetilde{f(\theta)}) - \mathrm{vec}(\widetilde{r}_{\mathrm{d}}) = \mathrm{vec}(I_f) + (J \otimes A)\cdot \mathrm{vec}(\widetilde{\Delta u}) + \mathrm{vec}(BJ^{\mathrm{T}}) - \mathrm{vec}(\widetilde{r}_{\mathrm{d}})$ signify the position error and $\mathrm{vec}(\widetilde{u}) = \mathrm{vec}(I_u) + (I \otimes E)\cdot \mathrm{vec}(\widetilde{\Delta u})$, with $\otimes$ denoting the Kronecker product, $\mathrm{vec}(\cdot)$ signifying the vectorization operation, and I being an identity matrix. Therefore, the MPC scheme for the trajectory tracking of redundant manipulators is formulated as

$$\min_{\widetilde{\Delta u}} \|\boldsymbol{\epsilon}\|_Q^2 + \|\mathrm{vec}(\widetilde{u})\|_{R_1}^2 + \|\mathrm{vec}(\widetilde{\Delta u})\|_{R_2}^2, \tag{4.9a}$$

$$\text{s.t.}\ \widetilde{\theta^-} \le \widetilde{\theta} \le \widetilde{\theta^+}, \tag{4.9b}$$

$$\widetilde{\dot{\theta}^-} \le \widetilde{u} \le \widetilde{\dot{\theta}^+}, \tag{4.9c}$$

$$\widetilde{\ddot{\theta}^-} \le \frac{\widetilde{\Delta u}}{\tau} \le \widetilde{\ddot{\theta}^+}, \tag{4.9d}$$

$$f(\boldsymbol{\theta}(k+N|k)) = \boldsymbol{r}_{\mathrm{d}}(k+N), \tag{4.9e}$$

where positive definite and symmetrical matrices $Q \in \mathbb{R}^{mN\times mN}$, $R_1 \in \mathbb{R}^{pN_u\times pN_u}$, and $R_2 \in \mathbb{R}^{pN_u\times pN_u}$ are separately the weight matrices for position error, input, and input increment; $\|\cdot\|_Q$, $\|\cdot\|_{R_1}$, and $\|\cdot\|_{R_2}$ symbolize Euclidean norms of vectors, i.e., $\|\boldsymbol{\epsilon}\|_Q = \sqrt{\boldsymbol{\epsilon}^{\mathrm{T}} Q \boldsymbol{\epsilon}}$. Notably, (4.9b)–(4.9d) represent the state constraint, input constraint, and input increment constraint, respectively, while (4.9e) enforces the terminal equality constraint that ensures the stability of the constructed MPC scheme. Further, the cost function (4.9a) can be simplified as

$$\Psi = \frac{1}{2}\mathrm{vec}(\widetilde{\Delta u})^{\mathrm{T}} \cdot W \cdot \mathrm{vec}(\widetilde{\Delta u}) + \boldsymbol{q}^{\mathrm{T}} \cdot \mathrm{vec}(\widetilde{\Delta u}),$$

with

$$W = 2[(J \otimes A)^{\mathrm{T}} Q (J \otimes A) + (I \otimes E)^{\mathrm{T}} R_1 (I \otimes E) + R_2],$$

$$\boldsymbol{q} = 2(J \otimes A)^{\mathrm{T}} Q[\mathrm{vec}(I_f) + \mathrm{vec}(BJ^{\mathrm{T}}) - \mathrm{vec}(\widetilde{r}_{\mathrm{d}})] + 2(I \otimes E)^{\mathrm{T}} R_1 \cdot \mathrm{vec}(I_u).$$

Note that constant items independent of $\widetilde{\Delta u}$ in the original cost function (4.9a) are omitted. Besides, constraints (4.9b)–(4.9d) can be combined in the following form:

$$\begin{bmatrix} A \\ -A \\ E \\ -E \\ I \\ -I \end{bmatrix} \widetilde{\Delta u} \leq \begin{bmatrix} \widetilde{\theta^+} - I_\theta - B \\ -\widetilde{\theta^-} + I_\theta + B \\ \widetilde{\dot{\theta}^+} - I_u \\ -\widetilde{\dot{\theta}^-} + I_u \\ \tau \widetilde{\ddot{\theta}^+} \\ -\tau \widetilde{\ddot{\theta}^-} \end{bmatrix}.$$

Let

$$M' = \begin{bmatrix} A \\ -A \\ E \\ -E \\ I \\ -I \end{bmatrix}, \ h' = \begin{bmatrix} \widetilde{\theta^+} - I_\theta - B \\ -\widetilde{\theta^-} + I_\theta + B \\ \widetilde{\dot{\theta}^+} - I_u \\ -\widetilde{\dot{\theta}^-} + I_u \\ \tau \widetilde{\ddot{\theta}^+} \\ -\tau \widetilde{\ddot{\theta}^-} \end{bmatrix},$$

and thus constraints (4.9b)–(4.9d) can be together converted into $(I \otimes M')\text{vec}(\widetilde{\Delta u}) \leq \text{vec}(h')$. Defining $\boldsymbol{c} = [0, 0, \ldots, 0, 1] \in \mathbb{R}^{1\times N}$ and combing (4.7) and (4.8), one can get

$$f^{\mathrm{T}}(\boldsymbol{\theta}(k+N|k)) = f^{\mathrm{T}}(\boldsymbol{\theta}(k)) + [\boldsymbol{c}A\widetilde{\Delta u} + N\tau\, \boldsymbol{u}^{\mathrm{T}}(k-1)]J^{\mathrm{T}}.$$

Therefore, constraint (4.9e) is transformed into

$$(J \otimes (\boldsymbol{c}A))\text{vec}(\widetilde{\Delta u}) = \boldsymbol{r}_{\mathrm{d}}(k+N) - f(\boldsymbol{\theta}(k)) - N\tau\, J\boldsymbol{u}(k-1).$$

Let $\boldsymbol{x} = \text{vec}(\widetilde{\Delta u})$, $M = I \otimes M'$, $\boldsymbol{h} = \text{vec}(h')$, $F = (J \otimes (\boldsymbol{c}A))$, $\boldsymbol{g} = \boldsymbol{r}_{\mathrm{d}}(k+N) - f(\boldsymbol{\theta}(k)) - N\tau\, J\boldsymbol{u}(k-1)$, and the constructed MPC scheme (4.9) can be expressed as the following QP problem:

$$\min_{\boldsymbol{x}} \frac{1}{2}\boldsymbol{x}^{\mathrm{T}} W \boldsymbol{x} + \boldsymbol{q}^{\mathrm{T}}\boldsymbol{x}, \tag{4.10a}$$

$$\text{s.t. } F\boldsymbol{x} = \boldsymbol{g}, \tag{4.10b}$$

$$M\boldsymbol{x} \leq \boldsymbol{h}, \tag{4.10c}$$

where $\boldsymbol{x} \in \mathbb{R}^{pN_u}$, $W \in \mathbb{R}^{pN_u \times pN_u}$, $\boldsymbol{q} \in \mathbb{R}^{pN_u}$, $F \in \mathbb{R}^{m\times pN_u}$, $\boldsymbol{g} \in \mathbb{R}^m$, $M \in \mathbb{R}^{p(2N+4N_u)\times pN_u}$, and $\boldsymbol{h} \in \mathbb{R}^{p(2N+4N_u)}$. Through solving (4.10) at sampling instant k, the future input increment $\widetilde{\Delta u}$ can be obtained, with its first element $\Delta\boldsymbol{u}(k)$ being applied as the control input increment over the time interval $[k\tau, (k+1)\tau)$. At each subsequent sampling instant $k+1$, the above procedure is repeated until the trajectory-tracking task ends.

4.3 ZDNN Model and Theoretical Analysis

This section presents a ZDNN approach for solving QP problem (4.10), achieving both high computational effectiveness and numerical precision. Additionally, theoretical analysis demonstrates the stability and global convergence of the constructed ZDNN model.

As the QP problem (4.10) is strictly convex, it admits a unique optimal solution characterized by the following Karush–Kuhn–Tucker (KKT) conditions:

$$\begin{cases} W\boldsymbol{x}^* + \boldsymbol{q} + F^{\mathrm{T}}\boldsymbol{\lambda}^* + M^{\mathrm{T}}\boldsymbol{\mu}^* = \mathbf{0}, & (4.11a) \\ F\boldsymbol{x}^* - \boldsymbol{g} = \mathbf{0}, & (4.11b) \\ \boldsymbol{\mu}^* \geq 0, \boldsymbol{h} - M\boldsymbol{x}^* \geq \mathbf{0}, & (4.11c) \\ \boldsymbol{\mu}^{*\mathrm{T}}(\boldsymbol{h} - M\boldsymbol{x}^*) = 0, & (4.11d) \end{cases}$$

of which $\boldsymbol{x}^*$ is the optimal solution to QP problem (4.10), and $\boldsymbol{\lambda}^* \in \mathbb{R}^m$ and $\boldsymbol{\mu}^* \in \mathbb{R}^{p(2N+4N_u)}$ separately symbolize the Lagrangian multipliers corresponding to constraints (4.10b) and (4.10c). For (4.11c) and (4.11d), one can replace them with an NCP function satisfying $\varphi(\boldsymbol{a}, \boldsymbol{b}) = \mathbf{0} \Leftrightarrow \boldsymbol{a} \geq \mathbf{0}, \boldsymbol{b} \geq \mathbf{0}, \boldsymbol{a} \circ \boldsymbol{b} = \mathbf{0}$, where $\boldsymbol{a} \in \mathbb{R}^n$, $\boldsymbol{b} \in \mathbb{R}^n$, and symbol $\circ$ represents the Hadamard product. The following NCP function is considered (Chen et al. 2000):

$$\varphi_\eta(\boldsymbol{a}, \boldsymbol{b}) = \eta\,(\boldsymbol{a} + \boldsymbol{b} - \sqrt{\boldsymbol{a} \circ \boldsymbol{a} + \boldsymbol{b} \circ \boldsymbol{b}}) + (1 - \eta)\boldsymbol{a}_+ \circ \boldsymbol{b}_+,$$

where $\eta \in (0, 1)$ is an arbitrarily fixed parameter, $\boldsymbol{a}_+ = \max\{\mathbf{0}, \boldsymbol{a}\}$, and $\boldsymbol{b}_+ = \max\{\mathbf{0}, \boldsymbol{b}\}$. Subsequently, KKT condition (4.11) is rewritten with the help of the NCP function:

$$\begin{cases} W\boldsymbol{x}^* + \boldsymbol{q} + F^{\mathrm{T}}\boldsymbol{\lambda}^* + M^{\mathrm{T}}\boldsymbol{\mu}^* = \mathbf{0}, \\ F\boldsymbol{x}^* - \boldsymbol{g} = \mathbf{0}, \\ \varphi_\eta(\boldsymbol{h} - M\boldsymbol{x}^*, \boldsymbol{\mu}^*) = \mathbf{0}. \end{cases}$$

On the strength of the above analyses, solving QP problem (4.10) is equivalent to solving the following equation:

$$\begin{bmatrix} W & F^{\mathrm{T}} & M^{\mathrm{T}} \\ F & 0 & 0 \\ -\eta M & 0 & \eta I \end{bmatrix} \begin{bmatrix} \boldsymbol{x} \\ \boldsymbol{\lambda} \\ \boldsymbol{\mu} \end{bmatrix} + \begin{bmatrix} \boldsymbol{q} \\ -\boldsymbol{g} \\ \eta \boldsymbol{h} + \boldsymbol{l} \end{bmatrix} = \mathbf{0}, \tag{4.12}$$

with $\boldsymbol{l} = (1 - \eta)(\boldsymbol{h} - M\boldsymbol{x})_+ \circ \boldsymbol{\mu}_+ - \eta\sqrt{(\boldsymbol{h} - M\boldsymbol{x}) \circ (\boldsymbol{h} - M\boldsymbol{x}) + \boldsymbol{\mu} \circ \boldsymbol{\mu}}$. Let

$$P = \begin{bmatrix} W & F^{\mathrm{T}} & M^{\mathrm{T}} \\ F & 0 & 0 \\ -\eta M & 0 & \eta I \end{bmatrix}, \boldsymbol{y} = \begin{bmatrix} \boldsymbol{x} \\ \boldsymbol{\lambda} \\ \boldsymbol{\mu} \end{bmatrix}, \boldsymbol{v} = \begin{bmatrix} \boldsymbol{q} \\ -\boldsymbol{g} \\ \eta \boldsymbol{h} + \boldsymbol{l} \end{bmatrix},$$

and (4.12) can be reformulated as

$$P\boldsymbol{y} + \boldsymbol{v} = \boldsymbol{0}. \tag{4.13}$$

As a result, solving QP problem (4.10) is transformed to solve Eq. (4.13).

The subsequent focuses on developing a ZDNN model to solve Eq. (4.13) with stability and convergence guaranteed. First, an error function $\boldsymbol{e} = P\boldsymbol{y} + \boldsymbol{v}$ is defined to monitor the solving process. As $\boldsymbol{e}$ converges to zero, the numerical solution $\boldsymbol{y}$ asymptotically approaches the theoretical solution $\boldsymbol{y}^*$. With this aim, the formula $\dot{\boldsymbol{e}} = -\gamma \Upsilon(\boldsymbol{e})$ is employed to facilitate error zeroing, where $\gamma > 0 \in \mathbb{R}$ is a scaling index controlling the convergence rate of $\boldsymbol{e}$ and $\Upsilon(\cdot)$ signifies an activation function expediting the convergence of $\boldsymbol{e}$, with odd and monotonically increasing properties. The ZDNN model for solving (4.13) is then formulated as follows:

$$S\dot{\boldsymbol{y}} = -\gamma \Upsilon(P\boldsymbol{y} + \boldsymbol{v}), \tag{4.14}$$

where $\Upsilon_i(e_i) = \exp(\kappa e_i) - \exp(-\kappa e_i)$,

$$S = \begin{bmatrix} W & F^{\mathrm{T}} & M^{\mathrm{T}} \\ F & 0 & 0 \\ -\eta M + \eta D_1 M - (1-\eta) D_3 M & 0 & \eta I - \eta D_2 + (1-\eta) D_4 \end{bmatrix},$$

and

$$\begin{aligned}
&\kappa = 3,\ D_1 = \Lambda((\boldsymbol{h} - M\boldsymbol{x}) \oslash \boldsymbol{w}),\ D_2 = \Lambda(\boldsymbol{\mu} \oslash \boldsymbol{w}),\\
&D_3 = \Lambda(\partial(\boldsymbol{h} - M\boldsymbol{x})_+ \circ \boldsymbol{\mu}_+),\ D_4 = \Lambda((\boldsymbol{h} - M\boldsymbol{x})_+ \circ \partial\boldsymbol{\mu}_+),\\
&\boldsymbol{w} = \sqrt{(\boldsymbol{h} - M\boldsymbol{x}) \circ (\boldsymbol{h} - M\boldsymbol{x}) + \boldsymbol{\mu} \circ \boldsymbol{\mu}},\\
&\partial z_+ = \begin{cases} 1, & z > 0, \\ [0, 1], & z = 0, \\ 0, & z < 0, \end{cases}
\end{aligned}$$

with e_i standing for the ith element of vector $\boldsymbol{e}$, and $\Lambda(\cdot)$ and $\oslash$ denoting the diagonal matrix and the Hadamard division, respectively. Ulteriorly, the stability and convergence of ZDNN model (4.14) for solving (4.13) are guaranteed by the following theorem.

Theorem 4.1 *The actual solution* $\boldsymbol{y}$ *calculated by ZDNN model (4.14) converges to the theoretical solution* $\boldsymbol{y}^*$ *to (4.13).*

Proof 4.1 Aiming at the design formula, a Lyapunov function is designed as $\chi = \boldsymbol{e}^{\mathrm{T}}\boldsymbol{e}/2$ that is evidently positive definite. Taking its time derivative generates

$$\dot{\chi} = \boldsymbol{e}^{\mathrm{T}}\dot{\boldsymbol{e}} = -\gamma \boldsymbol{e}^{\mathrm{T}}\Upsilon(\boldsymbol{e}).$$

In view of the odd and monotonically increasing properties of the activation function $\Upsilon(\cdot)$, one can infer that $\dot{\chi} \le 0$. Through Lyapunov stability analysis, we rigorously establish the asymptotic convergence of $\boldsymbol{e}$ to zero over time, demonstrating that $\boldsymbol{e}^* = \mathbf{0}$ constitutes a stable equilibrium point of the dynamical system. What is noteworthy is that the equilibrium point $\boldsymbol{e}^*$ corresponds to the theoretical solution $\boldsymbol{y}^*$ to (4.13). Consequently, the conclusion is concluded that the actual solution $\boldsymbol{y}$ solved by ZDNN model (4.14) converges to the theoretical solution $\boldsymbol{y}^*$ to (4.13). The proof is completed. □

4.4 Simulation and Experiment Results Based on Manipulator

This section implements the constructed MPC scheme (4.9), solved via the ZDNN model (4.14), for trajectory tracking of a redundant manipulator to demonstrate its efficiency and robustness. Additionally, several existing methods for addressing the trajectory-tracking problem of redundant manipulators are introduced.

The MAN scheme considering joint limits and error feedbacks without considering the elimination of joint-angle drift phenomenon is as follows (Jin and Zhang 2015):

$$
\begin{aligned}
&\min_{\ddot{\theta}} \frac{1}{2}\ddot{\boldsymbol{\theta}}^{\mathrm{T}}\ddot{\boldsymbol{\theta}}, && \text{(4.15a)}\\
&\text{s.t. } J(\boldsymbol{\theta})\ddot{\boldsymbol{\theta}} = \ddot{\boldsymbol{r}}_{\mathrm{d}} - \dot{J}(\boldsymbol{\theta})\dot{\boldsymbol{\theta}} + \alpha_{\mathrm{v}}\boldsymbol{\epsilon}_{\mathrm{v}} + \alpha_{\mathrm{p}}\boldsymbol{\epsilon}_{\mathrm{p}}, && \text{(4.15b)}\\
&\quad\ \boldsymbol{\theta}^- \leq \boldsymbol{\theta} \leq \boldsymbol{\theta}^+, && \text{(4.15c)}\\
&\quad\ \dot{\boldsymbol{\theta}}^- \leq \dot{\boldsymbol{\theta}} \leq \dot{\boldsymbol{\theta}}^+, && \text{(4.15d)}\\
&\quad\ \ddot{\boldsymbol{\theta}}^- \leq \ddot{\boldsymbol{\theta}} \leq \ddot{\boldsymbol{\theta}}^+, && \text{(4.15e)}
\end{aligned}
$$

of which $\boldsymbol{\epsilon}_{\mathrm{v}} = \dot{\boldsymbol{r}}_{\mathrm{d}} - J(\boldsymbol{\theta})\dot{\boldsymbol{\theta}}$ and $\boldsymbol{\epsilon}_{\mathrm{p}} = \boldsymbol{r}_{\mathrm{d}} - f(\boldsymbol{\theta})$ separately stand for velocity error and position error of the end-effector of the redundant manipulator with $\boldsymbol{r}_{\mathrm{d}}$ being the desired trajectory, and positive α_{v} and α_{p} denote the corresponding feedback gains. The aforementioned scheme is addressed using the piecewise-linear projection equation-based neural network (PLPENN) model introduced in Jin and Zhang (2015). Additionally, the MVN scheme is employed as a comparison scheme:

$$
\begin{aligned}
&\min_{\dot{\theta}} \frac{1}{2}\dot{\boldsymbol{\theta}}^{\mathrm{T}}\dot{\boldsymbol{\theta}}, && \text{(4.16a)}\\
&\text{s.t. } J(\boldsymbol{\theta})\dot{\boldsymbol{\theta}} = \dot{\boldsymbol{r}}_{\mathrm{d}}, && \text{(4.16b)}\\
&\quad\ \dot{\boldsymbol{\theta}}^- \leq \dot{\boldsymbol{\theta}} \leq \dot{\boldsymbol{\theta}}^+, && \text{(4.16c)}
\end{aligned}
$$

which does not consider the error feedback and is solved by a discrete perturbation-immunity neural network (DPINN) model (Ma et al. 2023). It is noted that in both schemes mentioned above, joint limits are not handled directly. Instead, they are managed indirectly by incorporating them into the joint acceleration or joint velocity levels. For the presented MPC scheme (4.9), a simplified dual neural network (SDNN) model introduced in Pan and Wang (2008) is employed for comparison.

4.4.1 Without Sudden External Interference

Initially, the feasibility, effectiveness, and accuracy of the presented MPC scheme (4.9) solved by the ZDNN model (4.14) are demonstrated using the PUMA 560 manipulator, which is also employed in Chaps. 2 and 3, under an environment without sudden external disturbances. The initial state of the manipulator is set as $\boldsymbol{\theta}(0) = [0, -\pi/4, \pi/4, \pi/6, 0, 0]^{\mathrm{T}}$ rad, and the joint limits are listed below:

$$
\begin{aligned}
\boldsymbol{\theta}^{+} &= \frac{\pi}{180}[160, 45, 225, 170, 100, 266]^{\mathrm{T}}\ \text{rad},\\
\boldsymbol{\theta}^{-} &= \frac{\pi}{180}[-160, -225, -45, -110, -100, -266]^{\mathrm{T}}\ \text{rad},\\
-\dot{\boldsymbol{\theta}}^{-} &= \dot{\boldsymbol{\theta}}^{+} = [1, 1, 1, 1, 1, 1]^{\mathrm{T}}\ \text{rad/s},\\
-\ddot{\boldsymbol{\theta}}^{-} &= \ddot{\boldsymbol{\theta}}^{+} = [10, 10, 10, 10, 10, 10]^{\mathrm{T}}\ \text{rad/s}^2.
\end{aligned}
$$

The sampling interval τ is set as 0.01 s, and the task duration T is 10 s. Without loss of generality, other parameters are simultaneously given: $\eta = 0.95$, $\gamma = 100$, $N = 5$, $N_u = 5$, $Q = 100I$, $R_1 = 0.1I$, $R_2 = 0.1I$. The manipulator is driven to track a flower-shaped trajectory, with the initial position of the desired trajectory coinciding with that of the manipulator's end-effector.

The simulation results obtained using the presented MPC scheme (4.9) solved by the ZDNN model (4.14) are illustrated in Figs. 4.1 and 4.2. The desired trajectory to be tracked and the end-effector trajectory of the PUMA 560 manipulator are shown in Fig. 4.1a, where their coincidence is clearly visible. Figure 4.1b illustrates the trajectory-tracking process of the manipulator, demonstrating reasonable configurations in three-dimensional space. Figure 4.1c displays the tracking errors stabilizing at the order of 10^{-4} m, which attests to the high effectiveness and accuracy of the presented MPC scheme (4.9) solved by ZDNN model (4.14) in the trajectory tracking of the redundant manipulator. Figure 4.2a–c illustrates the joint angle, joint velocity, and joint acceleration of the manipulator while tracking the flower-shaped trajectory. It can be observed that none of the joint parameters exceed the joint limits. To sum up, the desired trajectory-tracking task is successfully accomplished driven by the MPC scheme (4.9) solved by the ZDNN model (4.14), thereby demonstrating its feasibility, effectiveness, and accuracy.

Furthermore, different parameters have different effects on the results, and Fig. 4.3 portrays simulation results with $Q = 100I$, $R_1 = 0$, $R_2 = 0.1I$ and Fig. 4.4 portrays simulation results with $Q = 20000I$, $R_1 = 0.1I$, $R_2 = 0.1I$. Note that other

Fig. 4.1 Simulation results for the PUMA 560 manipulator executing the trajectory-tracking task with the aid of MPC scheme (4.9) solved by ZDNN model (4.14). **a** Actual trajectory and desired trajectory. **b** Motion process. **c** Position error

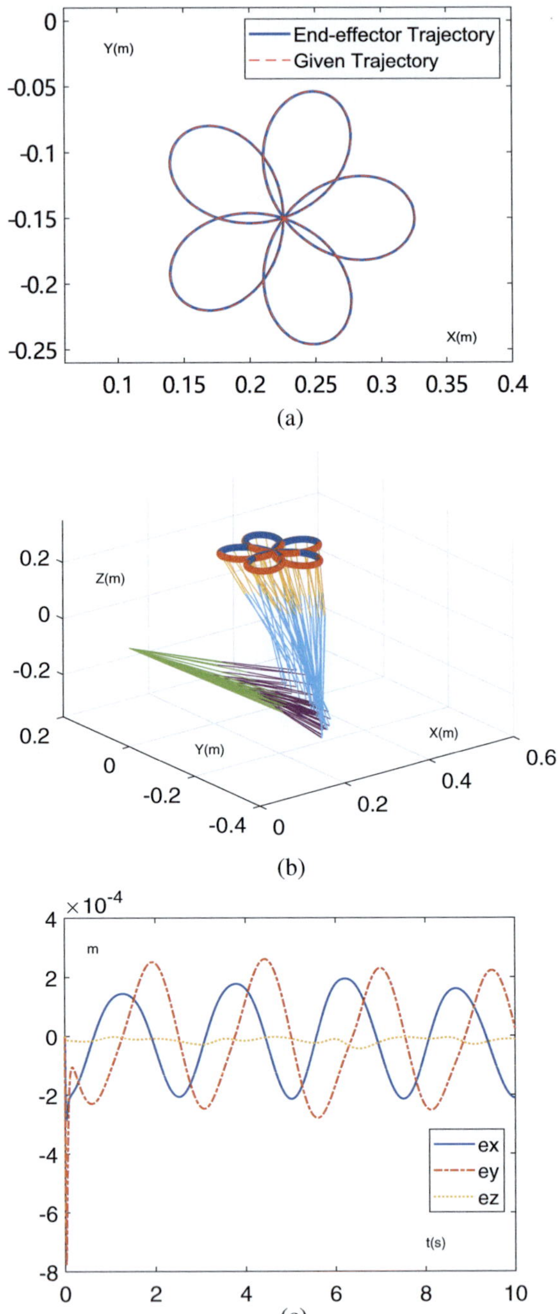

Fig. 4.2 Joint information for the PUMA 560 manipulator executing the trajectory-tracking task with the aid of MPC scheme (4.9) solved by ZDNN model (4.14). **a** Joint angle. **b** Joint velocity. **c** Joint acceleration

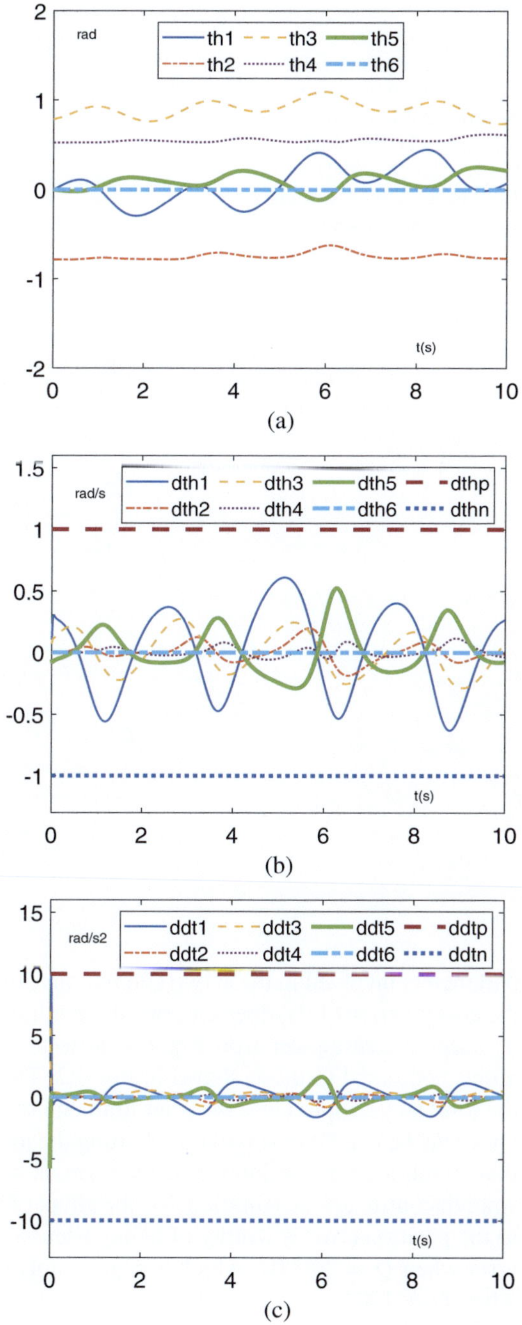

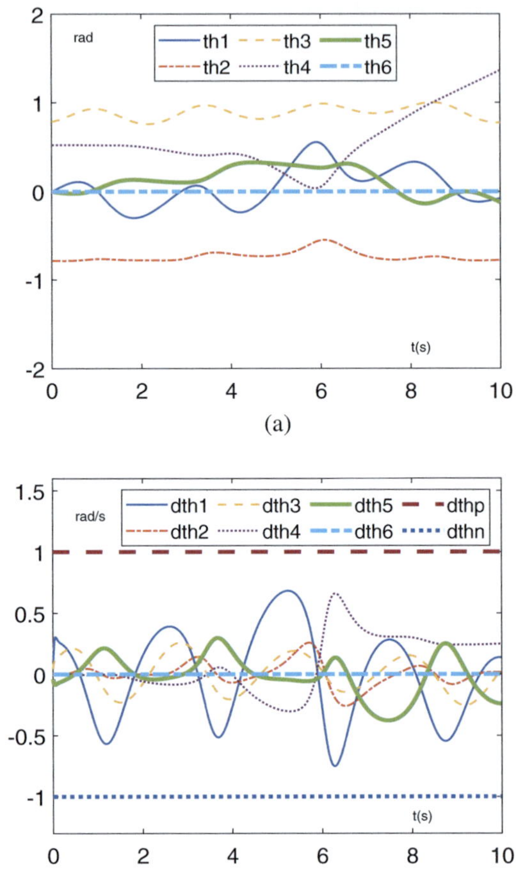

Fig. 4.3 Simulation results for the PUMA 560 manipulator executing the trajectory-tracking task with the aid of MPC scheme (4.9) solved by ZDNN model (4.14), with $Q = 100I$, $R_1 = 0$, $R_2 = 0.1I$. **a** Joint angle. **b** Joint velocity

parameters involved in these two simulations are the same as before. When $R_1 = 0$, the cost function (4.9a) does not consider minimizing the input norm, i.e., joint velocity norm. As can be seen from Fig. 4.3, the joint angles exhibit more pronounced variations compared to those shown in Fig. 4.2. There exists hidden trouble that when the joint angle approaches the joint limit and remains at the joint limit, the manipulator may be caught in singularity. Letting the joint angle change in a gentle manner, that is, minimizing the joint velocity norm, is significant for achieving a stable and smooth control effect. Additionally, the effect of the weight matrix Q corresponding to the position error is worthy of being discussed. Figure 4.4 illustrates the position error when $Q = 20000I$, which is significantly smaller than that shown in Fig. 4.1c when $Q = 100I$.

Comparison results obtained using the MPC scheme (4.9) with the SDNN model (Pan and Wang 2008) are presented in Fig. 4.5. Note that the scaling index controlling the convergence rate of the SDNN model (Pan and Wang 2008) is denoted by γ. When

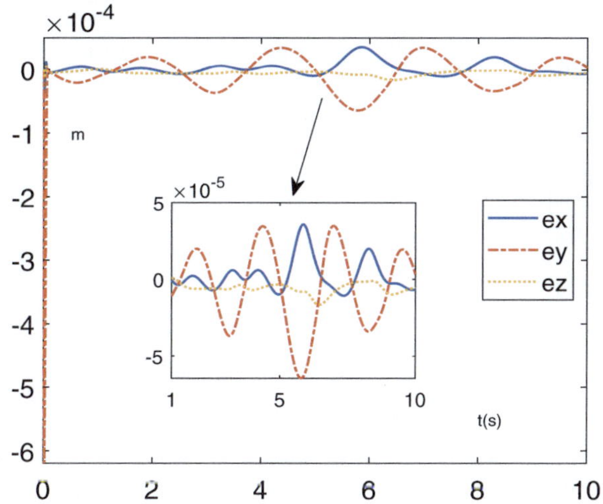

Fig. 4.4 Simulation results for the PUMA 560 manipulator executing the trajectory-tracking task with the aid of MPC scheme (4.9) solved by ZDNN model (4.14), with $Q = 20000I$, $R_1 = 0.1I$, $R_2 = 0.1I$

$\gamma = 100$, the same as that of the ZDNN model (4.14) in Fig. 4.1, the SDNN model (Pan and Wang 2008) fails to effectively solve the MPC scheme (4.9). This results in disordered tracking performance, as shown in Fig. 4.5a and b. When $\gamma = 5 \times 10^4$, the end-effector trajectory shown in Fig. 4.5c only marginally aligns with the desired trajectory, exhibiting a noticeable position error. In comparison with ZDNN model (4.14) equipped with $\gamma = 100$, SDNN model (Pan and Wang 2008) equipped with $\gamma = 5 \times 10^4$ still produces a larger position error, which underscores the effectiveness and superiority of the presented ZDNN model (4.14) in solving the MPC scheme (4.9).

4.4.2 With Sudden External Interference

In this subsection, the sudden external interference is considered for testing the quick-response capacity and robustness of MPC scheme (4.9) solved by ZDNN model (4.14). To be specific, when $t = 2$ s, the joint angle $\boldsymbol{\theta}$ of the PUMA 560 manipulator is increased by $[0.03, 0.03, 0.03, 0.03, 0.03, 0]^{\mathrm{T}}$ rad. Meanwhile, comparative simulations are conducted using the previously mentioned comparison schemes, with the results illustrated in Figs. 4.6, 4.7, and 4.8.

Figures 4.6 and 4.7 depict the simulation results obtained using the presented MPC scheme (4.9) solved by the ZDNN model (4.14) under sudden external interference, with $N = 8$ and $N_u = 8$. From Fig. 4.6a and b, it is evident that the end-effector of the PUMA 560 manipulator significantly deviates from the desired trajectory when the external interference is suddenly applied to the joint angle, despite the small magnitude of the interference. However, thanks to the quick-response capability and inherent robustness of the MPC scheme (4.9), the end-effector quickly returns to the

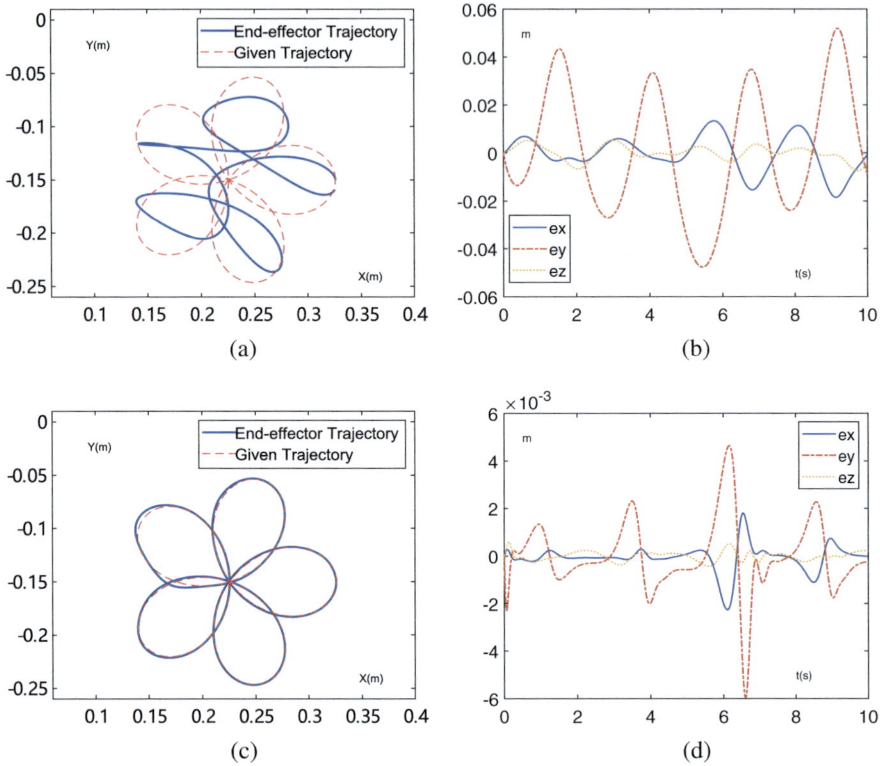

Fig. 4.5 Simulation results for the PUMA 560 manipulator executing the trajectory-tracking task with the aid of MPC scheme (4.9) solved by SDNN model (Pan and Wang 2008). **a** Actual trajectory and desired trajectory, $\gamma = 100$. **b** Position error, $\gamma = 100$. **c** Actual trajectory and desired trajectory, $\gamma = 5 \times 10^4$. **d** Position error, $\gamma = 5 \times 10^4$

desired trajectory. It is revealed that the MPC scheme (4.9) is capable of effectively handling abrupt situations that occur during the control process in a timely manner. Additionally, the tracking errors observed in Fig. 4.6c stabilize again at the order of 10^{-4} m after 0.3 s, indicating the high efficiency and quick-response capability of MPC scheme (4.9) solved by ZDNN model (4.14). Similar to Fig. 4.2a–c, the joint information of the PUMA 560 manipulator shown in Fig. 4.7a–c remain within the joint limits, ensuring the safe operation of the manipulator even when impacted by sudden external interference.

As for the comparative MAN scheme (4.15) solved by the PLPENN model (Jin and Zhang 2015), the related simulation results are presented in Figs. 4.8 and 4.9. From Fig. 4.8a, it can be observed that when subjected to sudden external interference, the PUMA 560 manipulator driven by the MAN scheme (4.15) returns to the desired trajectory more slowly compared to the one driven by the MPC scheme (4.9). Moreover, despite taking more response time than the MPC scheme (4.9), the MAN scheme

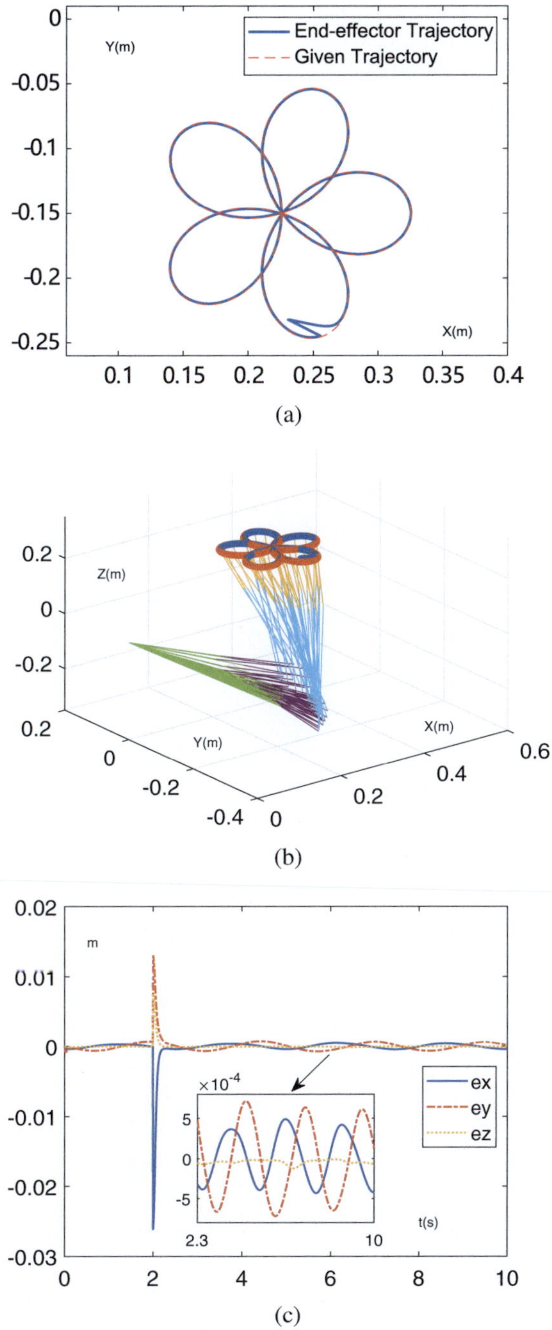

Fig. 4.6 Simulation results for trajectory-tracking task of the PUMA 560 manipulator with sudden external interference, synthesized by MPC scheme (4.9) solved by ZDNN model (4.14). **a** Actual trajectory and desired trajectory. **b** Motion process. **c** Position error

Fig. 4.7 Joint information for trajectory-tracking task of the PUMA 560 manipulator with sudden external interference, synthesized by MPC scheme (4.9) solved by ZDNN model (4.14). **a** Joint angle. **b** Joint velocity. **c** Joint acceleration

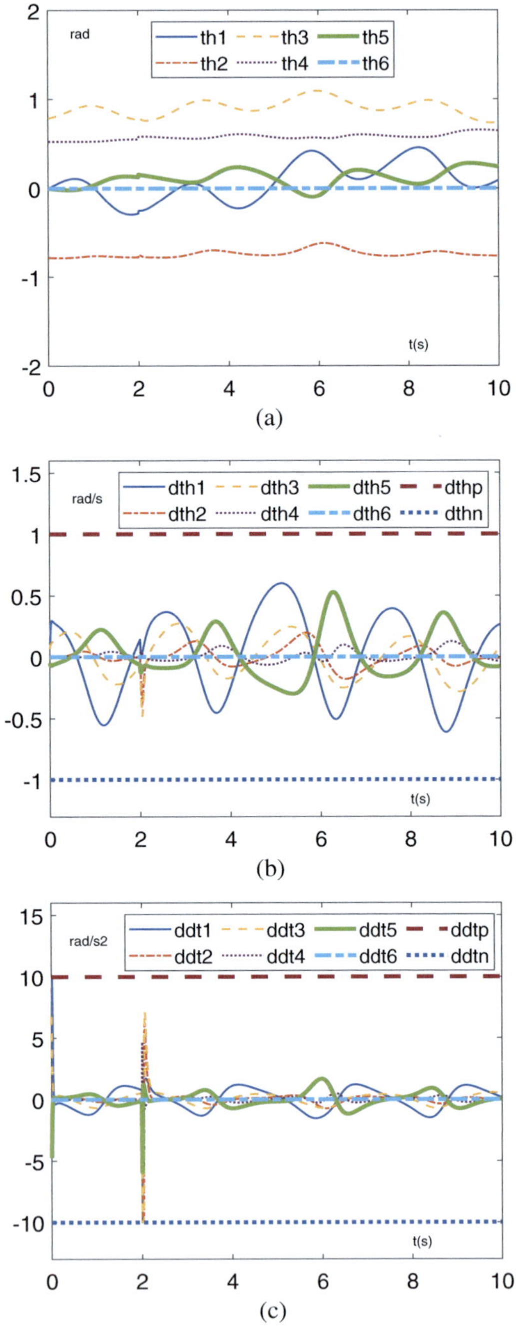

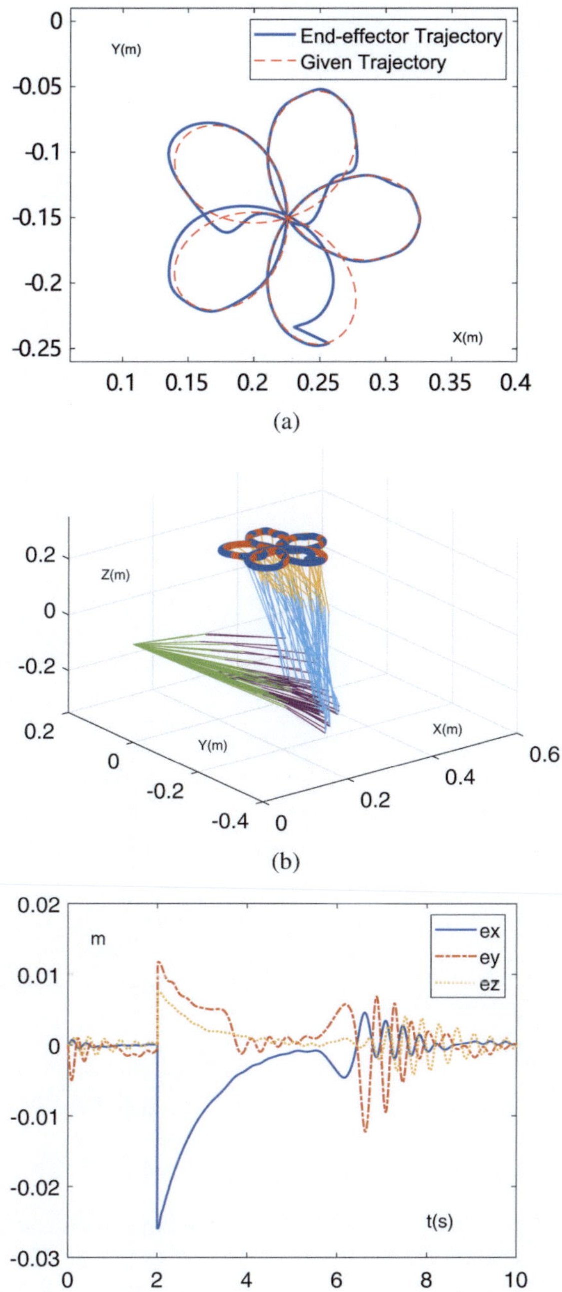

Fig. 4.8 Simulation results for trajectory-tracking task of the PUMA 560 manipulator with sudden external interference, synthesized by MAN scheme (4.15) solved by PLPENN model (Jin and Zhang 2015). **a** Actual trajectory and desired trajectory. **b** Motion process. **c** Position error

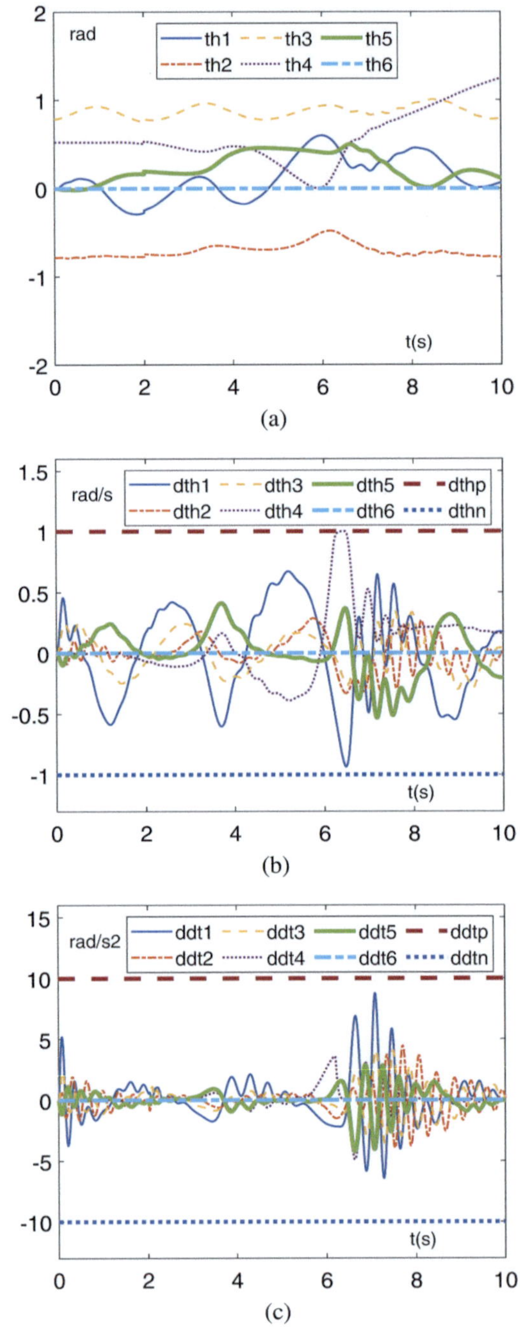

Fig. 4.9 Joint information for trajectory-tracking task of the PUMA 560 manipulator with sudden external interference, synthesized by MAN scheme (4.15) solved by PLPENN model (Jin and Zhang 2015). **a** Joint angle. **b** Joint velocity. **c** Joint acceleration

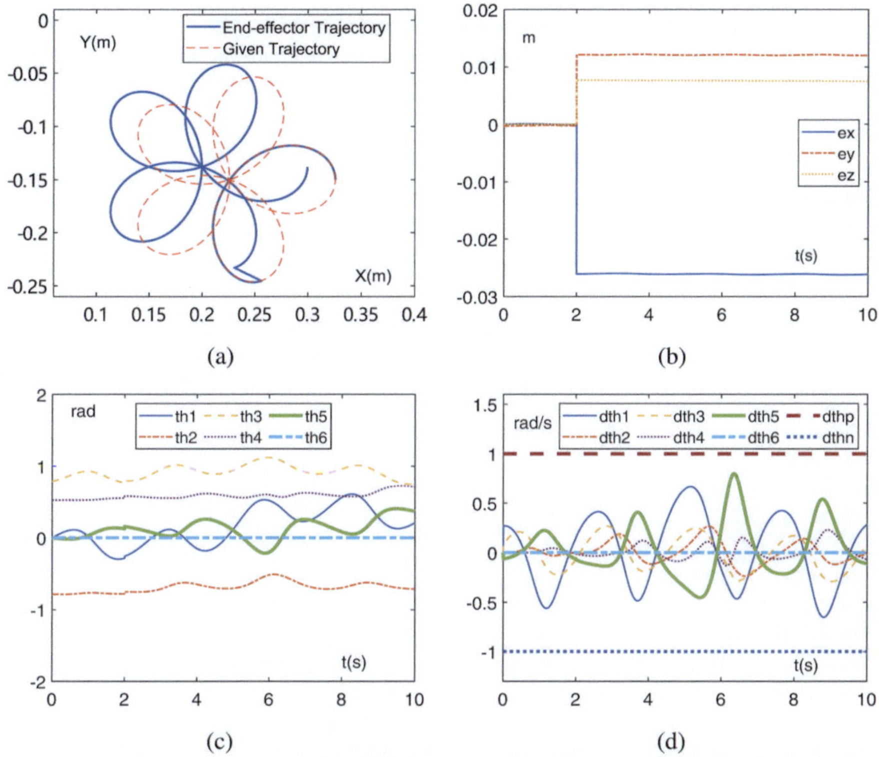

Fig. 4.10 Simulation results for trajectory-tracking task of the PUMA 560 manipulator with sudden external interference, synthesized by MVN scheme (4.16) solved by DPINN model (Ma et al. 2023). **a** Actual trajectory and desired trajectory. **b** Position error. **c** Joint angle. **d** Joint velocity

(4.15) exhibits lower accuracy. This indicates that the presented MPC scheme (4.9) solved by the ZDNN model (4.14) demonstrates stronger robustness, higher efficiency, and quicker response capability in handling sudden external interference compared to the MAN scheme (4.15) solved by the PLPENN model (Jin and Zhang 2015). Furthermore, due to the absence of minimizing the joint velocity norm in the performance index of the MAN scheme (4.15), the joint velocities shown in Fig. 4.9b are intuitively larger than those in Fig. 4.7b, and even approach the joint velocity limit.

In addition, as depicted in Fig. 4.10, the MVN scheme (4.16) solved by the DPINN model (Ma et al. 2023) exhibits notably inferior performance when the manipulator encounters sudden external interference during the trajectory-tracking task. From Fig. 4.10a and b, it is evident that the MVN scheme (4.16) fails to effectively address the significant tracking errors caused by the interference, resulting in the trajectory of the end-effector deviating entirely from the desired one. This is because the MVN scheme (4.16) lacks error feedback, rendering it unable to handle sudden external interference.

Table 4.1 Comparisons of various control schemes for trajectory tracking

	Joint limits level	Joint limits manner	Minimizing joint velocity norm or joint acceleration norm	Initial position	Returning to the desired trajectory
The presented scheme (4.9)	Multiple	Directly	Both	Any	√
Scheme in Zhang et al. (2022)	Single	Directly	Velocity	Any	√
Scheme in Bhattacharya et al. (2023)	None	NA⋆	Velocity	Any	√
Scheme in Ma et al. (2023)	Two	Indirectly	Velocity	Limited†	×
Scheme in Jin and Zhang (2015)	Multiple	Indirectly	Acceleration	Any	√
Scheme in Xiao and Zhang (2016)	Multiple	Indirectly	Acceleration	Limited†	×
Scheme in Zhang et al. (2017)	Two	Indirectly	Velocity	Limited†	×
Scheme in Xiao and Zhang (2013)	Multiple	Indirectly	Acceleration	Any	√
Scheme in Zhang et al. (2019)	Multiple	Indirectly	Velocity	Any	√
Scheme in Tan et al. (2024)	Multiple	Directly	Velocity	Any	√

⋆ "NA" denotes that joint limits are not considered
† "Limited" denotes that the initial position of the end-effector is required to be on the desired trajectory

To summarize, the simulation results shown in Figs. 4.1–4.10 confirm the high efficiency, quick-response capability, and strong robustness of the MPC scheme (4.9) solved by the ZDNN model (4.14), and further indicate its superiority over the comparison schemes and solvers. Additionally, further comparison schemes are considered. As indicated in Table 4.1, the presented scheme (4.9) is the only one capable of directly addressing joint limits at multiple levels. Additionally, scheme (4.9) uniquely minimizes both the joint velocity norm and the joint acceleration norm, whereas other schemes minimize only one of these. Furthermore, owing to a lack of error feedback, the schemes in Ma et al. (2023), Xiao and Zhang (2016), Zhang et al. (2017) have a requirement for the initial position of the end-effector of the manipulator and cannot help it return to the desired trajectory once it deviates.

4.5 Conclusions

In this chapter, an MPC scheme has been developed for trajectory tracking of redundant manipulators, with joint limits explicitly incorporated at three levels. Furthermore, the scheme aims to simultaneously minimize both the joint velocity and acceleration norms. Regarding the online solution for the MPC scheme, a ZDNN model has been devised, utilizing the technique of converting inequality constraints into equality constraints. The ZDNN model has been employed to solve the MPC scheme for the control of PUMA 560 manipulator, with simulations conducted under sudden external interference. Overall, the presented MPC scheme, solved by the ZDNN model, has been verified through simulation results to exhibit high efficiency, quick-response capability, and strong robustness, outperforming other existing schemes.

References

Ahmed AA, Koh BK, Lee YI (2017) A comparison of finite control set and continuous control set model predictive control schemes for speed control of induction motors. IEEE Trans Industr Inf 14(4):1334–1346

Arcari E, Minniti MV, Scampicchio A, Carron A, Farshidian F, Hutter M, Zeilinger MN (2023) Bayesian multi-task learning MPC for robotic mobile manipulation. IEEE Robotics and Automation Letters 8(6):3222–3229

Bhattacharya D, Chan YP, Shang S, Chan YS, Tan Y, Lau D (2023) Tri-space operational control of redundant multilink and hybrid cable-driven parallel robots using an iterative-learning-based reactive approach. IEEE Trans Control Syst Technol 31(6):2465–2483

Chen B, Chen X, Kanzow C (2000) A penalized Fischer-Burmeister NCP-function. Math Program 88:211–216

Cursi F, Bai W, Yeatman EM, Kormushev P (2022) Optimization of surgical robotic instrument mounting in a macro-micro manipulator setup for improving task execution. IEEE Trans Rob 38(5):2858–2874

Faulwasser T, Weber T, Zometa P, Findeisen R (2017) Implementation of nonlinear model predictive path-following control for an industrial robot. IEEE Trans Control Syst Technol 25(4):1505–1511

Garcia GA, Keshmiri SS, Stastny T (2015) Robust and adaptive nonlinear model predictive controller for unsteady and highly nonlinear unmanned aircraft. IEEE Trans Control Syst Technol 23(4):1620–1627

Gold T, Völz A, Graichen K (2023) Model predictive interaction control for robotic manipulation tasks. IEEE Trans Rob 39(1):76–89

Gulbudak O, Gokdag M (2021) Finite control set model predictive control approach of nine switch inverter-based drive systems: design, analysis, and validation. ISA Trans 110:283–304

Howell TA, Le Cleach S, Singh S, Florence P, Manchester Z, Sindhwani V (2022) Trajectory optimization with optimization-based dynamics. IEEE Robot Autom Lett 7(3):6750–6757

Jin L, Zhang YN (2015) G2-type SRMPC scheme for synchronous manipulation of two redundant robot arms. IEEE Transactions on Cybernetics 45(2):153–164

Jin L, Liufu Y, Lu H, Zhang Z (2020) Saturation-allowed neural dynamics applied to perturbed time-dependent system of linear equations and robots. IEEE Trans Industr Electron 68(10):9844–9854
Jin L, Wei L, Li S (2023) Gradient-based differential neural-solution to time-dependent nonlinear optimization. IEEE Trans Autom Control 68(1):62–627
Jin L, Liu L, Wang X, Shang M, Wang FY (2024) Physical-informed neural network for MPC-based trajectory tracking of vehicles with noise considered. IEEE Trans Intell Veh 9(3):4493–4503
Jin L, Su Z, Fu D, Xiao X (2024) Coevolutionary neural solution for nonconvex optimization with noise tolerance. IEEE Trans Neural Netw Learn Syst 35(12):17571–17581
Karimi M, Ahmadi M (2022) A reinforcement learning approach in assignment of task priorities in kinematic control of redundant robots. IEEE Robot Autom Lett 7(2):850–857
Li S, He J, Li Y, Rafique MU (2016) Distributed recurrent neural networks for cooperative control of manipulators: a game-theoretic perspective. IEEE Trans Neural Netw Learn Syst 28(2):415–426
Li S, Wang H, Rafique MU (2017) A novel recurrent neural network for manipulator control with improved noise tolerance. IEEE Trans Neural Netw Learn Syst 29(5):1908–1918
Li S, Zhou M, Luo X (2018) Modified primal-dual neural networks for motion control of redundant manipulators with dynamic rejection of harmonic noises. IEEE Trans Neural Netw Learn Syst 29(10):4791–4801
Li K, Wang J, Li S, Deng B, Yu H (2022) Latent characteristics and neural manifold of brain functional network under acupuncture. IEEE Trans Neural Syst Rehabil Eng 30:758–769
Lima PF, Nilsson M, Trincavelli M, Mårtensson J, Wahlberg B (2017) Spatial model predictive control for smooth and accurate steering of an autonomous truck. IEEE Trans Intell Veh 2(4):238–250
Liu M, Chen L, Du X, Jin L, Shang M (2021) Activated gradients for deep neural networks. IEEE Trans Neural Netw Learn Syst 34(4):2156–2168
Liu M, Li Y, Chen Y, Qi Y, Jin L (2024) A distributed competitive and collaborative coordination for multirobot systems. IEEE Trans Mob Comput 23(2):11436–11448
Ma B, Xie Z, Zhan B, Jiang Z, Liu Y, Liu H (2023) Actual shape-based obstacle avoidance synthesized by velocity-acceleration minimization for redundant manipulators: An optimization perspective. IEEE Trans Syst Man Cybern Syst 53(10):6460–6474
Pan Y, Wang J (2008) Nonlinear model predictive control using a recurrent neural network. In: 2008 IEEE international joint conference on neural networks, pp 2296–2301
Peng G, Yang C, He W, Chen CP (2019) Force sensorless admittance control with neural learning for robots with actuator saturation. IEEE Trans Industr Electron 67(4):3138–3148
Ren X, Guo J, Chen S, Zhang M, Zhang Z (2024) A distributed varying-parameter recurrent neural network for solving the motion generation problem of a multimanipulator collaborative system. IEEE Trans Syst Man Cybern Syst 54(8):4918–4928
Rubagotti M, Sangiovanni B, Nurbayeva A, Incremona GP, Ferrara A, Shintemirov A (2023) Shared control of robot manipulators with obstacle avoidance: a deep reinforcement learning approach. IEEE Control Syst Mag 43(1):44–63
Sleiman JP, Farshidian F, Minniti MV, Hutter M (2021) A unified MPC framework for whole-body dynamic locomotion and manipulation. IEEE Robot Autom Lett 6(3):4688–4695
Tan J, Shang M, Jin L (2024) Metaheuristic-based RNN for manipulability optimization of redundant manipulators. IEEE Trans Industr Inf 20(4):6489–6498
Tian G, Tan J, Li B, Duan G (2024) Optimal fully actuated system approach-based trajectory tracking control for robot manipulators. IEEE Trans Cybern 54(12):7469–7478
Van Duijkeren N, Faulwasser T, Pipeleers G (2018) Dual-objective NMPC: considering economic costs near manifolds. IEEE Trans Autom Control 64(9):3788–3795
Wang C, Tnunay H, Zuo Z, Lennox B, Ding Z (2019) Fixed-time formation control of multi-robot systems: Design and experiments. IEEE Trans Industr Electron 66(8):6292–6301
Wang X, Wang D, Du M, Song K, Ni Y, Li Y (2024) A two-layer trajectory tracking control scheme of manipulator based on ELM-SMC for autonomous robotic vehicle. IEEE Trans Autom Sci Eng 21(3):2337–2348

Xiao L, Zhang YN (2013) Acceleration-level repetitive motion planning and its experimental verification on a six-link planar robot manipulator. IEEE Trans Control Syst Technol 21(3):906–914

Xiao L, Zhang YN (2016) Dynamic design, numerical solution and effective verification of acceleration-level obstacle-avoidance scheme for robot manipulators. Int J Syst Sci 47(4):932–945

Xie Z, Jin L, Luo X, Zhou M, Zheng Y (2024) A biobjective scheme for kinematic control of mobile robotic arms with manipulability optimization. IEEE/ASME Trans Mechatron 29(2):1534–1545

Xu Z, Li S, Zhou X, Zhou S, Cheng T, Guan Y (2020) Dynamic neural networks for motion-force control of redundant manipulators: An optimization perspective. IEEE Trans Industr Electron 68(2):1525–1536

Yan J, Jin L, Yuan Z, Liu Z (2022) RNN for receding horizon control of redundant robot manipulators. IEEE Trans Industr Electron 69(2):1608–1619

Yan J, Jin L, Luo X, Li S (2024) Modified RNN for solving comprehensive Sylvester equation with TDOA application. IEEE Trans Neural Netw Learn Syst 35(9):12553–12563

Yan J, Su Z, Ma X, Jin L (2025) An obstacle avoidance scheme for manipulators aided by noise-tolerant neural dynamics. IEEE Trans Industr Inf 21(1):703–712

Yang C, Jiang Y, He W, Na J, Li Z, Xu B (2018) Adaptive parameter estimation and control design for robot manipulators with finite-time convergence. IEEE Trans Industr Electron 65(10):8112–8123

Yang C, Peng G, Cheng L, Na J, Li Z (2019) Force sensorless admittance control for teleoperation of uncertain robot manipulator using neural networks. IEEE Trans Syst Man Cybern Syst 51(5):3282–3292

Yang C, Huang D, He W, Cheng L (2020) Neural control of robot manipulators with trajectory tracking constraints and input saturation. IEEE Trans Neural Netw Learn Syst 32(9):4231–4242

Yang X, Zhao Z, Ma B, Xu Z, Zhao J, Liu H (2023) Kinematic and dynamic manipulability optimizations of redundant manipulators based on RNN model. IEEE Trans Industr Inf 20(4):5763–5773

Zhang Z, Zheng L, Yu J, Li Y, Yu Z (2017) Three recurrent neural networks and three numerical methods for solving a repetitive motion planning scheme of redundant robot manipulators. IEEE/ASME Trans Mechatron 22(3):1423–1434

Zhang YY, Li S, Zhou X (2019) Recurrent-neural-network-based velocity-level redundancy resolution for manipulators subject to a joint acceleration limit. IEEE Trans Industr Electron 66(5):3573–3582

Zhang Z, Chen S, Liang J (2022) Discrete-time circadian rhythms neural network for perturbed redundant robot manipulators tracking problem with periodic noises. IEEE Trans Industr Inf 18(1):242–251

Chapter 5
Cerebellum-Inspired MPC with Discrete DNN

5.1 Introduction

Robotics has witnessed rapid advancement in recent years, driving extensive related research efforts across both industrial (Omisore et al. 2022; Lillo et al. 2021; Yan et al. 2024a) and academic domains (Wang et al. 2023; Renda et al. 2022; Liu et al. 2021). As a representative category of robotic systems, manipulators serve a vital function across numerous application domains (Tong et al. 2023; Reister et al. 2022; Liu et al. 2024; Cao et al. 2023). Manipulators with kinematic redundancy, those possessing more degrees of freedom (DOFs) than required (Obregón-Flores et al. 2021; Chen et al. 2023; Fiore et al. 2023), are widely favored for their capacity to perform sophisticated tasks (Jin et al. 2024a; Picotti et al. 2023; Sabelhaus et al. 2021). A common approach to addressing trajectory tracking in redundant manipulators involves solving inverse kinematics at either the velocity (Na et al. 2023; Abadia et al. 2021; Yan et al. 2025) or acceleration level (Dahlin and Karayiannidis 2021; Ma et al. 2023). Liu et al. present a motion control method utilizing quadratic programming (QP) at the velocity level for redundant manipulators, incorporating the trajectory-tracking task as an equality constraint within the framework (Yan et al. 2025). A kinematic control method at the acceleration level for redundant manipulators is introduced in Dahlin and Karayiannidis (2021), which addresses the joint drift and incorporates constraints on the joint acceleration. Nevertheless, these methods essentially realize the tracking control at the velocity or acceleration levels. Although certain methods incorporate feedback compensation at the position level, they remain unable to address sudden issues at this level in a timely manner (Baek et al. 2022). Moreover, another limitation of these methods is their inability to directly handle joint constraints, requiring the transformation of constraints at various levels to the variable level (velocity or acceleration). Following these transformations, the initial feasible region for the joint angles could be consequently narrowed (Zhang et al. 2023a).

M. Liu et al., *Dynamic Neural Networks for Motion Control of Redundant Manipulators*, Intelligent Control and Learning Systems 21,
https://doi.org/10.1007/978-981-96-9144-9_5

Model predictive control (MPC), which originates from industry in the previous century, has found successful applications in robot control (Capolei et al. 2020), autonomous driving (Khaitan et al. 2021), and other areas (Lu et al. 2022; Jin et al. 2024b), owing to its exceptional abilities, including constraints handling and the performance enhancement. In general, the MPC method requires making predictions based on the system model and optimizing performance over the prediction horizon. Moreover, system constraints are frequently considered. To achieve safe autonomous driving under uncertain conditions, Khaitan et al. implement a robust tube MPC method to guarantee that the vehicle adheres to a safe trajectory, even in the presence of noise and model uncertainties (Khaitan et al. 2021). In Li et al. (2022a), an event-triggered MPC method is explored and utilized for trajectory-tracking control of a wheeled mobile robot, considering system constraints while reducing computational costs through the event-triggered mechanism and an adaptive prediction horizon. These outstanding research results motivate us to apply the MPC method to the motion control of redundant manipulators. It is important to highlight that certain limitations of the existing motion control methods for redundant manipulators, as discussed earlier, can also be addressed by efficient MPC methods.

Generally, an MPC method involves two key components. The first component is to forecast the system's future output using the system model. The second component involves solving an optimization problem that incorporates control objectives and prediction data to enhance system performance. Future information prediction forms the foundation of MPC, and the system model employed for this prediction can be derived either through mathematical modeling or data-driven methods. Various neural network models have become widely used to capture the dynamic properties of unknown systems in recent years (Chen et al. 2024; Cai et al. 2024). Among these, the echo state network (ESN), introduced by Jaeger et al., is a recurrent neural network (RNN) that is simple and quick to train, excelling at processing and forecasting time series (Jaeger and Haas 2004). As a result, several studies have emerged that focus on designing model predictive controllers using the ESN (Jordanou et al. 2022a, b; Schwedersky et al. 2022; Pan and Wang 2012). Jordanou et al. employ the ESN for an electrical submersible to investigate a nonlinear MPC approach and a practical nonlinear MPC (PNMPC) method after linearization, both of which successfully control the electrical submersible pump (Jordanou et al. 2022b). Additionally, an adaptive MPC method utilizing the ESN is presented for controlling permanent magnet synchronous motors, where the ESN model is trained online using a recursive least-squares method with an adaptive directional forgetting factor (Schwedersky et al. 2022). Recently, research on MPC using ESN has been steadily growing. However, its application to motion control of redundant manipulators remains limited, which is exactly the focus of this chapter.

Furthermore, existing research indicates that the ESN exhibits similarities with the cerebellum and can serve as a potential computational model (Yamazaki and Tanaka 2009). In the cerebellum, the granule layer composed of granular cells exhibits reservoir computing characteristics similar to the ESN's reservoir and can project the input signal into a high-dimensional space to produce a complex output signal. Consequently, several studies on ESN-based cerebellum models have been designed

(Zhang et al. 2023b; Tan et al. 2023; Tanaka et al. 2022). In Zhang et al. (2023b), an ESN-based cerebellum prediction network is employed to forecast the position of a musculoskeletal robot under the original control signals, while a cerebellum correction network adjusts these control signals. In Tan et al. (2023), an ESN-based cerebellum model is developed to act as a controller and to fine-tune a multi-layer perceptron for controlling a model-unknown manipulator. However, these results do not consider performance optimization or system constraints. In contrast to the previous studies, this chapter aims to design an ESN-based cerebellum model as the process model for the redundant manipulator with unknown structural information and further develop an MPC algorithm that incorporates both performance optimization and system constraints for trajectory tracking.

As mentioned earlier, previous studies identify similarities between the cerebellum circuitry and the ESN. This resemblance motivates the development of a cerebellum-inspired model utilizing the ESN, which is used in place of the kinematic model of the redundant manipulator to replicate its input–output behavior. Inspired by the PNMPC strategy (Plucenio et al. 2007), an MPC framework is established by performing a linearization of the cerebellum-based model along the reference trajectory. This trajectory-based linearization enables the retention of most nonlinear dynamics of the robotic system, ensuring the linearized representation closely approximates the original nonlinear prediction model (Yan et al. 2024b; Wang and Tan 2023; Hoxha et al. 2016). To enhance the prediction accuracy and tracking performance, this work introduces an online training approach for the cerebellum model along with a correction algorithm based on ND. The cerebellum model is trained online by continuously updating the state and output matrices, which enables the real-time adjustment of the output weight matrix. An ND-based correction algorithm is developed using the neural dynamics approach to further reduce both the prediction error of the cerebellum model and the inaccuracies introduced by the linearization process. The ND approach is good at processing complex problems effectively (Thomas et al. 2022; Kalidindi et al. 2019; Jin et al. 2023) and has been applied in various fields (Honerkamp et al. 2024; Wang and Santos 2023; Bazaraa et al. 2006; Niddam et al. 2024). Moreover, the framework incorporates joint constraints across three hierarchical levels of the redundant manipulator to guarantee the operational safety. For solving the presented MPC scheme, a discrete dynamic neural network (DDNN) model is designed. Unlike the cerebellum model, the ND-based correction algorithm and the DDNN model do not require training (Kim et al. 2021; Cai et al. 2024; Zhang et al. 2023c), and its parameters are determined at the time of design (Jin et al. 2024c; Li et al. 2022b; Yang et al. 2024; Wei et al. 2021). In conclusion, the approach is encapsulated as the cerebellum-inspired MPC (CIMPC) algorithm.

The remainder of this chapter is structured as follows: Section 5.2 presents the cerebellum-inspired model utilizing the ESN. The online training algorithm, the ND-based correction algorithm, and the MPC scheme for the redundant manipulator are presented in Sect. 5.3. Section 5.4 presents the DDNN model and summarizes the presented method as the CIMPC algorithm. Section 5.5 offers the theoretical analyses on the convergence of both the ND-based correction algorithm and the

DDNN model. Section 5.6 presents and discusses the simulation and experimental results. Section 5.7 provides the conclusion of this chapter. The main contributions are as follows:

1. The CIMPC algorithm, which uses the cerebellum-inspired model, is introduced for trajectory tracking of the redundant manipulator with unknown structural information.
2. An online training algorithm for the cerebellum model is introduced to enhance its prediction accuracy and further optimize the tracking performance of the CIMPC algorithm.
3. An ND-based correction algorithm is designed to adjust the prediction error of the cerebellum model and the error introduced by the linearization process.

5.2 Cerebellum-Inspired Model

This section presents the cerebellum-inspired model based on the ESN, which is used to identify the unknown dynamic system and the associated training process.

5.2.1 *Model*

The ESN model excels at modeling nonlinear dynamic systems and demands minimal computationally efficient training. Previous studies have demonstrated that the ESN shares similarities with the cerebellum and holds promise as a computational model (Yamazaki and Tanaka 2009). The cerebellum-inspired model utilizing the ESN includes mossy fibers, granular cells, parallel fibers, and Purkinje cells. Mossy fibers serve as the main input fibers of the cerebellar cortex, transmitting the input $\boldsymbol{u}(k) \in \mathbb{R}^m$ into the cerebellum at time k. Granular cells receive the input $\boldsymbol{u}(k)$ from mossy fibers and establish synaptic connections with other granular cells and Purkinje cells. The signals $\boldsymbol{c}(k) \in \mathbb{R}^n$ and $\boldsymbol{c}(k+1) \in \mathbb{R}^n$ in granular cells at times k and $k+1$ are expressed as

$$\boldsymbol{c}(k+1) = (1-a)\boldsymbol{c}(k) + a\, g(W_{\text{in}}\boldsymbol{u}(k) + W_{\text{gc}}\boldsymbol{c}(k) + W_{\text{fb}}\boldsymbol{y}(k) + \boldsymbol{w}_{\text{bias}}), \quad (5.1)$$

where a represents the leakage rate; $g(\cdot) : \mathbb{R}^n \to \mathbb{R}^n$ is a nonlinear activation function and usually selected as $g(\cdot) = \tanh(\cdot)$; the weight matrices $W_{\text{in}} \in \mathbb{R}^{n\times m}$, $W_{\text{gc}} \in \mathbb{R}^{n\times n}$, and $W_{\text{fb}} \in \mathbb{R}^{n\times l}$ represent the connections from mossy fibers to granular cells, between granular cells, and from Purkinje cells to granular cells, respectively; $\boldsymbol{y}(k) \in \mathbb{R}^l$ represents output at time k; and $\boldsymbol{w}_{\text{bias}} \in \mathbb{R}^n$ signifies the bias term. The axons of granular cells merge beneath the cerebellar cortex to create parallel fibers, with the state $\boldsymbol{x}(k+1) \in \mathbb{R}^n$ expressed as

$$\boldsymbol{x}(k+1) = \boldsymbol{c}(k+1). \tag{5.2}$$

Purkinje cells receive synaptic inputs from parallel fibers and serve as the efferent neurons within the cerebellar cortex. The output generated by Purkinje cells is expressed as

$$\boldsymbol{y}(k+1) = W_{\text{out}}\boldsymbol{x}(k+1), \tag{5.3}$$

of which $W_{\text{out}} \in \mathbb{R}^{l\times n}$ denotes the weight matrix.

5.2.2 *Training*

For cerebellum model (5.1)–(5.3), only the output weight matrix W_{out} needs to be trained, while the other weight matrices are randomly initialized and remain fixed. The training procedure for the cerebellum model (5.1)–(5.3) consists of the following steps:

1. Randomly generates weight matrices W_{in}, W_{gc}, and W_{fb}, and bias term $\boldsymbol{w}_{\text{bias}}$, where W_{gc} requires to be scaled so that its spectral radius is less than 1 (Jordanou et al. 2022a). Set the initial signal $\boldsymbol{c}(0)$ in granular cells and the state $\boldsymbol{x}(0)$ in parallel fibers to the zero vector, and choose an appropriate leaking rate a.
2. Collect the training data sampled from the unknown system, which includes input vectors $\boldsymbol{u}(0), \boldsymbol{u}(1), \cdots, \boldsymbol{u}(k_{\text{w}}), \cdots, \boldsymbol{u}(k_{\text{L}}-1)$ and output vectors $\boldsymbol{y}(0), \boldsymbol{y}(1), \cdots, \boldsymbol{y}(k_{\text{w}}), \cdots, \boldsymbol{y}(k_{\text{L}})$. In this context, k_{w} refers to the flushing time, indicating that the initial memory of granular cells will be cleared after time k_{w}, while k_{L} represents the size of the training data. The training data at each time step are fed to granular cells through (5.1) to obtain the corresponding signal $\boldsymbol{c}(1), \cdots, \boldsymbol{c}(k_{\text{w}}), \cdots, \boldsymbol{c}(k_{\text{L}})$. Then, state vectors $\boldsymbol{x}(1), \cdots, \boldsymbol{x}(k_{\text{w}}), \cdots, \boldsymbol{x}(k_{\text{L}})$ in the parallel fibers are obtained through (5.2). Discard state vectors and training data at time $k = 0, 1, \cdots, k_{\text{w}}$, and construct the state matrix as $X = [\boldsymbol{x}(k_{\text{w}}+1), \boldsymbol{x}(k_{\text{w}}+2), \cdots, \boldsymbol{x}(k_{\text{L}})] \in \mathbb{R}^{n\times(k_{\text{L}}-k_{\text{w}})}$ and output matrix as $Y = [\boldsymbol{y}(k_{\text{w}}+1), \boldsymbol{y}(k_{\text{w}}+2), \cdots, \boldsymbol{y}(k_{\text{L}})] \in \mathbb{R}^{l\times(k_{\text{L}}-k_{\text{w}})}$.
3. Using the state matrix X and the output matrix Y, the ridge regression method described in Bishop (2006) is employed to compute the output weight matrix:

$$W_{\text{out}} = YX^{\text{T}}(XX^{\text{T}} + \lambda I)^{-1}, \tag{5.4}$$

of which I is an identity matrix with the appropriate dimension; λ symbolizes the Tikhonov regularization parameter; superscript ^{T} stands for the transpose operation; superscript $^{-1}$ denotes the inversion operation.

It is important to note that only the output weight matrix W_{out} requires training in the cerebellum model (5.1)–(5.3), which reduces the problem to a linear regression task. As a result, the training of the cerebellum model (5.1)–(5.3) is relatively straightforward and computationally efficient.

5.3 Cerebellum-Inspired MPC

This section presents the use of the cerebellum model (5.1)–(5.3) for identifying the kinematics of a redundant manipulator when its structural information is unknown. Following the linearization of the cerebellum model (5.1)–(5.3), an MPC algorithm is presented, utilizing the previously introduced online training and ND-based correction algorithms.

5.3.1 Preliminaries

The kinematic model of the redundant manipulator is analogous to (8.2) and expressed by the following equation:

$$\boldsymbol{r}(t) = f(\boldsymbol{\theta}(t)), \tag{5.5}$$

where $\boldsymbol{r}(t) \in \mathbb{R}^p$ represents the position of the end-effector, $\boldsymbol{\theta}(t) \in \mathbb{R}^s$ represents the joint angle with $s > p$, and $f(\cdot) : \mathbb{R}^s \to \mathbb{R}^p$ denotes a nonlinear function that models the mapping from the joint angles to the position of the end-effector. The velocity-level equation of the above kinematics is analogous to (7.2) and can be expressed as

$$\dot{\boldsymbol{r}}(t) = \Gamma(\boldsymbol{\theta}(t))\dot{\boldsymbol{\theta}}(t), \tag{5.6}$$

of which $\dot{\boldsymbol{r}}(t) \in \mathbb{R}^p$ and $\dot{\boldsymbol{\theta}}(t) \in \mathbb{R}^s$ represent the velocity of the end-effector and the joint velocity, respectively, and $\Gamma(\boldsymbol{\theta}(t)) = \partial f(\boldsymbol{\theta}(t))/\partial\boldsymbol{\theta}(t) \in \mathbb{R}^{p\times s}$ denotes the Jacobian matrix.

In this work, we employ the MPC method to achieve trajectory tracking of the redundant manipulator. At each time step, it is necessary to forecast the future output of the system, i.e., the position of the end-effector, and the following cost function is formulated:

$$\Upsilon(k) = \sum_{i=1}^{N} \|\boldsymbol{r}(k+i|k) - \boldsymbol{r}_{\mathrm{d}}(k+i|k)\|_Q^2,$$

where $N \in \mathbb{N}^+$ denotes the prediction horizon; $\boldsymbol{r}(k+i|k)$ represents the output at time $k+i$ predicted at time k; $\boldsymbol{r}_{\mathrm{d}}(k+i|k)$ denotes the desired output at time $k+i$; $Q \in \mathbb{R}^{rN\times rN}$ is a positive definite and symmetric weight matrix; and $\|\cdot\|_Q$ denotes the weighted 2-norm, i.e., $\|\boldsymbol{a}\|_Q = \sqrt{\boldsymbol{a}^{\mathrm{T}}Q\boldsymbol{a}}$ with $\boldsymbol{a} \in \mathbb{R}^{rN}$. In order to predict the future output, the input vector defined below is selected and fed into the granular cells of the cerebellum model through (5.1):

$$\boldsymbol{u}(k) = [\boldsymbol{\theta}^{\mathrm{T}}(k), \mathrm{z}^{\mathrm{T}}(k)]^{\mathrm{T}},$$

where $z(k) = \tau\dot{\boldsymbol{\theta}}(k) \in \mathbb{R}^s$ represents the increment of the joint angle serving as the control input of the redundant manipulator; τ denotes the sampling interval. The output vector generated by the cerebellum model (5.1)–(5.3) corresponds to the end-effector's position:

$$\boldsymbol{y}(k) = \boldsymbol{r}(k).$$

At each time step, the current output $\boldsymbol{y}(k)$ and joint angle $\boldsymbol{\theta}(k)$ are known and measurable from the redundant manipulator, whereas the current control input $z(k)$ remains unknown and must be computed.

5.3.2 Processing of Cerebellum Model

In the previous subsection, the cerebellum model (5.1)–(5.3) is employed as a predictive model to estimate the future output of the redundant manipulator, eliminating the need for the kinematic equations (5.5) and (5.6). It should be noted that the cerebellum model (5.1)–(5.3) is inherently nonlinear, which introduces the significant complexity into the subsequent MPC formulation and derivation. To address this issue, we introduce the concept of the free response and forced response to reformulate the prediction model (Plucenio et al. 2007). The processed predicted output $\boldsymbol{y}(k+i|k)$ is expressed as

$$\boldsymbol{y}(k+i|k) = \sum_{j=0}^{N_u-1} E(i,j)\,(z(k+j|k) - z(k-1)) + \hat{\boldsymbol{y}}(k+i|k),\ i = 1, 2, \cdots, N, \tag{5.7}$$

with

$$E(i,j) = \left.\frac{\partial \boldsymbol{y}(k+i|k)}{\partial z(k+j|k)}\right|_{z(k+j|k)=z(k-1)},$$

where $N_u \in \mathbb{N}_+$ signifies the control horizon; $z(k+j|k)$ represents the future control input; and $\hat{\boldsymbol{y}}(k+i|k)$ is obtained by feeding the last control input $z(k-1)$ to cerebellum model (5.1)–(5.3). The process is detailed below:

$$\begin{aligned}
\hat{\boldsymbol{x}}(k+1|k) =& (1-a)\boldsymbol{x}(k) \\
&+ a\,g(W_{\text{in}}[\boldsymbol{\theta}^{\mathrm{T}}(k), z^{\mathrm{T}}(k-1)]^{\mathrm{T}} + W_{\text{gc}}\boldsymbol{x}(k) + W_{\text{fb}}\boldsymbol{y}(k) + \boldsymbol{w}_{\text{bias}}), \\
\hat{\boldsymbol{y}}(k+1|k) =& W_{\text{out}}\hat{\boldsymbol{x}}(k+1|k), \\
\hat{\boldsymbol{\theta}}(k+1|k) =& \boldsymbol{\theta}(k) + z(k-1), \\
\hat{\boldsymbol{x}}(k+2|k) =& (1-a)\hat{\boldsymbol{x}}(k+1|k) + a\,g(W_{\text{in}}[\hat{\boldsymbol{\theta}}^{\mathrm{T}}(k+1|k), z^{\mathrm{T}}(k-1)]^{\mathrm{T}} \\
&+ W_{\text{gc}}\hat{\boldsymbol{x}}(k+1|k) + W_{\text{fb}}\hat{\boldsymbol{y}}(k+1|k) + \boldsymbol{w}_{\text{bias}}), \\
\hat{\boldsymbol{y}}(k+2|k) =& W_{\text{out}}\hat{\boldsymbol{x}}(k+2|k),
\end{aligned}$$

$$
\begin{aligned}
&\vdots\\
\hat{\boldsymbol{\theta}}(k+N-1|k) =&\hat{\boldsymbol{\theta}}(k+N-2)+z(k-1),\\
\hat{\boldsymbol{x}}(k+N|k) =&(1-a)\hat{\boldsymbol{x}}(k+N-1|k)+a\,g(W_{\text{in}}[\hat{\boldsymbol{\theta}}^{\text{T}}(k+N-1|k),\boldsymbol{z}^{\text{T}}(k-1)]^{\text{T}}\\
&+W_{\text{gc}}\hat{\boldsymbol{x}}(k+N-1|k)+W_{\text{fb}}\hat{\boldsymbol{y}}(k+N-1|k)+\boldsymbol{w}_{\text{bias}}),\\
\hat{\boldsymbol{y}}(k+N|k) =&W_{\text{out}}\hat{\boldsymbol{x}}(k+N|k).
\end{aligned}
$$

For clarity, we introduce a new variable as follows:

$$
\begin{aligned}
\hat{\boldsymbol{\xi}}(i) =&W_{\text{in}}[\hat{\boldsymbol{\theta}}^{\text{T}}(k+i-1|k),\boldsymbol{z}^{\text{T}}(k-1)]^{\text{T}}+W_{\text{gc}}\hat{\boldsymbol{x}}(k+i-1|k)\\
&+W_{\text{fb}}\hat{\boldsymbol{y}}(k+i-1|k)+\boldsymbol{w}_{\text{bias}}.
\end{aligned}
$$

Furthermore, by applying the chain rule of differentiation, it can be obtained as follows:

$$
\frac{\partial \boldsymbol{y}(k+i|k)}{\partial z(k+j|k)}=\frac{\partial \boldsymbol{y}(k+i|k)}{\partial \boldsymbol{x}(k+i|k)}\frac{\partial \boldsymbol{x}(k+i|k)}{\partial \boldsymbol{u}(k+j|k)}\frac{\partial \boldsymbol{u}(k+j|k)}{\partial z(k+j|k)},
$$

with

$$
\frac{\partial \boldsymbol{y}(k+i|k)}{\partial \boldsymbol{x}(k+i|k)}=W_{\text{out}},\ \frac{\partial \boldsymbol{u}(k+j|k)}{\partial z(k+j|k)}=[0_{s\times s},I_{s\times s}]^{\text{T}}.
$$

Depending on the values of i and j, the partial derivative $\partial \boldsymbol{x}(k+i|k)/\partial \boldsymbol{u}(k+j|k)$ can be classified into the following cases:

$$
\left.\frac{\partial \boldsymbol{x}(k+i|k)}{\partial \boldsymbol{u}(k+j|k)}\right|_{z(k+j|k)=z(k-1)}=\begin{cases}G(i), & i-j=1,\\ H(i)\dfrac{\partial \boldsymbol{x}(k+i-1|k)}{\partial \boldsymbol{u}(k+j|k)}, & i-j>1,\\ 0, & i-j<1,\end{cases}
$$

with

$$
\begin{aligned}
G(i)=&\left.\frac{\partial \boldsymbol{x}(k+i|k)}{\partial \boldsymbol{u}(k+i-1|k)}\right|_{z(k+j|k)=z(k-1)}=a\,\frac{\partial g(\hat{\boldsymbol{\xi}}(i))}{\partial \hat{\boldsymbol{\xi}}(i)}W_{\text{in}},\\
H(i)=&\left.\frac{\partial \boldsymbol{x}(k+i|k)}{\partial \boldsymbol{x}(k+i-1|k)}\right|_{z(k+j|k)=z(k-1)}=(1-a)I+a\,\frac{\partial g(\hat{\boldsymbol{\xi}}(i))}{\partial \hat{\boldsymbol{\xi}}(i)}(W_{\text{gc}}+W_{\text{fb}}W_{\text{out}}).
\end{aligned}
$$

Note that $g(\cdot)=\tanh(\cdot)$, and thus we have $\partial g(\hat{\boldsymbol{\xi}}(i))/\partial \hat{\boldsymbol{\xi}}(i)=1-\tanh^2(\hat{\boldsymbol{\xi}}(i))$. The following variables are defined:

$$\overline{\boldsymbol{y}} = \begin{bmatrix} \boldsymbol{y}(k+1|k) \\ \boldsymbol{y}(k+2|k) \\ \vdots \\ \boldsymbol{y}(k+N|k) \end{bmatrix} \in \mathbb{R}^{rN}, \overline{\hat{\boldsymbol{y}}} = \begin{bmatrix} \hat{\boldsymbol{y}}(k+1|k) \\ \hat{\boldsymbol{y}}(k+2|k) \\ \vdots \\ \hat{\boldsymbol{y}}(k+N|k) \end{bmatrix} \in \mathbb{R}^{rN}, \overline{\boldsymbol{r}}_{\mathrm{d}} = \begin{bmatrix} \boldsymbol{r}_{\mathrm{d}}(k+1|k) \\ \boldsymbol{r}_{\mathrm{d}}(k+2|k) \\ \vdots \\ \boldsymbol{r}_{\mathrm{d}}(k+N|k) \end{bmatrix} \in \mathbb{R}^{rN},$$

$$\overline{z} = \begin{bmatrix} z(k|k) \\ z(k+1|k) \\ \vdots \\ z(k+N_u-1|k) \end{bmatrix} \in \mathbb{R}^{sN_u}, \Delta\overline{z} = \begin{bmatrix} \Delta z(k|k) \\ \Delta z(k+1|k) \\ \vdots \\ \Delta z(k+N_u-1|k) \end{bmatrix} \in \mathbb{R}^{sN_u},$$

$$\overline{z}_{k-1} = \begin{bmatrix} z(k-1) \\ z(k-1) \\ \vdots \\ z(k-1) \end{bmatrix} \in \mathbb{R}^{sN_u}, C = \begin{bmatrix} I & 0 & \cdots & 0 \\ I & I & \cdots & 0 \\ \vdots & \vdots & \ddots & \vdots \\ I & I & \cdots & I \end{bmatrix} \in \mathbb{R}^{sN_u \times sN_u},$$

$$E = \begin{bmatrix} E(1,0) & E(1,1) & \cdots & E(1,N_u-1) \\ E(2,0) & E(2,1) & \cdots & E(2,N_u-1) \\ \vdots & \vdots & \ddots & \vdots \\ E(N,0) & E(N,1) & \cdots & E(N,N_u-1) \end{bmatrix} \in \mathbb{R}^{rN \times sN_u},$$

where $\Delta z(k+i|k) = z(k+i|k) - z(k+i-1|k)$ represents the control input increment with $i = 0, 1, \cdots, N_u - 1$. Based on the previous derivation, the predicted output $\overline{\boldsymbol{y}}$ is given by

$$\overline{\boldsymbol{y}} = EC\Delta\overline{z} + \overline{\hat{\boldsymbol{y}}}. \tag{5.8}$$

Besides, the relationship between the future control input and the future control input increment is formulated as

$$\overline{z} = C\Delta\overline{z} + \overline{z}_{k-1}.$$

5.3.3 Online Training and Correction

Typically, the ESN model is employed as a prediction model following offline training, as seen in Jordanou et al. (2022a). However, the cerebellum model (5.1)–(5.3), when trained offline, exhibits the suboptimal performance in prediction tasks in our previous work. Therefore, to enhance the prediction accuracy of cerebellum model (5.1)–(5.3), this study focuses on its online training approach. It is important to highlight that the training of cerebellum model (5.1)–(5.3) primarily involves updating the output weight matrix W_{out}, and therefore the online training process focuses on the update of W_{out}.

Apart from the parameters within the cerebellum model (5.1)–(5.3), the quality of the training data also plays a crucial role in determining its prediction accuracy. In other words, examining the training algorithm (5.4) for the output weight matrix

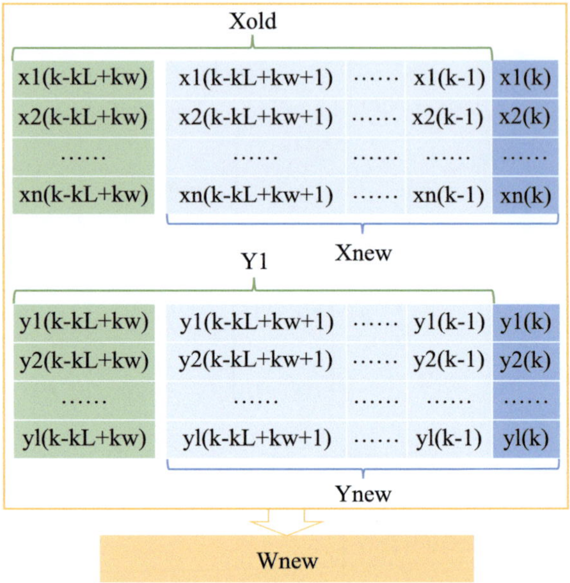

Fig. 5.1 Update mechanism for state matrix X, output matrix Y, and output weight matrix W_{out} of cerebellum model (5.1)–(5.3)

W_{out} reveals that W_{out} is strongly dependent on the state matrix X and output matrix Y, both of which are derived from the training data. Hence, by updating the state matrix X and output matrix Y in real time, the process of updating the output weight matrix W_{out} can be further enhanced. The update mechanism for the state matrix X, output matrix Y, and output weight matrix W_{out} of the cerebellum model (5.1)–(5.3) is illustrated in Fig. 5.1. Building upon the update mechanism depicted in Fig. 5.1, this study introduces an online training method for the cerebellum model (5.1)–(5.3), which is summarized in Algorithm 1.

Algorithm 1 Online Training of Cerebellum Model (5.1)–(5.3)

1. Feed the last input $\boldsymbol{u}(k-1)$ to the granular cells via (5.1) and further obtain the current state $\boldsymbol{x}(k)$ in the parallel fibers using (5.2).
2. Obtain the current output $\boldsymbol{y}(k)$ from the redundant manipulator through measurement.
3. Generate a new state X_{new} and a new output Y_{new}:
   a. Add $\boldsymbol{x}(k)$ and $\boldsymbol{y}(k)$ to X and Y, respectively.
   b. Discard the first state $\boldsymbol{x}(k-k_{\text{L}}+k_{\text{w}})$ in X and the first output $\boldsymbol{y}(k-k_{\text{L}}+k_{\text{w}})$ in Y.
4. Adopt ridge regression algorithm (5.4) to update output weight matrix W_{out} with X_{new} and Y_{new}:

$$W_{\text{out}} = Y_{\text{new}} X_{\text{new}}^{\text{T}} (X_{\text{new}} X_{\text{new}}^{\text{T}} + \lambda I)^{-1}.$$

Additionally, considering the prediction error of cerebellum model (5.1)–(5.3) and the error induced by the linearization operation described in the previous subsection, it is essential to include a correction term in Eq. (5.8) to adjust the predicted output $\overline{\boldsymbol{y}}$. The predicted output derived from Eq. (5.8) at time $k-1$ is represented as $\boldsymbol{y}_{\mathrm{li}}(k|k-1)$, while the predicted output incorporating the correction term is denoted as $\boldsymbol{y}(k|k-1)$, which is expressed as

$$\boldsymbol{y}(k|k-1) = \boldsymbol{y}_{\mathrm{li}}(k|k-1) + \boldsymbol{\sigma}(k-1),$$

with $\boldsymbol{\sigma}(k-1) \in \mathbb{R}^p$ standing for the correction term at time $k-1$. The purpose of introducing the correction term $\boldsymbol{\sigma}(k-1)$ is to ensure that the predicted output $\boldsymbol{y}(k|k-1)$ closely matches the actual output $\boldsymbol{y}(k)$, which is measured from the redundant manipulator. Therefore, the ND approach is utilized to compute the suitable correction term. Initially, an error function is defined as $\boldsymbol{\varsigma}(k-1) = \boldsymbol{y}(k) - \boldsymbol{y}(k|k-1)$. Secondly, an evolution formula is presented to ensure that the error $\boldsymbol{\varsigma}$ converges to zero, causing the predicted output $\boldsymbol{y}(k|k-1)$ to approach the actual output $\boldsymbol{y}(k)$. The evolution formula is chosen as $\dot{\boldsymbol{\varsigma}} = -\varpi\boldsymbol{\varsigma}$ with $\varpi > 0 \in \mathbb{R}$, where the symbol $\cdot$ represents the operation of taking the time derivative. Thirdly, by injecting the error function into the evolution formula, we obtain the following:

$$\dot{\boldsymbol{\sigma}}(k-1) = \dot{\boldsymbol{y}}(k) - \dot{\boldsymbol{y}}_{\mathrm{li}}(k|k-1) + \varpi(\boldsymbol{y}(k) - \boldsymbol{y}(k|k-1)),$$

which is further discretized using the Euler formula to calculate the current correction term as follows:

$$\begin{aligned}\boldsymbol{\sigma}(k) =& \boldsymbol{\sigma}(k-1) + \boldsymbol{y}(k) - \boldsymbol{y}(k-1) - \boldsymbol{y}_{\mathrm{li}}(k|k-1) + \boldsymbol{y}_{\mathrm{li}}(k-1|k-2) \\ &+ \tau\varpi(\boldsymbol{y}(k) - \boldsymbol{y}(k|k-1)).\end{aligned} \tag{5.9}$$

Finally, the correction term, which is calculated by the above ND-based correction algorithm (5.9), is added to Eq. (5.8), resulting in the modified predicted output:

$$\overline{\boldsymbol{y}} = EC\Delta\overline{\boldsymbol{z}} + \overline{\hat{\boldsymbol{y}}} + \overline{\boldsymbol{\sigma}}, \tag{5.10}$$

where $\overline{\boldsymbol{\sigma}} = [\boldsymbol{\sigma}(k)^{\mathrm{T}}, \boldsymbol{\sigma}(k)^{\mathrm{T}}, \cdots, \boldsymbol{\sigma}(k)^{\mathrm{T}}]^{\mathrm{T}} \in \mathbb{R}^{rN}$. In this work, the aforementioned equation is employed to predict the future output of the redundant manipulator, eliminating the need for its explicit model.

5.3.4 MPC Scheme

A key advantage of MPC is its ability to directly handle system constraints. For kinematic control of the redundant manipulator, constraints encompass joint limits at various levels, such as joint angle, joint velocity, and joint acceleration. By reviewing the definition of control input $\boldsymbol{z} = \tau\dot{\boldsymbol{\theta}}$, the joint angle, joint velocity, and joint acceleration are expressed as follows:

$$\overline{\boldsymbol{\theta}} = CC\Delta\overline{z} + C\overline{z}_{k-1} + \overline{\boldsymbol{\theta}_k}, \quad \overline{\dot{\boldsymbol{\theta}}} = \frac{1}{\tau}(C\Delta\overline{z} + \overline{z}_{k-1}), \quad \overline{\ddot{\boldsymbol{\theta}}} = \frac{1}{\tau^2}\Delta\overline{z},$$

with

$$\overline{\boldsymbol{\theta}} = \begin{bmatrix} \boldsymbol{\theta}(k+1|k) \\ \boldsymbol{\theta}(k+2|k) \\ \vdots \\ \boldsymbol{\theta}(k+N_u|k) \end{bmatrix} \in \mathbb{R}^{sN_u}, \overline{\dot{\boldsymbol{\theta}}} = \begin{bmatrix} \dot{\boldsymbol{\theta}}(k|k) \\ \dot{\boldsymbol{\theta}}(k+1|k) \\ \vdots \\ \dot{\boldsymbol{\theta}}(k+N_u-1|k) \end{bmatrix} \in \mathbb{R}^{sN_u},$$

$$\overline{\ddot{\boldsymbol{\theta}}} = \begin{bmatrix} \ddot{\boldsymbol{\theta}}(k|k) \\ \ddot{\boldsymbol{\theta}}(k+1|k) \\ \vdots \\ \ddot{\boldsymbol{\theta}}(k+N_u-1|k) \end{bmatrix} \in \mathbb{R}^{sN_u}, \overline{\boldsymbol{\theta}_k} = \begin{bmatrix} \boldsymbol{\theta}(k) \\ \boldsymbol{\theta}(k) \\ \vdots \\ \boldsymbol{\theta}(k) \end{bmatrix} \in \mathbb{R}^{sN_u}.$$

Finally, the MPC framework for the redundant manipulator with unknown structural information is formulated as

$$\min_{\Delta\overline{z}} \|\overline{\boldsymbol{y}} - \overline{\boldsymbol{r}}_{\mathrm{d}}\|_Q^2 + \|\overline{z}\|_{V_1}^2 + \|\Delta\overline{z}\|_{V_2}^2, \tag{5.11a}$$

$$\text{s.t. } \overline{\boldsymbol{\theta}^-} \le \overline{\boldsymbol{\theta}} \le \overline{\boldsymbol{\theta}^+}, \tag{5.11b}$$

$$\overline{\dot{\boldsymbol{\theta}}^-} \le \overline{\dot{\boldsymbol{\theta}}} \le \overline{\dot{\boldsymbol{\theta}}^+}, \tag{5.11c}$$

$$\overline{\ddot{\boldsymbol{\theta}}^-} \le \overline{\ddot{\boldsymbol{\theta}}} \le \overline{\ddot{\boldsymbol{\theta}}^+}, \tag{5.11d}$$

with

$$\overline{\boldsymbol{\theta}^-} = \begin{bmatrix} \boldsymbol{\theta}^- \\ \boldsymbol{\theta}^- \\ \vdots \\ \boldsymbol{\theta}^- \end{bmatrix} \in \mathbb{R}^{sN_u}, \overline{\dot{\boldsymbol{\theta}}^-} = \begin{bmatrix} \dot{\boldsymbol{\theta}}^- \\ \dot{\boldsymbol{\theta}}^- \\ \vdots \\ \dot{\boldsymbol{\theta}}^- \end{bmatrix} \in \mathbb{R}^{sN_u}, \overline{\ddot{\boldsymbol{\theta}}^-} = \begin{bmatrix} \ddot{\boldsymbol{\theta}}^- \\ \ddot{\boldsymbol{\theta}}^- \\ \vdots \\ \ddot{\boldsymbol{\theta}}^- \end{bmatrix} \in \mathbb{R}^{sN_u},$$

$$\overline{\boldsymbol{\theta}^+} = \begin{bmatrix} \boldsymbol{\theta}^+ \\ \boldsymbol{\theta}^+ \\ \vdots \\ \boldsymbol{\theta}^+ \end{bmatrix} \in \mathbb{R}^{sN_u}, \overline{\dot{\boldsymbol{\theta}}^+} = \begin{bmatrix} \dot{\boldsymbol{\theta}}^+ \\ \dot{\boldsymbol{\theta}}^+ \\ \vdots \\ \dot{\boldsymbol{\theta}}^+ \end{bmatrix} \in \mathbb{R}^{sN_u}, \overline{\ddot{\boldsymbol{\theta}}^+} = \begin{bmatrix} \ddot{\boldsymbol{\theta}}^+ \\ \ddot{\boldsymbol{\theta}}^+ \\ \vdots \\ \ddot{\boldsymbol{\theta}}^+ \end{bmatrix} \in \mathbb{R}^{sN_u},$$

where $Q \in \mathbb{R}^{rN\times rN}$, $V_1 \in \mathbb{R}^{sN_u\times sN_u}$, and $V_2 \in \mathbb{R}^{sN_u\times sN_u}$ represent positive definite and symmetric weight matrices; $\boldsymbol{\theta}^-$, $\dot{\boldsymbol{\theta}}^-$, and $\ddot{\boldsymbol{\theta}}^-$ represent the lower bounds of the joint angle, joint velocity, and joint acceleration, respectively; and $\boldsymbol{\theta}^+$, $\dot{\boldsymbol{\theta}}^+$, and $\ddot{\boldsymbol{\theta}}^+$ represent the upper bounds of the joint angle, joint velocity, and joint acceleration, respectively. In the MPC scheme described in Eq. (5.11), the cost function incorporates not only the position error but also the control input and its increment, aiming to ensure smooth control.

5.4 DDNN Model

This part focuses on the design of the DDNN model to address the MPC scheme in Eq. (5.11). To begin with, the MPC scheme in Eq. (5.11) is reformulated as a QP problem:

$$\min_{\Delta\overline{z}} \frac{1}{2}\Delta\overline{z}^{\mathrm{T}} K \Delta\overline{z} + \boldsymbol{b}^{\mathrm{T}}\Delta\overline{z}, \tag{5.12a}$$

$$\text{s.t. } J\Delta\overline{z} \leq \boldsymbol{d}, \tag{5.12b}$$

where

$$K = 2(C^{\mathrm{T}}E^{\mathrm{T}}QEC + C^{\mathrm{T}}V_1C + V_2) \in \mathbb{R}^{\kappa\times\kappa},$$
$$\boldsymbol{b} = 2C^{\mathrm{T}}E^{\mathrm{T}}Q(\hat{\overline{y}} - \overline{\boldsymbol{r}}_{\mathrm{d}}) + 2C^{\mathrm{T}}V_1\overline{z}_{k-1}) \in \mathbb{R}^{\kappa},$$
$$J = \begin{bmatrix} CC \\ -CC \\ C \\ -C \\ I \\ -I \end{bmatrix} \in \mathbb{R}^{\iota\times\kappa}, \boldsymbol{d} = \begin{bmatrix} \overline{\boldsymbol{\theta}}^{+} - \overline{\boldsymbol{\theta}_k} - C\overline{z}_{k-1} \\ -\overline{\boldsymbol{\theta}}^{-} + \overline{\boldsymbol{\theta}_k} + C\overline{z}_{k-1} \\ \tau\overline{\dot{\boldsymbol{\theta}}}^{+} - \overline{z}_{k-1} \\ -\tau\overline{\dot{\boldsymbol{\theta}}}^{-} + \overline{z}_{k-1} \\ \tau^2\overline{\ddot{\boldsymbol{\theta}}}^{+} \\ -\tau^2\overline{\ddot{\boldsymbol{\theta}}}^{-} \end{bmatrix} \in \mathbb{R}^{\iota},$$

with $\iota = 6sN_u$ and $\kappa = sN_u$. In contrast to cost function (5.11a,) cost function (5.12a) omits constants that are independent of $\Delta\overline{z}$. Since QP problem (5.12) is a convex optimization problem, its optimal solution $\Delta\overline{z}^*$ corresponds to the optimal solution to the following Karush–Kuhn–Tucker (KKT) conditions:

$$\begin{cases} K\Delta\overline{z}^* + \boldsymbol{b} + J^{\mathrm{T}}\boldsymbol{\mu}^* = \boldsymbol{0}, & (5.13a) \\ \boldsymbol{\mu}^* \geq \boldsymbol{0}, \boldsymbol{d} - J\Delta\overline{z}^* \geq \boldsymbol{0}, & (5.13b) \\ \boldsymbol{\mu}^{*\mathrm{T}}(\boldsymbol{d} - J\Delta\overline{z}^*) = 0, & (5.13c) \end{cases}$$

of which $\boldsymbol{\mu}^* \in \mathbb{R}^{\iota}$ signifies the Lagrangian multiplier. It can be found that Eqs. (5.13b) and (5.13c) form a nonlinear complementarity problem (NCP), which can be reformulated using an appropriate NCP function, such as the one constructed in Chen et al. (2000):

$$\varphi_\gamma(\boldsymbol{\alpha}, \boldsymbol{\beta}) = \gamma\,(\boldsymbol{\alpha} + \boldsymbol{\beta} - \sqrt{\boldsymbol{\alpha} \circ \boldsymbol{\alpha} + \boldsymbol{\beta} \circ \boldsymbol{\beta}}) + (1-\gamma)\boldsymbol{\alpha}_+ \circ \boldsymbol{\beta}_+ = \mathbf{0},$$
$$\Longleftrightarrow \boldsymbol{\alpha} \geq \mathbf{0}, \boldsymbol{\beta} \geq \mathbf{0}, \boldsymbol{\alpha}^{\mathrm{T}}\boldsymbol{\beta} = 0,$$

where $\boldsymbol{\alpha} \in \mathbb{R}^{\iota}$; $\boldsymbol{\beta} \in \mathbb{R}^{\iota}$; $\gamma \in (0, 1)$; $\circ$ represents the Hadamard product; $\boldsymbol{\alpha}_+ = \max\{\mathbf{0}, \boldsymbol{\alpha}\}$; $\boldsymbol{\beta}_+ = \max\{\mathbf{0}, \boldsymbol{\beta}\}$. Hence, KKT conditions (5.13) are converted into

$$\begin{cases} K\overline{\Delta z}^* + \boldsymbol{b} + J^{\mathrm{T}}\boldsymbol{\mu}^* = \mathbf{0}, \\ \varphi_\gamma(\boldsymbol{d} - J\Delta\bar{z}^*, \boldsymbol{\mu}^*) = \mathbf{0}, \end{cases}$$

which can be further expressed in a compact form as follows:

$$\Psi\boldsymbol{\eta} + \boldsymbol{\varphi}(\boldsymbol{\eta}) = \mathbf{0}, \tag{5.14}$$

where

$$\Psi = \begin{bmatrix} K & J^{\mathrm{T}} \\ -\gamma J & \gamma I \end{bmatrix} \in \mathbb{R}^{(\kappa+\iota)\times(\kappa+\iota)}, \boldsymbol{\eta} = \begin{bmatrix} \Delta z \\ \boldsymbol{\mu} \end{bmatrix} \in \mathbb{R}^{\kappa+\iota}, \boldsymbol{\varphi}(\boldsymbol{\eta}) = \begin{bmatrix} \boldsymbol{b} \\ \gamma\boldsymbol{d} + \boldsymbol{\epsilon} \end{bmatrix} \in \mathbb{R}^{\kappa+\iota},$$
$$\boldsymbol{\epsilon} = -\gamma\boldsymbol{\varepsilon} + (1-\gamma)(\boldsymbol{d} - J\Delta\bar{z})_+ \circ \boldsymbol{\mu}_+ \in \mathbb{R}^{\iota},$$
$$\boldsymbol{\varepsilon} = \sqrt{(\boldsymbol{d} - J\Delta\bar{z}) \circ (\boldsymbol{d} - J\Delta\bar{z}) + \boldsymbol{\mu} \circ \boldsymbol{\mu}} \in \mathbb{R}^{\iota}.$$

It should be noted that the initial κ components of the solution to the nonlinear equation (5.14) represent the optimal solution of the MPC scheme (5.11). To address the nonlinear equation (5.14), a DDNN model is subsequently constructed. Initially, an error function is introduced as $\boldsymbol{\xi} = \Psi\boldsymbol{\eta} + \boldsymbol{\varphi}(\boldsymbol{\eta})$. Once the error function converges to zero, the resulting solution $\boldsymbol{\eta}$ corresponds to the theoretical solution to Eq. (5.14). Therefore, to drive the error $\boldsymbol{\xi}$ toward zero, the following evolution law is adopted: $\dot{\boldsymbol{\xi}} = -\nu\boldsymbol{\xi}$ with $\nu > 0 \in \mathbb{R}$. Combining the error function and evolution formula generates the following model:

$$\dot{\boldsymbol{\eta}} = -\nu(\Psi + \Phi(\boldsymbol{\eta}))^{\dagger}(\Psi\boldsymbol{\eta} + \boldsymbol{\varphi}(\boldsymbol{\eta})),$$

where

$$\Phi(\boldsymbol{\eta}) = \begin{bmatrix} 0 & 0 \\ \gamma\Lambda_1 J - (1-\gamma)\Lambda_3 J & -\gamma\Lambda_2 + (1-\gamma)\Lambda_4 \end{bmatrix},$$
$$\Lambda_1 = \mathrm{diag}(\boldsymbol{\chi}), \Lambda_3 = \mathrm{diag}(\partial(\boldsymbol{d} - J\Delta\bar{z})_+ \circ \boldsymbol{\mu}_+), \boldsymbol{\chi} \in \mathbb{R}^{\iota},$$
$$\Lambda_2 = \mathrm{diag}(\boldsymbol{\upsilon}), \Lambda_4 = \mathrm{diag}((\boldsymbol{d} - J\Delta\bar{z})_+ \circ \partial\boldsymbol{\mu}_+), \boldsymbol{\upsilon} \in \mathbb{R}^{\iota},$$

$$\chi_i = \begin{cases} \dfrac{(\boldsymbol{d} - J\Delta\bar{z})_i}{\varepsilon_i}, & \varepsilon_i \neq 0, \\ [0, 0.5], & \varepsilon_i = 0, \end{cases}, \upsilon_i = \begin{cases} \dfrac{\mu_i}{\varepsilon_i}, & \varepsilon_i \neq 0, \\ [0, 0.5], & \varepsilon_i = 0, \end{cases}, \partial\omega_+ = \begin{cases} 1, & \omega > 0, \\ [0, 1], & \omega = 0, \\ 0, & \omega < 0. \end{cases}$$

Moreover, χ_i, $(\boldsymbol{d} - J\Delta\overline{\mathbf{z}})_i$, ε_i, υ_i, and μ_i denote the ith component of the vectors $\boldsymbol{\chi}$, $\boldsymbol{d} - J\Delta\overline{\mathbf{z}}$, $\boldsymbol{\varepsilon}$, $\boldsymbol{\upsilon}$, and $\boldsymbol{\mu}$, respectively; diag$(\cdot)$ denotes the diagonal matrix; and superscript † denotes the pseudoinverse operation. By applying the Euler formula for discretizing the above model, the following DDNN model is obtained:

$$\boldsymbol{\eta}_{i+1} = \boldsymbol{\eta}_i - \rho(\Psi + \Phi(\boldsymbol{\eta}_i))^\dagger(\Psi\boldsymbol{\eta}_i + \boldsymbol{\varphi}(\boldsymbol{\eta}_i)), \tag{5.15}$$

of which $\rho = \nu\delta$ with δ symbolizing the step size.

The first s elements of the solution $\Delta\overline{\mathbf{z}}$ obtained from the ND solver in (5.15) are used to compute the current control input $\mathbf{z}(k|k)$, where the current control input increment $\Delta\mathbf{z}(k|k)$ is determined. Subsequently, the current control input $\mathbf{z}(k|k)$ is processed to compute the joint velocity $\dot{\boldsymbol{\theta}}(k)$, which is then fed to the redundant manipulator prior to the arrival of the next time step.

By integrating cerebellum model (5.1)–(5.3), the online training algorithm, the ND-based correction algorithm (5.9), the MPC scheme (5.11), and the DDNN model (5.15), the CIMPC algorithm introduced in this chapter can be concisely presented in Algorithm 2.

Algorithm 2 CIMPC Algorithm

1. The initial output weight matrix W_{out} and the other weight matrices of the cerebellum model (5.1)–(5.3) are obtained through offline training, as outlined in the training method described in Sect. 5.2.2.
2. Start with initial states: $\boldsymbol{\theta}(0)$ and $\boldsymbol{r}(0)$. Set parameters: N, N_u, Q, V_1, and V_2.
3. for $k = 1, 2, \ldots$ do:

   a. Update W_{out} by Algorithm 1.
   b. Calculate the free response and forced response of cerebellum model (5.1)–(5.3) by (5.7).
   c. Calculate the correction term by ND-based correction algorithm (5.9).
   d. Predict the future output $\overline{\mathbf{y}}$ by (5.10).
   e. Update MPC scheme (5.11) by $\overline{\mathbf{y}}$ and $\overline{\boldsymbol{r}_\text{d}}$.
   f. Solve MPC scheme (5.11) by (5.15), and obtain the control input increment $\Delta\mathbf{z}(k|k)$.
   g. Calculate the joint velocity $\dot{\boldsymbol{\theta}}(k)$ by $\Delta\mathbf{z}(k|k)$ and $\mathbf{z}(k-1)$.

   end.

5.5 Theoretical Analyses

In this section, the convergence of the ND-based correction algorithm (5.9) and the DDNN model (5.15) is demonstrated through theoretical analyses.

5.5.1 Convergence of Correction Algorithm

Initially, the following theorem establishes the convergence of the ND-based correction algorithm (5.9).

Theorem 5.1 *Driven by ND-based correction algorithm (5.9), provided that* $0 < \tau\varpi < 2$, *the residual error* $\lim_{k\to\infty} \|\boldsymbol{\varsigma}(k)\|_2$ *converges to* $o(\tau^2)$.

Proof Correction algorithm (5.9) is rewritten as follows:

$$
\begin{aligned}
\boldsymbol{\sigma}(k) =& \boldsymbol{\sigma}(k-1) + \boldsymbol{y}(k) - \boldsymbol{y}(k-1) - \boldsymbol{y}_{\text{li}}(k|k-1) + \boldsymbol{y}_{\text{li}}(k-1|k-2) \\
&+ \tau\varpi(\boldsymbol{y}(k) - \boldsymbol{y}(k|k-1)) + \boldsymbol{o}(\tau^2),
\end{aligned}
$$

where $\boldsymbol{o}(\tau^2)$ denotes the error resulting from the Euler formula $\dot{\boldsymbol{\sigma}}(k-1) = (\boldsymbol{\sigma}(k) - \boldsymbol{\sigma}(k-1))/\tau + \boldsymbol{o}(\tau)$. By applying the Euler formula again, we obtain

$$
\begin{aligned}
\tau\dot{\boldsymbol{y}}(k) &= \boldsymbol{y}(k) - \boldsymbol{y}(k-1) + \boldsymbol{o}(\tau^2) = \boldsymbol{y}(k+1) - \boldsymbol{y}(k) + \boldsymbol{o}(\tau^2), \\
\tau\dot{\boldsymbol{y}}_{\text{li}}(k|k-1) &= \boldsymbol{y}_{\text{li}}(k|k-1) - \boldsymbol{y}_{\text{li}}(k-1|k-2) + \boldsymbol{o}(\tau^2) \\
&= \boldsymbol{y}_{\text{li}}(k+1|k) - \boldsymbol{y}_{\text{li}}(k|k-1) + \boldsymbol{o}(\tau^2).
\end{aligned}
$$

By combining the results above, we can rewrite the ND-based correction algorithm (5.9) as

$$
\begin{aligned}
\boldsymbol{\sigma}(k) =& \boldsymbol{\sigma}(k-1) + \boldsymbol{y}(k+1) - \boldsymbol{y}(k) - \boldsymbol{y}_{\text{li}}(k+1|k) + \boldsymbol{y}_{\text{li}}(k|k-1) \\
&+ \tau\varpi(\boldsymbol{y}(k) - \boldsymbol{y}(k|k-1)) + \boldsymbol{o}(\tau^2).
\end{aligned}
$$

Taking into account $\boldsymbol{\varsigma}(k-1) = \boldsymbol{y}(k) - \boldsymbol{y}(k|k-1) = \boldsymbol{y}(k) - \boldsymbol{y}_{\text{li}}(k|k-1) - \boldsymbol{\sigma}(k-1)$, the above equation is further converted as

$$
\boldsymbol{\varsigma}(k) = (1 - \tau\varpi)\boldsymbol{\varsigma}(k-1) + \boldsymbol{o}(\tau^2), \tag{5.16}
$$

whose jth subsystem is as follows:

$$
\varsigma_j(k) = (1 - \tau\varpi)\varsigma_j(k-1) + o(\tau^2).
$$

The following results can be obtained through iteration:

$$
\begin{aligned}
\varsigma_j(k) &= (1 - \tau\varpi)\varsigma_j(k-1) + o(\tau^2) \\
&= (1 - \tau\varpi)^2\varsigma_j(k-2) + o(\tau^2) \\
&\ \ \vdots \\
&= (1 - \tau\varpi)^k\varsigma_j(0) + o(\tau^2).
\end{aligned}
$$

If $\varsigma_j(0) = 0$, then $\varsigma_j(k) = o(\tau^2)$. Besides, provided that $0 < \tau\varpi < 2$, one can derive that $\lim_{k\to\infty}(1 - \tau\varpi)^k = 0$. Consequently, if $\varsigma_j(0) \neq 0$, then $\lim_{k\to\infty} \varsigma_j(k) = o(\tau^2)$. Combining the above analyses, one concludes that $\lim_{k\to\infty} \|\boldsymbol{\varsigma}(k)\|_2 = o(\tau^2)$.

5.5.2 Convergence of DDNN Model

Subsequently, the following theorem ensures the convergence of the DDNN model (5.15).

Theorem 5.2 *When solving nonlinear equation (5.14), provided that* $0 < \rho < 2$, *solution* $\boldsymbol{\eta}$ *obtained by DDNN model (5.15) converges to its theoretical solution* $\boldsymbol{\eta}^*$ *with a residual error of* $\boldsymbol{o}(\delta^2)$, *whose first* κ *elements constitute the optimal solution of MPC scheme (5.11).*

Proof The presented DDNN model (5.15) can be reformulated as

$$(\Psi + \Phi_i)(\boldsymbol{\eta}_{i+1} - \boldsymbol{\eta}_i) = -\rho(\Psi\boldsymbol{\eta}_i + \boldsymbol{\varphi}_i) + \boldsymbol{o}(\delta^2),$$

where $\Phi_i = \Phi(\boldsymbol{\eta}_i)$, $\boldsymbol{\varphi}_i = \boldsymbol{\varphi}(\boldsymbol{\eta}_i)$, and $\boldsymbol{o}(\delta^2)$ represents the discretization error introduced by applying the Euler formula. That is, based on Euler formula $\dot{\boldsymbol{\eta}}_i = (\boldsymbol{\eta}_{i+1} - \boldsymbol{\eta}_i)/\delta + \boldsymbol{o}(\delta)$, we know that $\boldsymbol{\eta}_{i+1} = \boldsymbol{\eta}_i + \delta\dot{\boldsymbol{\eta}}_i + \boldsymbol{o}(\delta^2)$. The above expression can be further reformulated as

$$\Psi\boldsymbol{\eta}_{i+1} - \Psi\boldsymbol{\eta}_i + \Phi_i(\boldsymbol{\eta}_{i+1} - \boldsymbol{\eta}_i) = -\rho(\Psi\boldsymbol{\eta}_i + \boldsymbol{\varphi}_i) + \boldsymbol{o}(\delta^2).$$

Following the Euler formula, the expression $\Phi_i(\boldsymbol{\eta}_{i+1} - \boldsymbol{\eta}_i)$ can be transformed into the form below:

$$\Phi_i(\boldsymbol{\eta}_{i+1} - \boldsymbol{\eta}_i) = \delta\Phi_i\dot{\boldsymbol{\eta}}_i + \boldsymbol{o}(\delta^2) = \delta\dot{\boldsymbol{\varphi}}_i + \boldsymbol{o}(\delta^2) = \boldsymbol{\varphi}_{i+1} - \boldsymbol{\varphi}_i + \boldsymbol{o}(\delta^2).$$

Therefore, DDNN model (5.15) is further converted as

$$\Psi\boldsymbol{\eta}_{i+1} - \Psi\boldsymbol{\eta}_i + \boldsymbol{\varphi}_{i+1} - \boldsymbol{\varphi}_i = -\rho(\Psi\boldsymbol{\eta}_i + \boldsymbol{\varphi}_i) + \boldsymbol{o}(\delta^2).$$

By revisiting the definition of the error function, it follows that

$$\boldsymbol{\xi}_{i+1} - \boldsymbol{\xi}_i = -\rho\boldsymbol{\xi}_i + \boldsymbol{o}(\delta^2).$$

That is

$$\boldsymbol{\xi}_{i+1} = (1 - \rho)\boldsymbol{\xi}_i + \boldsymbol{o}(\delta^2).$$

It is worth noting that the above system shares similarities with system (5.16), and therefore the subsequent analyses are omitted for brevity. It is straightforward to deduce that $\lim_{i\to\infty} |\boldsymbol{\xi}_i|_2 = o(\delta^2)$ under the condition that $0 < \rho < 2$. In other words, the solution $\boldsymbol{\eta}$ produced by the DDNN model (5.15) approaches the exact solution of the nonlinear equation (5.14), with a remaining error on the order of $\boldsymbol{o}(\delta^2)$. Furthermore, it is concluded that the obtained $\Delta\bar{z}$, i.e., the first κ elements of $\boldsymbol{\eta}$, converges to the optimal solution of the MPC scheme (5.11) with a residual error of $\boldsymbol{o}(\delta^2)$.

5.6 Simulations and Experiments

This section presents computer simulations and physical experiments conducted with the redundant manipulator to validate the effectiveness and superiority of the CIMPC algorithm.

5.6.1 Preliminaries

The control object chosen for this study is a 7-DOF manipulator, the Franka Emika Panda (referred to as the Franka manipulator), and its Denavit–Hartenberg (D-H) parameters and joint limits are provided in Baek et al. (2022). The corresponding parameters are configured as follows: $\tau = 0.01$ s, $N = 3$, $N_u = 3$, $Q = 10 \times I_{9\times 9}$, $V_1 = I_{21\times 21}$, $V_2 = I_{21\times 21}$, $\gamma = 0.95$, $\varpi = 100$, $\rho = 1$, $m = 14$, $n = 340$, $l = 3$, $k_{\mathrm{w}} = 40$, $k_{\mathrm{L}} = 400$, and $\lambda = 10^{-8}$. In general, the prediction accuracy of the cerebellum model (5.1)–(5.3) is positively correlated with the size of the granular layer. However, increasing the granular layer size leads to higher computational costs for the presented algorithm. To guarantee the real-time performance of the presented algorithm, all computations within a loop must be completed within the sampling interval. Therefore, the granular layer size is set to 340, enabling high-precision tracking while maintaining real-time control. The data for offline training are collected from the Franka manipulator by feeding it a set of joint velocities.

Besides the aforementioned parameters, the leaking rate a and the spectral radius of W_{gc} also influence the prediction performance of cerebellum model (5.1)–(5.3), subsequently affecting the tracking accuracy of the manipulator. Thus, it is essential to determine the appropriate leaking rate and spectral radius of W_{gc} for the Franka manipulator system. The integral of absolute error (IAE) is introduced as a metric to evaluate the tracking performance of the Franka manipulator:

$$\mathrm{IAE} = \sum_{k=1}^{N_T} \left(|e_{\mathrm{X}}(k)| + |e_{\mathrm{Y}}(k)| + |e_{\mathrm{Z}}(k)|\right), \tag{5.17}$$

where $\boldsymbol{e}(k) = [e_{\mathrm{X}}(k), e_{\mathrm{Y}}(k), e_{\mathrm{Z}}(k)]^{\mathrm{T}}$ is the position error of the end-effector, and N_T denotes the iterations of the loop, i.e., $N_T = T/\tau$, with $T = 20$ s symbolizing the task duration. Subsequently, computer simulations are conducted to apply the CIMPC algorithm with various leaking rates and spectral radii of W_{gc}. The leaking rate varies

Fig. 5.2 Integral of absolute error (IAE) on the trajectory tracking of the Franka manipulator with the aid of the CIMPC algorithm under various leaking rates and spectral radii of W_{gc}

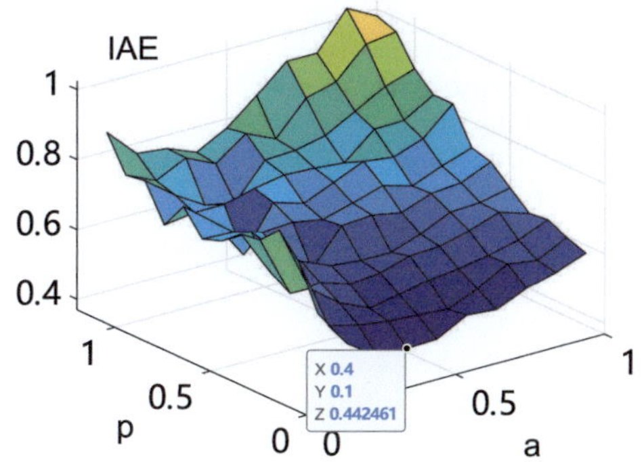

from 0.1 to 1, with increments of 0.1, while the spectral radius of W_{gc} ranges from 0.1 to 1.2 (Jordanou et al. 2022a), also with a step size of 0.1. Figure 5.2 displays the surface plot of IAE in relation to the leaking rate and spectral radius of W_{gc}, where a represents the leaking rate, and ϱ denotes the spectral radius of W_{gc}. The minimum IAE is achieved when $a = 0.4$ and $\varrho = 0.1$. Therefore, the leaking rate and spectral radius of W_{gc} are selected as 0.4 and 0.1, respectively.

5.6.2 Verification Results

The verification results for the Franka manipulator are presented in Figs. 5.3 and 5.4. In this task, the Franka manipulator is controlled using the CIMPC algorithm to track an epicycloid trajectory. Figure 5.3a shows the desired trajectory alongside the actual trajectory traced by the end-effector of the Franka manipulator. Additionally, the motion sequence of each link of the Franka manipulator is illustrated in Fig. 5.3b. Figure 5.4a depicts the actual and desired positions of the end-effector. In addition, the position error $\boldsymbol{e}$ of the end-effector is presented in Fig. 5.3c, which is divided into three components along the X-, Y-, and Z-axes, namely, e_X, e_Y, and e_Z. The tracking errors along the three axes are on the order of 10^{-4} m. Figure 5.4b and c illustrates the joint angles and joint velocities, respectively. According to the joint limits at various levels, the movement of each joint remains within the safe range. Thus, the CIMPC algorithm demonstrated its effectiveness on the Franka manipulator, and it operates without relying on its kinematic model.

Figure 5.5 illustrates the relative errors of the predicted output, both with and without the ND-based correction algorithm (5.9). The relative errors are defined as follows:

Fig. 5.3 Simulation results on the Franka manipulator for trajectory tracking with the aid of the CIMPC algorithm. **a** Actual trajectory and desired trajectory. **b** Motion process. **c** Position error

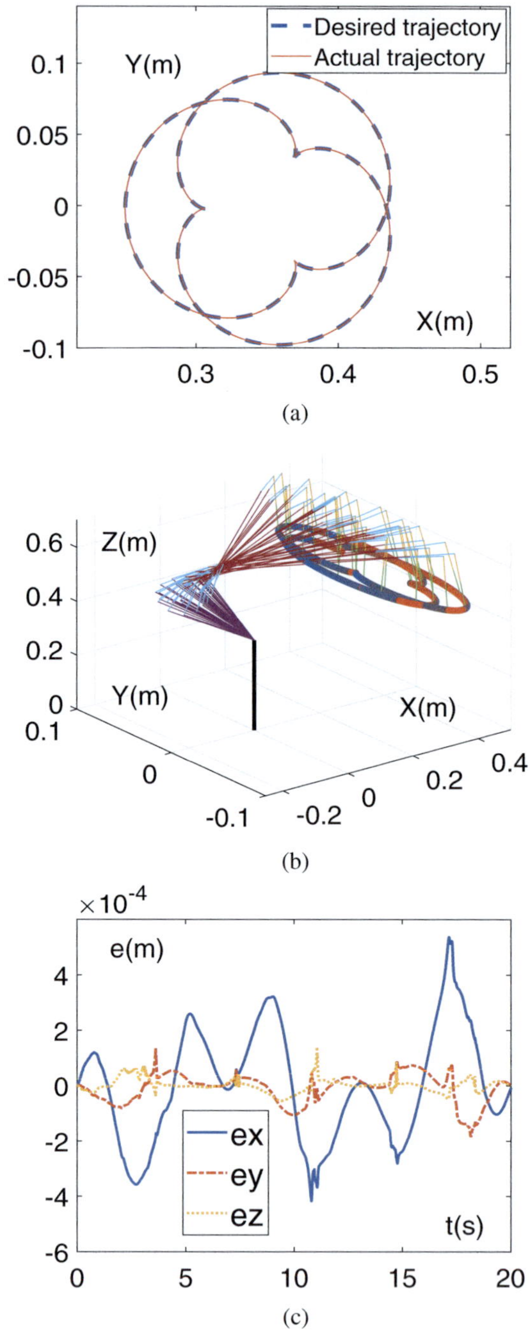

Fig. 5.4 State information of the Franka manipulator for trajectory tracking with the aid of the CIMPC algorithm. **a** Position of the end-effector. **b** Joint angle. **c** Joint velocity

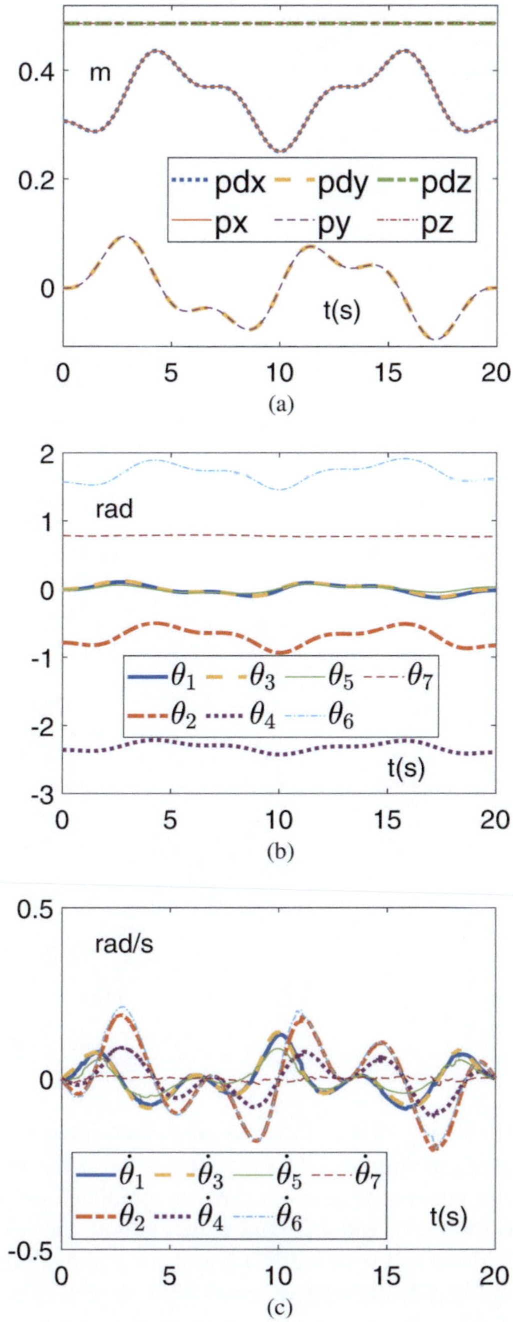

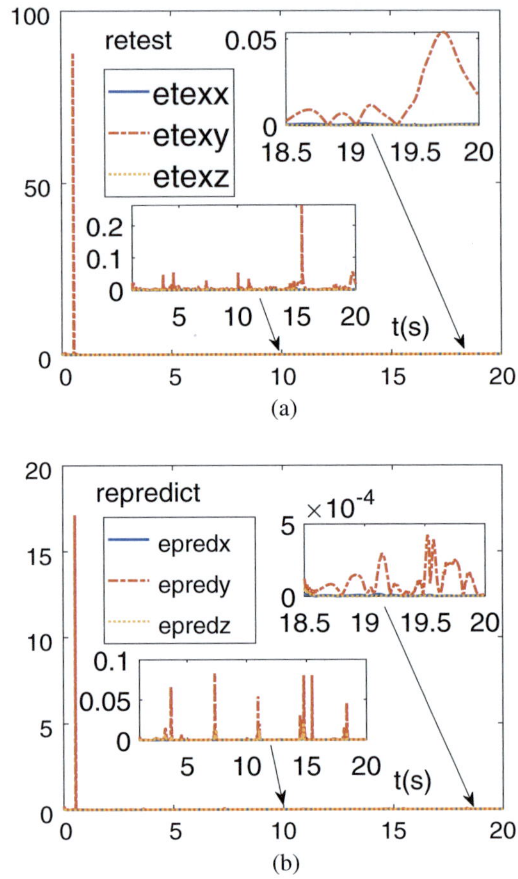

Fig. 5.5 Predicted output errors without or with algorithm (5.9). **a** Errors without correction. **b** Errors with correction

$$\boldsymbol{e}_{\mathrm{cere}} = \left| \frac{\boldsymbol{y}(k) - \boldsymbol{y}_{\mathrm{cere}}(k|k-1)}{\boldsymbol{y}(k)} \right| \times 100\%, \tag{5.18a}$$

$$\boldsymbol{e}_{\mathrm{predict}} = \left| \frac{\boldsymbol{y}(k) - \boldsymbol{y}_{\mathrm{predict}}(k|k-1)}{\boldsymbol{y}(k)} \right| \times 100\%, \tag{5.18b}$$

where $\boldsymbol{y}_{\mathrm{cere}}(k|k-1)$ is the predicted output of cerebellum model (5.1)–(5.3) computed at time $k-1$; $\boldsymbol{y}_{\mathrm{predict}}(k|k-1)$ is the predicted output according to (5.10) computed at time $k-1$, i.e., the predicted output with ND-based correction algorithm (5.9); and $\boldsymbol{e}_{\mathrm{cere}}$ and $\boldsymbol{e}_{\mathrm{predict}}$ denote the relative errors of the predicted output without and with the ND-based correction algorithm (5.9), respectively. As shown in Fig. 5.5, the relative errors with correction are significantly smaller compared to those without correction, demonstrating the necessity of the ND-based correction algorithm (5.9). Additionally, the IAE with the ND-based correction algorithm (5.9) is smaller than that without the correction algorithm.

Moreover, the experimental results for the Franka manipulator are shown in Fig. 5.6, illustrating the motion process of the Franka manipulator at various moments and the actual trajectory traced by its end-effector. As seen in Fig. 5.6, the Franka manipulator, guided by the CIMPC algorithm, effectively performs the trajectory-tracking task without depending on its kinematic model.

5.6.3 Comparison Results

To emphasize the advantages of the CIMPC algorithm, several existing algorithms are selected for comparison. The details are as follows:

1. *Linear MPC algorithm* (Baek et al. 2022): The algorithm is developed based on the kinematic model of the Franka manipulator, achieving satisfactory trajectory-tracking results, even though it utilizes linearization techniques. This algorithm is exemplified when the model of the redundant manipulator is available.
2. *ESN-based PNMPC (ESN-PNMPC) algorithm* (Jordanou et al. 2022a): The algorithm is developed by adopting the ESN model as the process model for the unknown system, which linearizes the ESN model by deriving its forced and free responses. The ESN model used in the algorithm is trained offline, and the trained weight matrices remain constant throughout the simulation. Additionally, an error correction filter is incorporated into the algorithm to adjust the predicted output, thereby mitigating the effects of interference and model mismatch.
3. *Adaptive PNMPC algorithm* (Schwedersky et al. 2022): The algorithm is developed using an online-trained ESN model, which serves as the process model for the unknown system. The output weight matrix of the ESN model is updated online using a recursive least-squares algorithm with an adaptive forgetting factor. Additionally, the linearization of the ESN model is achieved by utilizing its forced and free responses.

In a similar manner to the presented CIMPC algorithm, the three algorithms mentioned above are applied to the Franka manipulator for the same trajectory-tracking task. The parameter settings are as follows. For consistency, the prediction horizon N, control horizon N_u, and weight matrices Q, V_1, and V_2 are set to the same values as in Sect. 5.6.1. For the linear MPC algorithm (Baek et al. 2022), the remaining parameters are consistent with those provided in Baek et al. (2022). For both the ESN-PNMPC algorithm (Jordanou et al. 2022a) and the adaptive PNMPC algorithm (Schwedersky et al. 2022), the size of the reservoir is set equal to that of the granular layer in Sect. 5.6.1. Additionally, for the ESN-PNMPC algorithm (Jordanou et al. 2022a), the Tikhonov regularization parameter is set to be the same as in Sect. 5.6.1. The leaking rate and spectral radius of W_{gc} are chosen to be 0.1 and 0.6, respectively, which correspond to the minimum IAE for trajectory tracking. The parameters involved in the error correction filter are consistent with those in Jordanou et al. (2022a). For the adaptive PNMPC algorithm (Schwedersky et al. 2022), the spectral radius of W_{gc} is set to 0.1, which corresponds to the minimum

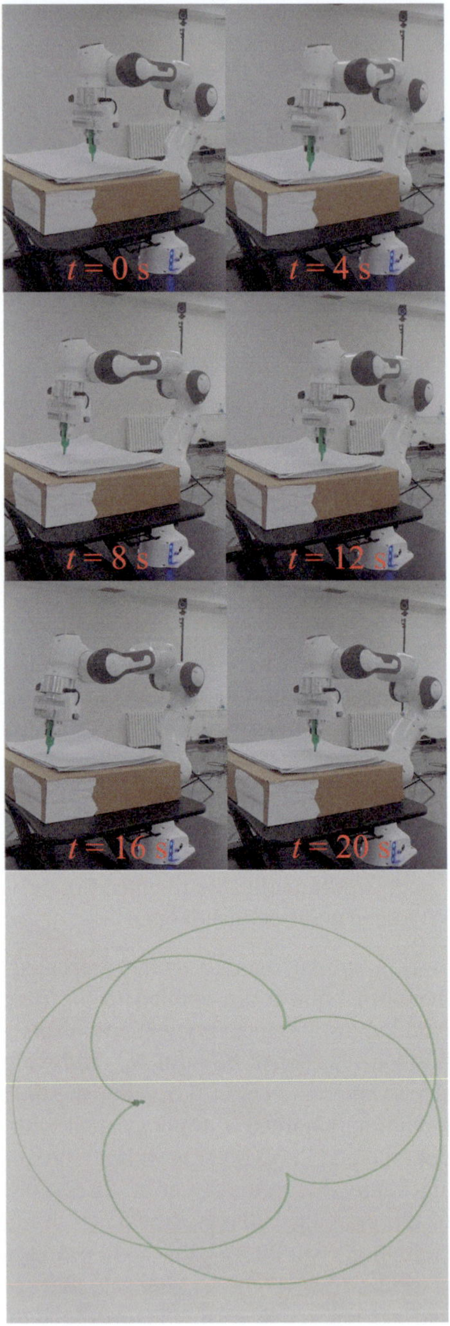

Fig. 5.6 Experimental results of the Franka manipulator for trajectory tracking with the aid of the CIMPC algorithm

IAE for trajectory tracking. Parameters used in the online training of the ESN are identical to those in Schwedersky et al. (2022).

The simulation results are shown in Fig. 5.7, where Fig. 5.7a illustrates the norm of the position error, and Fig. 5.7b shows the IAE over time. Due to the use of the kinematic model of the Franka manipulator, the linear MPC algorithm (Baek et al. 2022) yields the smallest position error. In the absence of the kinematic model, other algorithms perform slightly worse than the linear MPC algorithm (Baek et al. 2022). Notably, the presented CIMPC algorithm outperforms both the ESN-PNMPC algorithm (Jordanou et al. 2022a) and the adaptive PNMPC algorithm (Schwedersky et al. 2022). To be specific, the norm of the position error for the CIMPC algorithm is on the order of 10^{-4}, while those for the other two algorithms are on the order of 10^{-3}. The final IAE for the CIMPC algorithm is 0.44, whereas for the ESN-PNMPC algorithm (Jordanou et al. 2022a), it is 6.34, and for the adaptive PNMPC algorithm (Schwedersky et al. 2022), it is 1.36. Furthermore, the three components of IAE along the X-, Y-, and Z-axes, as well as the calculation time for each loop, are presented in Table 5.1. It can be observed that the ESN-PNMPC algorithm (Jordanou et al. 2022a) requires the shortest time to complete a loop, but it results in the largest IAE, whereas the other two algorithms take more time due to the online training of the ESN model and cerebellum model. In summary, in the absence of the kinematic model of the Franka manipulator, the presented CIMPC algorithm achieves the smallest position error and IAE, outperforming the comparison algorithms.

Subsequently, additional MPC algorithms based on ESN models are discussed and compared in Table 5.2. These MPC algorithms utilize ESN models as the process model of the unknown system, with only the cerebellum model (5.1)–(5.3) used in the presented CIMPC algorithm and the ESN model used in the adaptive PNMPC algorithm (Schwedersky et al. 2022) being trained online. In addition to the ESN-MPC algorithm (Wang et al. 2020), other algorithms incorporate error correction methods to counter interference and model mismatch, such as the ND-based correction algorithm (5.9) in this chapter and the error correction filter in Jordanou et al. (2022a). For the nonlinear ESN model, choosing an appropriate linearization method is crucial. The CIMPC algorithm introduced in this study, along with the ESN-PNMPC algorithm (Jordanou et al. 2022a), the adaptive PNMPC algorithm (Schwedersky et al. 2022), and the ESN-PNMPC algorithm (Jordanou et al. 2022b), all utilize a linearization technique for the process model along the trajectory. Specifically, this involves separating the predicted output into the forced response and the free response (Plucenio et al. 2007). Whereas the RNN-based MPC algorithm (Pan and Wang 2012) linearizes the process model around equilibrium points, which retains only a few of the nonlinear characteristics of the system, resulting in the lower prediction accuracy along the prediction horizon (Jordanou et al. 2022a). Finally, in contrast to the comparison algorithms, the presented CIMPC algorithm is utilized for trajectory tracking of the redundant manipulator, where the structural information is unknown.

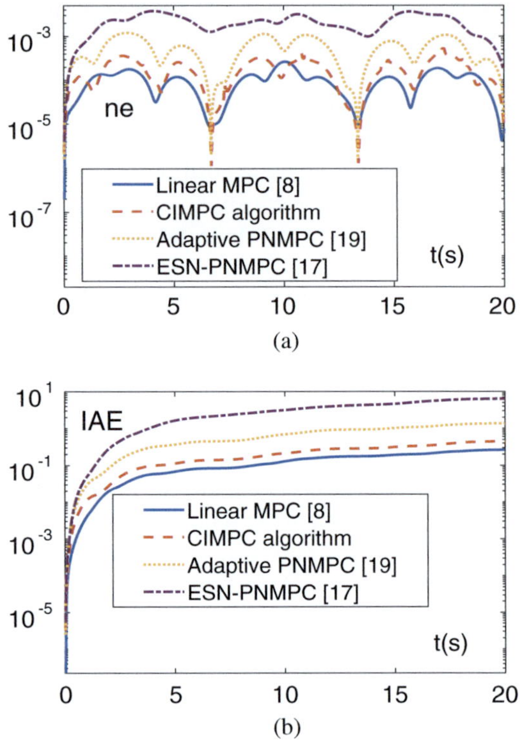

Fig. 5.7 Tracking error of the Franka manipulator for trajectory tracking with the aid of the CIMPC algorithm, linear MPC algorithm (Baek et al. 2022), ESN-PNMPC algorithm (Jordanou et al. 2022a), and adaptive PNMPC algorithm (Schwedersky et al. 2022), where linear MPC algorithm (Baek et al. 2022) utilizes the kinematic model of the Franka manipulator, while other algorithms do not. **a** Norm of the position error. **b** IAE over time

Table 5.1 IAE and calculation time under various algorithms

	IAE	IAE (X)	IAE (Y)	IAE (Z)	Calculation time for each loop (ms)
ESN-PNMPC (Jordanou et al. 2022a)	6.34	2.82	2.22	1.30	4.5
Adaptive PNMPC (Schwedersky et al. 2022)	1.36	1.02	0.28	0.06	6.3
CIMPC algorithm	0.44	0.33	0.08	0.03	5.3

The simulation and experimental results presented in this section demonstrate that the CIMPC algorithm is effective for trajectory tracking of the redundant manipulator, even when the structural information is unknown, and it outperforms the comparison algorithms.

Table 5.2 Comparisons under various MPC algorithms for the unknown system

	Process model	Training method	Error correction	Linearization method	Application object
CIMPC algorithm	ESN-based cerebellum	Online	√	Linearization along the trajectory	Redundant manipulator
RNN-based MPC (Pan and Wang 2012)	ESN	Offline	√	Linearization around equilibrium points	Vehicle
ESN-PNMPC (Jordanou et al. 2022a)	ESN	Offline	√	Linearization along the trajectory	Four-tank system and oil production platform
Adaptive PNMPC (Schwedersky et al. 2022)	ESN	Online	√	Linearization along the trajectory	Permanent magnet synchronous motor
ESN-PNMPC (Jordanou et al. 2022b)	ESN	Offline	√	Linearization along the trajectory	Electrical submersible pump
ESN-MPC (Wang et al. 2020)	ESN	Offline	×	Not mentioned	Dynamic thermal management

5.7 Conclusions

This chapter has introduced a CIMPC algorithm for trajectory tracking of a redundant manipulator with unknown structural information. In this algorithm, a cerebellum-inspired model based on the ESN has been utilized to substitute the kinematic model of the redundant manipulator. Additionally, an online training algorithm for the cerebellum model, along with an ND-based correction algorithm, has been presented and integrated into the CIMPC algorithm to enhance its performance. Moreover, a DDNN model has been introduced to implement the presented MPC scheme. Simulation and experimental results have demonstrated that the CIMPC algorithm has effectively controlled the redundant manipulator without relying on its structural information. Furthermore, in comparison with the other algorithms, the presented CIMPC algorithm has reduced the prediction error and further minimized the position error.

References

Abadia I, Naveros F, Ros E, Carrillo RR, Luque NR (2021) A cerebellar-based solution to the nondeterministic time delay problem in robotic control. Science Robotics 6(58):eabf2756

Baek S, Lee H, Han S (2022) Communication-efficient event-triggered time-delay control and its application to robot manipulators. IEEE Trans Industr Electron 69(9):9288–9297

Bazaraa MS, Sherali HD, Shetty CM (2006) Nonlinear programming: theory and algorithms. Wiley, Hoboken, NJ, USA

Bishop CM (2006) Pattern recognition and machine learning. Inf Sci Stat

Cai M, Wang Y, Wang S, Wang R, Tan M (2024) Autonomous manipulation of an underwater vehicle-manipulator system by a composite control scheme with disturbance estimation. IEEE Trans Autom Sci Eng 21(1):1012–1022

Cao R, Cheng L, Li H (2023) Passive model predictive impedance control for safe physical human-robot interaction. IEEE Trans Cogn Devel Syst 16(2):426–435

Capolei MC, Andersen NA, Lund HH, Falotico E, Tolu S (2020) A cerebellar internal models control architecture for online sensorimotor adaptation of a humanoid robot acting in a dynamic environment. IEEE Robot Autom Lett 5(1):80–87

Chen B, Chen X, Kanzow C (2000) A penalized Fischer-Burmeister NCP-function. Math Program 88:211–216

Chen B, Zhang H, Zhang F, Liu Y, Tan C, Yu H, Wang Y (2023) A multi-robot distributed collaborative region coverage search algorithm based on Glasius bio-inspired neural network. IEEE Trans Cogn Devel Syst 15(3):1449–1462

Chen L, Jin L, Shang M, Wang FY (2024) Enhancing representation power of deep neural networks with negligible parameter growth for industrial applications. IEEE Trans Syst Man Cybern Syst 54(11):6837–6848

Dahlin A, Karayiannidis Y (2021) Temporal coupling of dynamical movement primitives for constrained velocities and accelerations. IEEE Robot Autom Lett 6(2):2233–2239

Fiore MD, Meli G, Ziese A, Siciliano B, Natale C (2023) A general framework for hierarchical redundancy resolution under arbitrary constraints. IEEE Trans Rob 39(3):2468–2487

Honerkamp D, Büchner M, Despinoy F, Welschehold T, Valada A (2024) Language-grounded dynamic scene graphs for interactive object search with mobile manipulation. IEEE Robotics and Automation Letters 9(10):8298–8305

Hoxha E, Tempia F, Lippiello P, Miniaci MC (2016) Modulation, plasticity and pathophysiology of the parallel fiber-Purkinje cell synapse. Frontiers in Synaptic Neuroscience 8:35

Jaeger H, Haas H (2004) Harnessing nonlinearity: predicting chaotic systems and saving energy in wireless communication. Science 304(5667):78–80

Jin L, Wei L, Li S (2023) Gradient-based differential neural-solution to time-dependent nonlinear optimization. IEEE Trans Autom Control 68(1):620–627

Jin L, Zhao J, Chen L, Li S (2024a) Collective neural dynamics for sparse motion planning of redundant manipulators without Hessian matrix inversion. IEEE Trans Neural Netw Learn Syst 36(3):4326–4335

Jin L, Liu L, Wang X, Shang M, Wang FY (2024) Physical-informed neural network for MPC-based trajectory tracking of vehicles with noise considered. IEEE Trans Intell Veh 9(3):4493–4503

Jin L, Su Z, Fu D, Xiao X (2024c) Coevolutionary neural solution for nonconvex optimization with noise tolerance. IEEE Trans Neural Netw Learn Syst 35(12):17571–17581

Jordanou JP, Antonelo EA, Camponogara E (2022a) Echo state networks for practical nonlinear model predictive control of unknown dynamic systems. IEEE Trans Neural Netw Learn Syst 33(6):2615–2629

Jordanou JP, Osnes I, Hernes SB, Camponogara E, Antonelo EA, Imsland L (2022b) Nonlinear model predictive control of electrical submersible pumps based on echo state networks. Adv Eng Inf 52:101553

Kalidindi HT, Thuruthel TG, Laschi C, Falotico E (2019) Cerebellum-inspired approach for adaptive kinematic control of soft robots. In: 2nd IEEE international conference on soft robotics, pp 684–689

Khaitan S, Lin Q, Dolan JM (2021) Safe planning and control under uncertainty for self-driving. IEEE Trans Veh Technol 70(10):9826–9837

Kim H, Ohmura Y, Kuniyoshi Y (2021) Gaze-based dual resolution deep imitation learning for high-precision dexterous robot manipulation. IEEE Robot Autom Lett 6(2):1630–1637

Li P, Wang S, Yang H, Zhao H (2022a) Trajectory tracking and obstacle avoidance for wheeled mobile robots based on EMPC with an adaptive prediction horizon. IEEE Trans Cybern 52(12):13536–13545

Li Q, Zhang T, Li G, Li Z, Xia H, Su CY (2022b) Neural-dynamics optimization and repetitive learning control for robotic leg prostheses. IEEE/ASME Trans Mechatron 27(2):811–822

Lillo PD, Arrichiello F, Vito DD, Antonelli G (2021) BCI-controlled assistive manipulator: developed architecture and experimental results. IEEE Trans Cogn Devel Syst 13(1):91–104

Liu M, Chen L, Du X, Jin L, Shang M (2021) Activated gradients for deep neural networks. IEEE Trans Neural Netw Learn Syst 34(4):2156–2168

Liu M, Li Y, Chen Y, Qi Y, Jin L (2024) A distributed competitive and collaborative coordination for multirobot systems. IEEE Trans Mob Comput 23(2):11436–11448

Lu G, Xu W, Zhang F (2022) On-manifold model predictive control for trajectory tracking on robotic systems. IEEE Trans Industr Electron 70(9):9192–9202

Ma B, Xie Z, Zhan B, Jiang Z, Liu Y, Liu H (2023) Actual shape-based obstacle avoidance synthesized by velocity-acceleration minimization for redundant manipulators: An optimization perspective. IEEE Trans Syst Man Cybern Syst 53(10):6460–6474

Na X, Zhang M, Ren W, Han M (2023) Multistep-ahead chaotic time series prediction based on hierarchical echo state network with augmented random features. IEEE Trans Cogn Devel Syst 15(2):700–711

Niddam E, Dumon J, Cuvillon L, Durand S, Querry S, Hably A, Gangloff J (2024) Design of a suspended manipulator with aerial elliptic winding. IEEE Robot Autom Lett 9(9):7939–7946

Obregón-Flores J, Arechavaleta G, Becerra HM, Morales-Díaz A (2021) Predefined-time robust hierarchical inverse dynamics on torque-controlled redundant manipulators. IEEE Trans Rob 37(3):962–978

Omisore OM, Han S, Xiong J, Li H, Li Z, Wang L (2022) A review on flexible robotic systems for minimally invasive surgery. IEEE Trans Syst Man Cybern Syst 52(1):631–644

Pan Y, Wang J (2012) Model predictive control of unknown nonlinear dynamical systems based on recurrent neural networks. IEEE Trans Industr Electron 59(8):3089–3101

Picotti E, Bruschetta M, Mion E, Beghi A (2023) A nonlinear model-predictive contouring controller for shared control driving assistance in high-performance scenarios. IEEE Trans Syst Man Cybern Syst 53(1):204–215

Plucenio A, Pagano DJ, Bruciapaglia AH, Normey-Rico JE (2007) A practical approach to predictive control for nonlinear processes. IFAC Proc Vol 40(12):210–215

Reister F, Grotz M, Asfour T (2022) Combining navigation and manipulation costs for time-efficient robot placement in mobile manipulation tasks. IEEE Robot Autom Lett 7(4):9913–9920

Renda F, Armanini C, Mathew A, Boyer F (2022) Geometrically-exact inverse kinematic control of soft manipulators with general threadlike actuators' routing. IEEE Robot Autom Lett 7(3):7311–7318

Sabelhaus AP, Zhao H, Zhu EL, Agogino AK, Agogino AM (2021) Model-predictive control with inverse statics optimization for tensegrity spine robots. IEEE Trans Control Syst Technol 29(1):263–277

Schwedersky BB, Flesch RCC, Rovea SB (2022) Adaptive practical nonlinear model predictive control for echo state network models. IEEE Trans Neural Netw Learn Syst 33(6):2605–2614

Tan N, Yu P, Ni F (2023) A cerebellum-inspired network model and learning approaches for solving kinematic tracking control of redundant manipulators. IEEE Trans Cogn Devel Syst 15(1):150–162

Tanaka K, Minami Y, Tokudome Y, Inoue K, Kuniyoshi Y, Nakajima K (2022) Continuum-body-pose estimation from partial sensor information using recurrent neural networks. IEEE Robot Autom Lett 7(4):11244–11251
Thomas TL, Sikorski J, Ananthasuresh GK, Venkiteswaran VK, Misra S (2022) Design, sensing, and control of a magnetic compliant continuum manipulator. IEEE Trans Med Robot Bion 4(4):910–921
Tong Y, Liu J, Zhang X, Ju Z (2023) Four-criterion-optimization-based coordination motion control of dual-arm robots. IEEE Trans Cogn Devel Syst 15(2):794–807
Wang X, Santos VJ (2023) Gaze-based shared autonomy framework with real-time action primitive recognition for robot manipulators. IEEE Trans Neural Syst Rehabil Eng 31:4306–4317
Wang X, Tan N (2023) Cerebellum-inspired model-free tracking control and visual servoing of a rigid-flexible hybrid robotic endoscope with RCM constraints. IEEE Trans Industr Electron 70(12):12626–12635
Wang H, Guo X, Tan XD, Zhang C, Tang H, Yuan Y (2020) Leakage-aware predictive thermal management for multi-core systems using echo state network. IEEE Trans Comput Aided Des Integr Circuits Syst 39(7):1400–1413
Wang R, Su C, Yu H, Wang S (2023) Six-dimensional target pose estimation for robot autonomous manipulation: methodology and verification. IEEE Trans Cogn Devel Syst 15(1):186–197
Wei L, Jin L, Yang C, Chen K, Li W (2021) New noise-tolerant neural algorithms for future dynamic nonlinear optimization with estimation on Hessian matrix inversion. IEEE Trans Syst Man Cybern Syst 51(4):2611–2623
Yamazaki T, Tanaka S (2009) Computational models of timing mechanisms in the cerebellar granular layer. Cerebellum 8(4):423–432
Yan J, Liu M, Jin L (2024a) Cerebellum-inspired model predictive control for redundant manipulators with unknown structure information. IEEE Trans Cogn Devel Syst 16(3):1198–1210
Yan J, Jin L, Luo X, Li S (2024b) Modified RNN for solving comprehensive Sylvester equation with TDOA application. IEEE Trans Neural Netw Learn Syst 35(9):12553–12563
Yan J, Su Z, Ma X, Jin L (2025) An obstacle avoidance scheme for manipulators aided by noise-tolerant neural dynamics. IEEE Trans Industr Inf 21(1):703–712
Yang X, Zhao Z, Ma B, Xu Z, Zhao J, Liu H (2024) Kinematic and dynamic manipulability optimizations of redundant manipulators based on RNN model. IEEE Trans Industr Inf 20(4):5763–5773
Zhang F, Jin L, Luo X (2023a) Error-summation enhanced Newton algorithm for model predictive control of redundant manipulators. IEEE Trans Industr Electron 70(3):2800–2811
Zhang J, Chen J, Wu W, Qiao H (2023b) A cerebellum-inspired prediction and correction model for motion control of a musculoskeletal robot. IEEE Trans Cogn Devel Syst 15(3):1209–1223
Zhang M, Tong W, Li P, Hou Y, Xu X, Zhu L, Wu EQ (2023c) Robust neural dynamics method for redundant robot manipulator control with physical constraints. IEEE Trans Industr Inf 19(12):11721–11729

Part II
Posture-Considered Robot Control with DNN

Chapter 6
Motion/Force Control with Fuzzy DNN

6.1 Introduction

Recently, the advancement of robotic technologies has enabled their widespread adoption across various fields, with manipulators being the most prevalent implementation (Liu et al. 2024; Qiu et al. 2022). In manipulator control systems, ensuring task accuracy is always a guaranteed priority (Liao and Liu 2015; Lu et al. 2021; Wang et al. 2020). Therefore, numerous control schemes focus on motion control to maintain accurate end-effector trajectory tracking (Li et al. 2018; Tan et al. 2022; Zhang et al. 2022b). For example, a receding-horizon control scheme for minimizing the tracking error is developed in Yan et al. (2022). Besides, Van et al. design a controller for the fault-tolerance motion planning of manipulators and improve the tracking precision by integrating a hybrid fuzzy approximation algorithm and a disturbance observer (Van and Ge 2021). In order to realize the precise motion control of a surgical robot, a feedforward compensation term is incorporated into the control scheme in Baek et al. (2022). These schemes primarily concentrate on the position control for redundant manipulators, overlooking the force control. However, for tasks like polishing and grinding, relying solely on precise motion control makes it challenging to ensure that the task is executed smoothly. In such scenarios, motion-controlled manipulators primarily focus on minimizing position errors, potentially leading to unexpected contact forces that could damage the manipulator. Therefore, the contact force control becomes essential. Karami et al. develop a control method for redundant manipulators that simultaneously addresses the orientation, position, and force control (Karami et al. 2019). In Wang et al. (2022), a contact force model of a 7-DOF redundant manipulator based on orthogonal point-to-point impacts is analyzed from an impact dynamics perspective. Besides, the hybrid motion/force control is also an effective method to control both position error and contact force from the kinematic perspective. Fan et al. introduce a motion/force control scheme for motion planning of a manipulator with an uncertain structure, maintaining the contact force space through adjustments to the desired position (Yan et al. 2024a).

M. Liu et al., *Dynamic Neural Networks for Motion Control of Redundant Manipulators*, Intelligent Control and Learning Systems 21,
https://doi.org/10.1007/978-981-96-9144-9_6

Joint drift is defined as the deviation between the initial and final states after a cyclic motion. For redundant manipulators performing cyclic tasks, this deviation may accumulate over time, progressively affecting subsequent movement accuracy. One possible approach is to add a state adjustment session to restore the joint state before performing the next task. Although such an approach can reduce joint error, it requires an additional adjustment session, which is not conducive to practical applications. When the minimum joint drift serves as the control index, this session can be omitted, and the operation process can be simplified. In Zhang and Zhang (2013), many schemes for minimizing joint error and corresponding solution algorithms are summarized. Then, Xie et al. construct a projection-based scheme to decouple the joint error and position error, which further enhances the joint error accuracy (Xie et al. 2022).

Generally, the motion control of redundant manipulators primarily relies on inverse kinematic equation solutions (Chen et al. 2019; Kong et al. 2021; Katsikis et al. 2021). Among available methods, the pseudo-inverse approach provides a straightforward solution, computing joint data through the inverse of the Jacobian matrix (Ayusawa et al. 2023; Jin et al. 2023; Xiao et al. 2024). Then, from a practical point of view, many studies improve and extend the pseudo-inverse method by adding index terms (Chang and Lee 1989; Guo et al. 2021). However, these approaches often fail to address joint limit constraints fully. Optimization-based methods overcome this limitation by constructing such requirements as the objective function, equality constraints, and inequality constraints (Ramos and Koivo 2002; Yan et al. 2024b; Li et al. 2023). Then, the resulting manipulator control commands are derived through optimal solution computation (Cao et al. 2021; Jin et al. 2024a; Yilmaz et al. 2022a). In recent years, neural networks have increasingly been used for optimization-based control of redundant manipulators, capitalizing on their exceptional computing power (Li et al. 2017; Guo and Bürger 2022). For example, a dynamic neural network (DNN) is exploited to solve the joint error optimization scheme and estimate the Jacobian matrix of redundant manipulators (Xie et al. 2021). Besides, Zhang et al. design a circadian rhythms neural network to solve the hybrid torque and velocity optimization scheme of the redundant manipulator in Zhang et al. (2022a).

A fuzzy system is a nonlinear system that originates from the deterministic system, utilizing fuzzy logic and fuzzy reasoning as its core design approaches (Eyoh et al. 2018). An essential feature of the fuzzy system is that it does not rely on an exact mathematical model, but defines a fuzzy rule that maps the input space to the output space on an empirical basis, forming an invisible bridge between input and output. Structurally, it is a closed-loop structure with feedback channels (Jia et al. 2021a). Therefore, it demonstrates the superior robustness, adaptability, and fault tolerance, making it suitable for controlling linear and nonlinear systems with parameter uncertainties, such as robot control (Yilmaz et al. 2022b; Gold et al. 2023), multi-agent control (Zhang and Chen 2021), and event-triggered control (Li et al. 2021). Based on a fuzzy logic, Yilmaz et al. design a closed-loop error control system to implement manipulator tracking control subject to dynamic uncertainties (Yilmaz et al. 2022b). Moreover, a fuzzy system is designed in Santoso et al. (2021) for the flapping-wing aircraft control, which employs an adaptive method to deal with uncertainties and nonlinearities of the presented system. The fuzzy system provided in Dong et al. (2021) implements the model predictive control during the data transmission from the controller to the actuator. In recent decades, numerous advancements and applications have emerged that integrate fuzzy systems with neural networks (Jia et al. 2021b). For example, in Mai and Wang (2014), a neural network combining the fuzzy system is exploited to assist the motion/force control of mobile robots and achieve high tracking accuracy, when the dynamics parameters are uncertain. Furthermore, the fuzzy recurrent neural network introduced in Zhang and Yan (2020) focuses on addressing the joint drift optimization scheme for manipulators. Both theoretical analyses and simulations confirm the feasibility and effectiveness of integrating neural networks with fuzzy logic (Xiao et al. 2022). Currently, control methods for manipulators that integrate neural networks and fuzzy systems can improve the control accuracy and flexibility. However, these approaches face challenges in handling the noise interference, particularly polynomial noises. Therefore, it is essential to enhance not only flexibility but also the noise tolerance capacity.

Given the above problems, this chapter presents an acceleration-level motion/force control (AMFC) scheme, which contains a motion/force-based cyclic motion planning (MFCMP) strategy and an improved fuzzy dynamic neural network (FDNN) model. The remainder of the chapter is organized as follows: Section 6.2 outlines the modeling process of the motion/force control and constructs the MFCMP strategy. In Sect. 6.3, the FDNN model is presented, with a proof of its convergence. Section 6.4 provides enhancements to the robustness and convergence proof of the FDNN model. Section 6.5 presents the comparisons and verifications. The final section concludes this chapter. Before concluding this section, the main contributions are summarized as follows:

1. In order to realize the motion/force control of a redundant manipulator, this chapter constructs an MFCMP strategy at the acceleration level, which considers the orientation maintenance of the end-effector to ensure the smooth execution of the task.
2. The force control ensures that the end-effector of the redundant manipulator maintains a fixed contact force in the direction vertical to the contact surface. In other words, it allows the end-effector to produce a slight position deviation while remaining close to the desired position.
3. The FDNN model sets the convergent parameter in the traditional DNN model as adaptive to improve flexibility and accuracy. Then, a set of state variables is added to the FDNN model to suppress the interference of polynomial noises, and it is proved that the improved FDNN model is able to converge to the same result in a noise-free case.

6.2 Strategy Construction

The manipulator's kinematic equation follows a similar form to (4.1) and can be expressed as

$$\boldsymbol{r}_{\mathrm{p}}(t) = \Upsilon(\boldsymbol{\theta}(t)), \boldsymbol{r}_{\omega}(t) = \Upsilon_{\omega}(\boldsymbol{\theta}(t)).$$

Then, differentiating the kinematic equation yields

$$\dot{\boldsymbol{r}}_{\mathrm{p}}(t) = J(t)\dot{\boldsymbol{\theta}}(t), \dot{\boldsymbol{r}}_{\omega}(t) = J_{\mathrm{d}}(t)\dot{\boldsymbol{\theta}}(t).$$

Further, taking the time derivative again yields

$$\ddot{\boldsymbol{r}}_{\mathrm{p}}(t) = J(t)\ddot{\boldsymbol{\theta}}(t) + \dot{J}(t)\dot{\boldsymbol{\theta}}(t), \ddot{\boldsymbol{r}}_{\omega}(t) = J_{\mathrm{d}}(t)\ddot{\boldsymbol{\theta}}(t) + \dot{J}_{\mathrm{d}}(t)\dot{\boldsymbol{\theta}}(t),$$

where $\dot{J}(t)$ and $\dot{J}_{\mathrm{d}}(t)$ denote the time derivative of $J(t)$ and $J_{\mathrm{d}}(t)$, respectively. To simplify notation, variable t is omitted in the following.

The motion/force control strategy primarily regulates the force between the manipulator's end-effector and the contact surface, which depends on the deformation of the contact surface and the inherent stiffness factor. In this chapter, force control is implemented by adjusting position deviation. As illustrated in Fig. 6.1, a small position offset from the desired trajectory is introduced to deform the contact surface. Assume that the manipulator operates in 3D space; the contact surface lies on a horizontal plane; the deformation of the contact surface occurs solely along the Z-axis and forces F_{c} in the same direction. Then, the contact force is expressed as

$$F_{\mathrm{c}} = f_{\mathrm{s}}d, \tag{6.1}$$

where $f_{\mathrm{s}} \in \mathbb{R}^{+}$ is the stiffness factor of the contact surface, while d corresponds to its deformation. In the base coordinate system, the contact surface is

$$\boldsymbol{F}_{\mathrm{b}} = f_{\mathrm{s}}\mathfrak{R}^{\mathrm{T}}\widehat{\mathscr{G}}\mathfrak{R}(\boldsymbol{r}_{\mathrm{p}} - \boldsymbol{r}_{\mathrm{D}}), \tag{6.2}$$

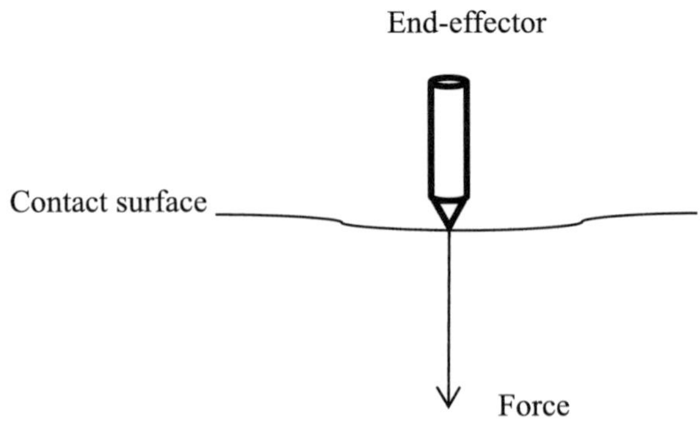

Fig. 6.1 Contact force modeling

where $\mathfrak{R} \in \mathbb{R}^{3\times 3}$ is the rotation matrix, serving to normalize the variables within the base coordinate system; $\widehat{\mathscr{G}}$ is a diagonal matrix with $[0, 0, 1]^{\mathrm{T}}$ as its diagonal elements, representing the force direction; $\boldsymbol{r}_{\mathrm{D}}$ denotes the desired position of the end-effector; and ^{T} signifies the transpose operator. Beyond the motion/force control problem, attention must also be given to the tracking error on the horizontal plane:

$$\boldsymbol{e}_{\mathrm{T}} = \mathfrak{R}^{\mathrm{T}}(\mathscr{I} - \widehat{\mathscr{G}})\mathfrak{R}(\boldsymbol{r}_{\mathrm{p}} - \boldsymbol{r}_{\mathrm{D}}), \tag{6.3}$$

where $\mathscr{I} - \widehat{\mathscr{G}}$ represents the direction of the tracking error $\boldsymbol{e}_{\mathrm{T}}$, with $\mathscr{I}$ being the unit matrix. For precise motion/force control of the manipulator, it is required that $\boldsymbol{F}_{\mathrm{b}} = \boldsymbol{F}_{\mathrm{d}}$ and $\boldsymbol{e}_{\mathrm{T}} = \boldsymbol{0}$, where $\boldsymbol{F}_{\mathrm{d}}$ denotes the expected force. Further combining these two conditions yields

$$\boldsymbol{\chi} = \mathscr{C}(\boldsymbol{r}_{\mathrm{p}} - \boldsymbol{r}_{\mathrm{D}}), \tag{6.4}$$

where

$$\boldsymbol{\chi} = \begin{bmatrix} \boldsymbol{F}_{\mathrm{d}} \\ \boldsymbol{0} \end{bmatrix} \in \mathbb{R}^{6}, \mathscr{C} = \begin{bmatrix} f_{\mathrm{s}}\mathfrak{R}^{\mathrm{T}}\widehat{\mathscr{G}}\mathfrak{R} \\ \mathfrak{R}^{\mathrm{T}}(\mathscr{I} - \widehat{\mathscr{G}})\mathfrak{R} \end{bmatrix} \in \mathbb{R}^{6\times 3}.$$

It should be noted that the orientation of the end-effector remains constant to ensure successful task execution, implying that the rotation matrix $\mathfrak{R}$ is fixed. Specifically, to maintain the orientation of the end-effector, setting the angular acceleration $\ddot{\boldsymbol{r}}_{\omega}(t)$ to $\boldsymbol{0}$ and adding a compensation for direction error lead to

$$J_{\mathrm{d}}\ddot{\boldsymbol{\theta}} + \dot{J}_{\mathrm{d}}\dot{\boldsymbol{\theta}} + b_{\mathrm{r}}(J_{\mathrm{d}}\dot{\boldsymbol{\theta}}) = \boldsymbol{0},$$

where $b_{\mathrm{r}}(J_{\mathrm{d}}\dot{\boldsymbol{\theta}})$ represents the compensation term with $b_{\mathrm{r}} > 0$ serves as the compensation coefficient.

Subsequently, the error function is defined as $\boldsymbol{e}_1 = \mathscr{C}(\boldsymbol{r}_{\mathrm{p}} - \boldsymbol{r}_{\mathrm{D}}) - \boldsymbol{\chi}$, its exponential convergence is guaranteed by applying $\dot{\boldsymbol{e}}_1 = -c_1\boldsymbol{e}_1$, resulting in

$$\mathscr{C}(J\dot{\boldsymbol{\theta}} - \dot{\boldsymbol{r}}_{\mathrm{D}}) - \dot{\boldsymbol{\chi}} + c_1\mathscr{C}(\boldsymbol{r}_{\mathrm{p}} - \boldsymbol{r}_{\mathrm{D}}) - c_1\boldsymbol{\chi} = \boldsymbol{0}, \tag{6.5}$$

where $c_1 > 0$; $\dot{\boldsymbol{r}}_{\mathrm{D}}$ is the end-effector's desired velocity; $\dot{\boldsymbol{\chi}}$ denotes the time derivative of $\boldsymbol{\chi}$. Once more, introduce a second error function $\boldsymbol{e}_2 = \mathscr{C}(J\dot{\boldsymbol{\theta}} - \dot{\boldsymbol{r}}_{\mathrm{D}}) - \dot{\boldsymbol{\chi}} + c_1\mathscr{C}(\boldsymbol{r}_{\mathrm{p}} - \boldsymbol{r}_{\mathrm{D}}) - c_1\boldsymbol{\chi}$ and employ the formula $\dot{\boldsymbol{e}}_2 = -c_2\boldsymbol{e}_2$, which yields

$$\begin{aligned} &\mathscr{C}(J\ddot{\boldsymbol{\theta}} + \dot{J}\dot{\boldsymbol{\theta}} - \ddot{\boldsymbol{r}}_{\mathrm{D}}) - \ddot{\boldsymbol{\chi}} + (c_1 + c_2)\mathscr{C}(J\dot{\boldsymbol{\theta}} - \dot{\boldsymbol{r}}_{\mathrm{D}}) \\ &-(c_1 + c_2)\dot{\boldsymbol{\chi}} + c_1c_2\mathscr{C}(\boldsymbol{r}_{\mathrm{p}} - \boldsymbol{r}_{\mathrm{D}}) - c_1c_2\boldsymbol{\chi} = \boldsymbol{0}, \end{aligned} \tag{6.6}$$

where $c_2 > 0$; $\ddot{\boldsymbol{r}}_{\mathrm{D}}$ is the desired acceleration of the end-effector; and $\ddot{\boldsymbol{\chi}}$ is the time derivative of $\dot{\boldsymbol{\chi}}$. Simplifying this expression results in

$$\mathscr{C}(J\ddot{\boldsymbol{\theta}} + \dot{J}\dot{\boldsymbol{\theta}} - \ddot{\boldsymbol{r}}_{\mathrm{D}}) + \eta = \boldsymbol{0}, \tag{6.7}$$

where $\boldsymbol{\eta} = (c_1 + c_2)\mathscr{C}(J\dot{\boldsymbol{\theta}} - \dot{\boldsymbol{r}}_{\mathrm{D}}) - (c_1 + c_2)\dot{\boldsymbol{\chi}} + c_1c_2\mathscr{C}(\boldsymbol{r}_{\mathrm{p}} - \boldsymbol{r}_{\mathrm{D}}) - c_1c_2\boldsymbol{\chi} - \ddot{\boldsymbol{\chi}}$. By this approach, the motion/force control objective is achieved when condition (6.7) is satisfied.

Based on the aforementioned foundational works, we present the following MFCMP strategy:

$$
\begin{aligned}
\min_{\ddot{\boldsymbol{\theta}}}\ & \frac{1}{2}\ddot{\boldsymbol{\theta}}(t)^{\mathrm{T}}\ddot{\boldsymbol{\theta}}(t) + \boldsymbol{\varrho}(t)^{\mathrm{T}}\ddot{\boldsymbol{\theta}}(t) \\
\text{s.t. } & J_{\mathrm{d}}\ddot{\boldsymbol{\theta}} + \dot{J}_{\mathrm{d}}\dot{\boldsymbol{\theta}} + b_{\mathrm{r}}(J_{\mathrm{d}}\dot{\boldsymbol{\theta}}) = \mathbf{0} \\
& \mathscr{C}(J\ddot{\boldsymbol{\theta}} + \dot{J}\dot{\boldsymbol{\theta}} - \ddot{\boldsymbol{r}}_{\mathrm{D}}) + \boldsymbol{\eta} = \mathbf{0} \\
& \ddot{\boldsymbol{\theta}} \in \Xi \\
\text{with } & \boldsymbol{\varrho}(t) = d_1\dot{\boldsymbol{\theta}}(t) + d_2(\boldsymbol{\theta}(t) - \boldsymbol{\theta}(0)),
\end{aligned}
\tag{6.8}
$$

where $d_1 > 0$; $d_2 > 0$; Ξ is the set of joint accelerations. As in Yan et al. (2024a), the constraint of joint accelerations is defined as $\underline{\Xi} < \ddot{\boldsymbol{\theta}} < \overline{\Xi}$, with

$$
\overline{\Xi} = \min\{\overline{\ddot{\boldsymbol{\theta}}}, \gamma_{\mathrm{p}}((\overline{\boldsymbol{\theta}} - \boldsymbol{\theta}) + \omega), \gamma_{\mathrm{v}}(\overline{\dot{\boldsymbol{\theta}}} - \dot{\boldsymbol{\theta}})\}
$$

$$
\underline{\Xi} = \max\{\underline{\ddot{\boldsymbol{\theta}}}, \gamma_{\mathrm{p}}((\underline{\boldsymbol{\theta}} - \boldsymbol{\theta}) - \omega), \gamma_{\mathrm{v}}(\underline{\dot{\boldsymbol{\theta}}} - \dot{\boldsymbol{\theta}})\},
$$

where $\gamma_{\mathrm{p}} > 0$, $\gamma_{\mathrm{v}} > 0$; $\omega > 0$ is used to adjust the safety region; $\overline{\ddot{\theta}}$ and $\underline{\ddot{\theta}}$, $\overline{\dot{\theta}}$ and $\underline{\dot{\theta}}$, as well as $\overline{\theta}$ and $\underline{\theta}$ represent the upper and lower bounds of joint accelerations, velocities, and angles, respectively.

Remark 6.1 Regarding MFCMP strategy (6.8) outlined above, three additional points need to be clarified. First, the objective function of the strategy is designed to minimize the joint drift $\boldsymbol{\theta}(t) - \boldsymbol{\theta}(0)$ at the joint acceleration level. Second, the first constraint equation is used to maintain the orientation of the end-effector, i.e., the elements of the rotation matrix remain unchanged (Yan et al. 2024a). Third, the second constraint equation in MFCMP strategy (6.8) is based on the requirements for the motion/force control. Specifically, it ensures that the end-effector follows the expected trajectory while also maintaining the contact force with the contact surface.

6.3 Strategy Solving

The next step is to focus on solving MFCMP strategy (6.8). A Lagrange function is defined as

$$
\begin{aligned}
\mathfrak{L}(\ddot{\boldsymbol{\theta}}, \boldsymbol{\lambda}_1, \boldsymbol{\lambda}_2) = & \frac{1}{2}\ddot{\boldsymbol{\theta}}^{\mathrm{T}}\ddot{\boldsymbol{\theta}} + \boldsymbol{\varrho}^{\mathrm{T}}\ddot{\boldsymbol{\theta}} + \boldsymbol{\lambda}_1^{\mathrm{T}}(J_{\mathrm{d}}\ddot{\boldsymbol{\theta}} + \dot{J}_{\mathrm{d}}\dot{\boldsymbol{\theta}} + b_{\mathrm{r}}J_{\mathrm{d}}\dot{\boldsymbol{\theta}}) \\
& + \boldsymbol{\lambda}_2^{\mathrm{T}}(\mathscr{C}(J\ddot{\boldsymbol{\theta}} + \dot{J}\dot{\boldsymbol{\theta}} - \ddot{\boldsymbol{r}}_{\mathrm{D}}) + \boldsymbol{\eta}),
\end{aligned}
\tag{6.9}
$$

where $\boldsymbol{\lambda}_1 \in \mathbb{R}^p$ and $\boldsymbol{\lambda}_2 \in \mathbb{R}^{p-q}$ are Lagrangian multipliers. From the Karush–Kuhn–Tucker conditions, it follows that

$$\begin{aligned}\frac{\partial \mathfrak{L}}{\ddot{\boldsymbol{\theta}}} &= \ddot{\boldsymbol{\theta}} + \boldsymbol{\varrho} + J_\mathrm{d}^\mathrm{T}\boldsymbol{\lambda}_1 + J^\mathrm{T}\mathscr{C}^\mathrm{T}\boldsymbol{\lambda}_2 = \mathbf{0}\\ \frac{\partial \mathfrak{L}}{\boldsymbol{\lambda}_1} &= J_\mathrm{d}\ddot{\boldsymbol{\theta}} + \dot{J}_\mathrm{d}\dot{\boldsymbol{\theta}} + b_\mathrm{r}J_\mathrm{d}\dot{\boldsymbol{\theta}} = \mathbf{0}\\ \frac{\partial \mathfrak{L}}{\boldsymbol{\lambda}_2} &= \mathscr{C}(J\ddot{\boldsymbol{\theta}} + \dot{J}\dot{\boldsymbol{\theta}} - \ddot{\boldsymbol{r}}_\mathrm{D}) + \boldsymbol{\eta} = \mathbf{0}.\end{aligned} \tag{6.10}$$

Thus, MFCMP strategy (6.8) yields an optimal solution through solving

$$\begin{aligned}&-\ddot{\boldsymbol{\theta}} + \aleph_\Xi(-\boldsymbol{\varrho} - J_\mathrm{d}^\mathrm{T}\boldsymbol{\lambda}_1 - J^\mathrm{T}\mathscr{C}^\mathrm{T}\boldsymbol{\lambda}_2) = \mathbf{0}\\ &J_\mathrm{d}\ddot{\boldsymbol{\theta}} + \dot{J}_\mathrm{d}\dot{\boldsymbol{\theta}} + b_\mathrm{r}J_\mathrm{d}\dot{\boldsymbol{\theta}} = \mathbf{0}\\ &\mathscr{C}(J\ddot{\boldsymbol{\theta}} + \dot{J}\dot{\boldsymbol{\theta}} - \ddot{\boldsymbol{r}}_\mathrm{D}) + \boldsymbol{\eta} = \mathbf{0},\end{aligned} \tag{6.11}$$

where

$$\aleph_\Xi(x_i) = \begin{cases}\overline{\Xi_i}, & x_i \geq \overline{\Xi_i}\\ x_i, & \underline{\Xi_i} < x_i < \overline{\Xi_i}\\ \underline{\Xi_i}, & x_i \leq \underline{\Xi_i}\end{cases}.$$

Consequently, solving MFCMP strategy (6.8) becomes to solving Eq. (6.11). The computational error is defined as $\boldsymbol{e}_q = [-\ddot{\boldsymbol{\theta}} + \aleph_\Xi(-\boldsymbol{\varrho} - J_\mathrm{d}^\mathrm{T}\boldsymbol{\lambda}_1 - J^\mathrm{T}\mathscr{C}^\mathrm{T}\boldsymbol{\lambda}_2), J_\mathrm{d}\ddot{\boldsymbol{\theta}} + \dot{J}_\mathrm{d}\dot{\boldsymbol{\theta}} + b_\mathrm{r}J_\mathrm{d}\dot{\boldsymbol{\theta}}, \mathscr{C}(J\ddot{\boldsymbol{\theta}} + \dot{J}\dot{\boldsymbol{\theta}} - \ddot{\boldsymbol{r}}_\mathrm{D}) + \boldsymbol{\eta}]^\mathrm{T}$.

6.3.1 Conventional DNN Model

Based on previous works (Jin et al. 2024b), we can develop a conventional DNN model to tackle MFCMP strategy (6.8) as follows:

$$\begin{aligned}\frac{\mathrm{d}\ddot{\boldsymbol{\theta}}}{\mathrm{d}t} &= \epsilon(-\ddot{\boldsymbol{\theta}} + \aleph_\Xi(-\boldsymbol{\varrho} - J_\mathrm{d}^\mathrm{T}\boldsymbol{\lambda}_1 - J^\mathrm{T}\mathscr{C}^\mathrm{T}\boldsymbol{\lambda}_2))\\ \frac{\mathrm{d}\boldsymbol{\lambda}_1}{\mathrm{d}t} &= \epsilon(J_\mathrm{d}\ddot{\boldsymbol{\theta}} + \dot{J}_\mathrm{d}\dot{\boldsymbol{\theta}} + b_\mathrm{r}J_\mathrm{d}\dot{\boldsymbol{\theta}})\\ \frac{\mathrm{d}\boldsymbol{\lambda}_2}{\mathrm{d}t} &= \epsilon(\mathscr{C}(J\ddot{\boldsymbol{\theta}} + \dot{J}\dot{\boldsymbol{\theta}} - \ddot{\boldsymbol{r}}_\mathrm{D}) + \boldsymbol{\eta}),\end{aligned} \tag{6.12}$$

where $\epsilon > 0$ is the scaling factor.

6.3.2 FDNN Model

The following content aims to introduce an FDNN model built upon the conventional DNN model (6.12) and discuss its convergence. Specifically, the original scaling factor ϵ in (6.12) is substituted with a fuzzy factor, a variable that can be dynamically adjusted based on the computational error $\boldsymbol{e}_q$. In order to determine the value of this fuzzy factor, a fuzzy approach must first be defined, where the computing error $\boldsymbol{e}_q$ serves as the input, the output is calculated by the fuzzy rules, and then the adjusted scaling factor is obtained.

The first step is fuzzification. The input subset $\Omega_{\mathrm{I}} = \{ZO, PS, PM, PB, PL\}$ (ZO means zero, PS means positive small, PM means positive medium, PB means positive big, PL means positive large) of the fuzzy strategy is obtained through fuzzifying the two norm of computational error e_q, as shown in Fig. 6.2. During the fuzzification, the S-type membership function (Smf), the general bell-type membership function (gbellmf), and the Z-type membership function (Zmf) are employed.

The second step involves constructing a fuzzy inference engine. The corresponding output subset is designated as Ω_{O}. The mapping relationships between the input subset and output subset are established as follows:

$$
\begin{aligned}
\Omega_{\mathrm{I}} = ZO &\Rightarrow \Omega_{\mathrm{O}} = ZO \\
\Omega_{\mathrm{I}} = PS &\Rightarrow \Omega_{\mathrm{O}} = PS \\
\Omega_{\mathrm{I}} = PM &\Rightarrow \Omega_{\mathrm{O}} = PM \\
\Omega_{\mathrm{I}} = PB &\Rightarrow \Omega_{\mathrm{O}} = PB \\
\Omega_{\mathrm{I}} = PL &\Rightarrow \Omega_{\mathrm{O}} = PL.
\end{aligned}
$$

The third step is the defuzzification. To obtain a precise output value, the defuzzification procedure is essential. In this chapter, the centroid method is applied to generate the output.

Let the output of the fuzzy system be μ, which is acquired through the above steps. Thus, the presented FDNN model is expressed as

$$
\begin{aligned}
\frac{\mathrm{d}\ddot{\boldsymbol{\theta}}}{\mathrm{d}t} &= -(1+\epsilon^{\natural})\epsilon(\ddot{\boldsymbol{\theta}} - \aleph_{\Xi}(-\boldsymbol{\varrho} - J_{\mathrm{d}}^{\mathrm{T}}\boldsymbol{\lambda}_1 - J^{\mathrm{T}}\mathscr{C}^{\mathrm{T}}\boldsymbol{\lambda}_2)) \\
\frac{\mathrm{d}\boldsymbol{\lambda}_1}{\mathrm{d}t} &= (1+\epsilon^{\natural})\epsilon(J_{\mathrm{d}}\ddot{\boldsymbol{\theta}} + \dot{J}_{\mathrm{d}}\dot{\boldsymbol{\theta}} + b_{\mathrm{r}}J_{\mathrm{d}}\dot{\boldsymbol{\theta}}) \\
\frac{\mathrm{d}\boldsymbol{\lambda}_2}{\mathrm{d}t} &= (1+\epsilon^{\natural})\epsilon(\mathscr{C}(J\ddot{\boldsymbol{\theta}} + \dot{J}\dot{\boldsymbol{\theta}} - \ddot{\boldsymbol{r}}_{\mathrm{D}}) + \boldsymbol{\eta}),
\end{aligned}
\tag{6.13}
$$

where $\epsilon^{\natural} = \alpha\mu$ with α being a scaling factor.

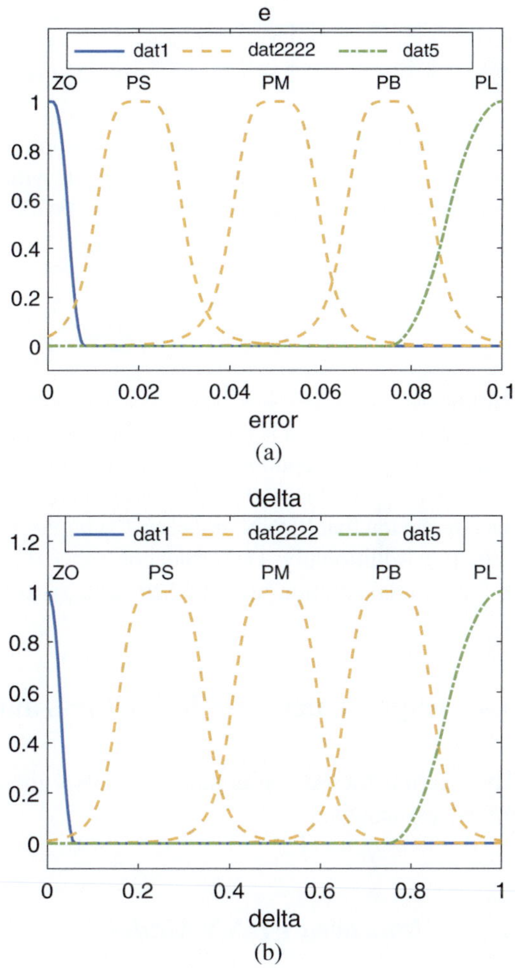

Fig. 6.2 Input and output in the system. **a** Input subset. **b** Output subset

6.3.3 Convergence Analysis

The next focus is to prove the convergence of FDNN model (6.13).

Theorem 6.1 *The FDNN model (6.13) is stable and globally convergent in the noise-free condition, while the first q elements of the solution asymptotically approach the optimal solution of MFCMP strategy (6.8).*

Proof Firstly, FDNN model (6.13) is reformulated using a canonical dynamic system as follows:

$$\frac{\mathrm{d}\hbar}{\mathrm{d}t} = -(1+\epsilon^{\natural})\epsilon(\hbar - \aleph_{\Lambda}(\hbar - G(\hbar))), \tag{6.14}$$

where

$$\hbar = \begin{bmatrix} \ddot{\boldsymbol{\theta}} \\ \boldsymbol{\lambda}_1 \\ \boldsymbol{\lambda}_2 \end{bmatrix}, G(\hbar) = \begin{bmatrix} \ddot{\boldsymbol{\theta}} + \boldsymbol{\varrho} + J_{\mathrm{d}}^{\mathrm{T}}\boldsymbol{\lambda}_1 + J^{\mathrm{T}}\mathscr{C}^{\mathrm{T}}\boldsymbol{\lambda}_2 \\ -J_{\mathrm{d}}\ddot{\boldsymbol{\theta}} - \dot{J}_{\mathrm{d}}\dot{\boldsymbol{\theta}} - b_{\mathrm{r}}J_{\mathrm{d}}\dot{\boldsymbol{\theta}} \\ -\mathscr{C}(J\ddot{\boldsymbol{\theta}} + \dot{J}\dot{\boldsymbol{\theta}} - \ddot{\boldsymbol{r}}_{\mathrm{D}}) - \boldsymbol{\eta} \end{bmatrix}.$$

Similarly, $\aleph_{\Lambda}$ follows an analogous definition to $\aleph_{\Xi}$, where $\underline{\hbar} = [\underline{\Xi}; -\boldsymbol{v}_{\infty}] \in \mathbb{R}^{2q}$ and $\overline{\hbar} = [\overline{\Xi}; \boldsymbol{v}_{\infty}] \in \mathbb{R}^{2q}$ serve as thresholds with $\boldsymbol{v}_{\infty}$ being a positive vector of large elements. Then, the gradient of $G(\hbar)$ can be expressed as

$$\nabla G(\hbar) = \begin{bmatrix} \mathscr{I} & J_{\mathrm{d}} & \mathscr{C}J \\ -J_{\mathrm{d}}^{\mathrm{T}} & \mathbf{0} & \mathbf{0} \\ -J^{\mathrm{T}}\mathscr{C}^{\mathrm{T}} & \mathbf{0} & \mathbf{0} \end{bmatrix}.$$

It follows that $\nabla G(\hbar) + \nabla^{\mathrm{T}} G(\hbar)$ is a semi-positive definite matrix, which indicates that the function $G(\hbar)$ is monotone. Moreover, it can be concluded that the dynamic system (6.14) is Lyapunov stable, converging to the equilibrium point described by $-\hbar + \aleph_{\Lambda}(\hbar - G(\hbar)) = \mathbf{0}$, as derived in Theorem 6.1 from Gao (2003). This leads to the conclusion that FDNN model (6.13) remains stable and converges to the solution of (6.11). Additionally, it is further confirmed that the first q element of $\hbar$ corresponds to the optimal solution for MFCMP strategy (6.8). The proof is completed. □

6.4 Improvements on Robustness and Proof

This section focuses on enhancing FDNN model (6.13) and conducting relevant robustness analyses.

6.4.1 Improved FDNN Model

The FDNN model (6.13) can be further refined to incorporate the capability of noise suppression. This chapter primarily considers the impact of polynomial noise and discusses an efficient noise suppression method. Let the noise be represented as $\mathfrak{N} = n_0 + n_1 t + n_2 t^2 + \cdots + n_k t^k$, where $n_0, n_1, n_2, \ldots,$ and n_k denote the corresponding coefficients. The FDNN model (6.13) is expressed as

$$\frac{\mathrm{d}\hbar}{\mathrm{d}t} = -(1 + \epsilon^{\natural})\epsilon(\hbar - \aleph_{\Lambda}(\hbar - G(\hbar))) + \mathfrak{N}. \tag{6.15}$$

A characteristic of polynomial noise lies in its becoming zero through successive differentiation, whereby the result of (6.15) should be zero after $k + 1$ derivatives. Based on this, we present the following improved FDNN model:

$$\begin{aligned}\frac{\mathrm{d}\hbar}{\mathrm{d}t} =& -(1+\epsilon^{\natural})\epsilon(\hbar - \aleph_{\Lambda}(\hbar - G(\hbar))) - b_0\beta_0 - b_1\beta_1 - \cdots - b_k\beta_k + \mathfrak{N}\\ \frac{\mathrm{d}\beta_0}{\mathrm{d}t} =& \dot{\hbar} + (1+\epsilon^{\natural})\epsilon(\hbar - \aleph_{\Lambda}(\hbar - G(\hbar)))\\ \frac{\mathrm{d}\beta_i}{\mathrm{d}t} =& \beta_{i-1},\ i = 1, 2, 3, \ldots, k,\end{aligned} \tag{6.16}$$

where $\beta_0,\ \beta_1,\ \beta_2, \ldots,$ and β_k are designed to learn the properties of noises, and $b_0,\ b_1,\ b_2, \ldots,$ and b_k are designed parameters with explanation provided in the following content.

Concerning the improved FDNN model (6.16), several critical aspects merit discussions. The presented improved FDNN model (6.16) is specifically developed to handle polynomial-noise interference, requiring adaptively adjusted order based on the noises. In reality, not only polynomial noises, it remains equally effective for time-varying noises that can be represented through Taylor expansion. Although the order of such noises is infinite following polynomial expansion, for slowly varying noises, a comparatively low-order polynomial model can effectively simulate them over a short time interval. Therefore, besides polynomial noises, the improved FDNN model (6.16) is capable of certain slowly changing time-varying noises.

6.4.2 Robustness Analysis

To demonstrate the convergence of the improved FDNN model (6.16), the following theorem is provided.

Theorem 6.2 *In the presence of polynomial noise $\mathfrak{N} = n_0 + n_1 t + n_2 t^2 + \cdots + n_k t^k$, the improved FDNN model (6.16) exhibits asymptotic stability and convergence. Specifically, the first q elements of the solution asymptotically converge to the globally optimal solution of MFCMP strategy (6.8), provided that all roots of $b_k + b_{k-1}x + b_{k-2}x^2 + \cdots + b_0 x^k + x^{k+1} = 0$ lie in the left half S-plane.*

Proof Firstly, by reformulating (6.16) as a linear system, we obtain

$$\begin{aligned}\frac{\mathrm{d}\beta_0}{\mathrm{d}t} &= -(b_0\beta_0 + b_1\beta_1 + \cdots + b_k\beta_k) + \mathfrak{N}\\ \frac{\mathrm{d}\beta_i}{\mathrm{d}t} &= \beta_{i-1},\ i = 1, 2, 3, \ldots, k\\ y &= \beta_0,\end{aligned} \tag{6.17}$$

where $\beta_0,\ \beta_1,\ \ldots,$ and β_k represent the state variables; $\mathfrak{N}$ and y represent input and output variables, respectively. To explore the stability and convergence of the linear system (6.17), it is analyzed from the frequency domain via Laplace transform. One gets that

$$\begin{aligned} s\varpi_0 - \beta_0(0) &= -(b_0\varpi_0 + b_1\varpi_1 + \cdots + b_k\varpi_k) + \mathscr{N} \\ s\varpi_i - \beta_i(0) &= \varpi_{i-1},\ i = 1, 2, 3, \ldots, k \\ \mathscr{Y} &= \varpi_0, \end{aligned} \tag{6.18}$$

where s represents the Laplace operator; $\varpi_0,\ \varpi_1,\ \ldots,$ and ϖ_k are obtained through the Laplace transform of $\beta_0,\ \beta_1,\ \ldots,$ and β_k; $\mathscr{Y}$ and $\mathscr{N}$ imply the Laplace transform of the output and input, respectively. From this, the output function $\mathscr{Y}$ and transfer function $\mathscr{H}$ of system (6.17) are

$$\begin{aligned} \mathscr{Y} =& \frac{s^k}{s^{k+1} + s^k b_0 + s^{k-1} b_1 + \cdots + s b_{k-1} + b_k} \\ & \times \left(\beta_0(0) - \frac{1}{s}\sum_{i=1}^{k} b_i\beta_i(0) - \frac{1}{s^2}\sum_{i=1}^{k-1} b_{i+1}\beta_i(0)\right. \\ & \left. - \cdots - \frac{1}{s^k} b_k\beta_1(0) + \mathscr{N}\right) \\ \mathscr{H} =& \frac{s^k}{s^{k+1} + s^k b_0 + s^{k-1} b_1 + \cdots + s b_{k-1} + b_k}. \end{aligned} \tag{6.19}$$

In general, a system is deemed stable if its transfer function meets the Routh–Hurwitz criterion (Franklin et al. 2001). Under the precondition provided in the theorem, the poles of the above transfer function $\mathscr{H}$ are all positioned in the left half S-plane, meaning that the system is stable. Furthermore, applying the final value theorem leads to the conclusion that

$$\begin{aligned} \lim_{t\to\infty} y &= \lim_{s\to 0} s\mathscr{Y} \\ &= \lim_{s\to 0} \frac{s^{k+1}\beta_0(0) - s^k \sum_{i=1}^{k} b_i\beta_i(0) - \cdots - s b_k \beta_1(0)}{s^{k+1} + s^k b_0 + s^{k-1} b_1 + \cdots + s b_{k-1} + b_k} \\ &\quad + \frac{s^{k+1}\mathscr{N}}{s^{k+1} + s^k b_0 + s^{k-1} b_1 + \cdots + s b_{k-1} + b_k} \\ &= \lim_{s\to 0} \frac{s^{k+1}\left(\frac{n_0}{s^1} + \frac{n_1}{s^2} + \cdots + \frac{n_k k!}{s^{k+1}}\right)}{s^{k+1} + s^k b_0 + s^{k-1} b_1 + \cdots + s b_{k-1} + b_k} \\ &= \frac{n_k k!}{b_k}. \end{aligned} \tag{6.20}$$

It suggests $\lim_{t\to\infty} y = n_k k!/b_k$, and further one gets

$$\lim_{t\to\infty} \frac{\mathrm{d}\beta_0}{\mathrm{d}t} = \lim_{t\to\infty} \dot{\hbar} + (1 + \epsilon^{\natural})\epsilon(\hbar - \aleph_{\Lambda}(\hbar - G(\hbar))) = 0. \tag{6.21}$$

Evidently, Eq. (6.21) is equivalent to (6.14). This equivalence demonstrates that the improved FDNN model (6.16) achieves asymptotic convergence, with its theoretical predictions maintaining full consistency with the noise-free condition. The proof is completed. □

6.5 Verifications and Comparisons

In this section, a series of comparisons and simulations are performed to demonstrate the effectiveness of the presented AMFC scheme (i.e., MFCPM strategy (6.8) aided with the improved FDNN model (6.16)).

6.5.1 Comparisons

Several common schemes to the manipulator control are summarized in Table 6.1, where they are compared with the presented AMFC scheme. In contrast, the scheme in Zhang et al. (2020) is designed only for motion and orientation control, while the remaining schemes focus solely on motion control. Additionally, the schemes in Yan et al. (2024a) and Xu et al. (2021) employ static convergent parameter and lack the ability to adapt noisy environments. Similarly, the scheme in Liu et al. (2023) utilizes a static convergent parameter and exhibits limited adaptability, functioning effectively only under bounded noise conditions. Furthermore, although the scheme in Zhang and Yan (2020) adopts an adaptive convergent parameter, its capacity for noise tolerance is restricted.

6.5.2 Simulation Verification

To demonstrate the effectiveness, robustness, and superiority of the present AMFC scheme, we perform simulations and compare it with other schemes under both noise-free and noisy cases. In the simulation, a 7-DOF manipulator, named Franka Emika Panda (abbreviated as Franka manipulator), is programmed to track a number "8" trajectory, and the configuration parameters are assigned as follows: $\epsilon = 1000$; $\alpha = 10$; $T = 10$ s; $c_1 = c_2 = d_1 = d_1 = b_\mathrm{r} = 10$; $f_\mathrm{s} = 2000$ N/m; $F_\mathrm{d} = [0, 0, 5]^\mathrm{T}$ N; $\gamma_\mathrm{p} = \gamma_\mathrm{v} = 20$; $\theta(0) = [0, -0.7854, 0, -2.3542, 0, 1.5716, 0.7854]^\mathrm{T}$ rad; $\dot{\theta}(0) = [0, 0, 0, 0, 0, 0, 0]^\mathrm{T}$ rad/s; $\ddot{\theta}(0) = [0, 0, 0, 0, 0, 0, 0]^\mathrm{T}$ rad/s^2; $\overline{\boldsymbol{\theta}} = [2.8, 1.7, 2.8, -0.07, 2.9, 3.8, 2.9]^\mathrm{T}$ rad; $\underline{\boldsymbol{\theta}} = [-2.9, -1.8, -2.9, -3.1, -2.9, -0.01, -2.9]^\mathrm{T}$ rad; $\overline{\dot{\boldsymbol{\theta}}} = -\underline{\dot{\boldsymbol{\theta}}} = [2.2, 2.2, 2.2, 2.2, 2.6, 2.6, 2.6]^\mathrm{T}$ rad/s; $\overline{\ddot{\boldsymbol{\theta}}} = -\underline{\ddot{\boldsymbol{\theta}}} = [15, 7.5, 10, 12.5, 15, 20, 20]^\mathrm{T}$ rad/s^2.

Table 6.1 Performance comparisons on different schemes for manipulators control

Schemes	Cyclic motion	Motion or force control	Orientation maintenance	Convergent parameter	Polynomial noises
AMFC scheme	Yes	Motion/force	Yes	Fuzzy-logic-based: $(1+\epsilon^{\natural})\epsilon$	Yes
Scheme in Yan et al. (2024a)	No	Motion/force	Yes	Static: ϵ	No
Scheme in Zhang and Yan (2020)	Yes	Motion	No	Fuzzy-logic-based: $\epsilon + \upsilon$ [♭]	Partial[◇]
Scheme in Xu et al. (2021)	No	Motion/force	No	Static: ϵ	No
Scheme in Zhang et al. (2020)	No	Motion	Yes	Static: ϵ	No
Scheme in Liu et al. (2023)	Yes	Motion	No	Static: ϵ	Partial[◇]
Scheme in Tan et al. (2024)	No	Motion	No	Static: ϵ	No

[◇] Partial indicates that the scheme can adapt to partial polynomial noises
[♭] υ Indicates the fuzzy control value, as detailed in Zhang and Yan (2020)

6.5.2.1 Noise-Free Case

First, the manipulator synthesized by the AMFC scheme (i.e., MFCPM strategy (6.8) combined with the improved FDNN model (6.16)) is simulated to execute the assigned task in a noise-free environment. The outcomes are illustrated in Figs. 6.3, 6.4, and 6.5. As revealed in Fig. 6.4a, the task is completed with high precision, as position errors $\triangledown R$ ($\triangledown R = \boldsymbol{r}_{\mathrm{p}} - \boldsymbol{r}_{\mathrm{D}}$) on the X-axis and Y-axis remain within the order of 10^{-4} m. The error on the Z-axis, linked to the contact force, stabilizes at 2.5×10^{-3} m throughout the task, aligning with predictions. Additionally, the joint data in Fig. 6.5a–c conform to expectations. To precisely characterize the orientation variation of the manipulator's end-effector, elements of rotation matrix $\mathfrak{R}$ are displayed in Fig. 6.4b. Evidently, these elements are almost constant, indicating that the orientation of the end-effector remains largely stable. Lastly, the contact force between the contact surface and the end-effector, as depicted in Fig. 6.4c, matches the preset value. Then, in the same scenario, the simulation results by employing MFCPM strategy (6.8) aided with DNN model (6.12) are presented in Fig. 6.6. Obviously, the improved FDNN model (6.16) is superior to DNN model (6.12) both in terms of accuracy, orientation control and motion/force control. Finally, joint drifts $|\nabla\theta| = |\theta(T) - \theta(0)|$ are collated in Table 6.2. It can be seen that the improved FDNN model (6.16) produces the smaller joint drift than DNN model (6.12).

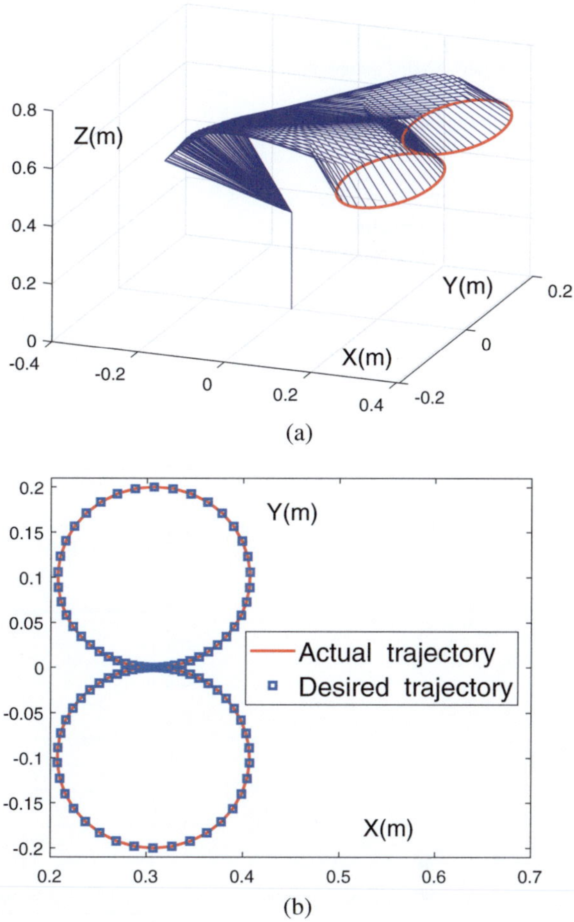

Fig. 6.3 The trajectory and the motion of the manipulator synthesized by the presented AMFC scheme (i.e., MFCMP strategy (6.8) aided with the improved FDNN model (6.16)) in tracking the trajectory of the number "8." **a** Motion process. **b** Actual trajectory and desired trajectory

6.5.2.2 Time-Varying Polynomial-Noise Cases

This part conducts simulations across various polynomial-noise cases ($\mathfrak{N} = 5$, $\mathfrak{N} = 5t + 5$, and $\mathfrak{N} = 5t^2 + 5t + 5$). Simulation results synthesized by the AMFC scheme are displayed in Fig. 6.7, and it can be seen that position errors in different noise cases align with those in the noise-free case, confirming the theoretical analyses. Besides, the performance of force control satisfies the previous results, as indicated by the position error along the Z-axis.

For comparison, scheme in Zhang and Yan (2020) and MFCPM strategy (6.8) aided with DNN model (6.12) are simulated in the same scenarios, and the corresponding position errors are revealed in Figs. 6.8 and 6.9. It is noticeable that the tracking accuracy decreases compared with the AMFC scheme, and even the phenomenon of error divergence occurs.

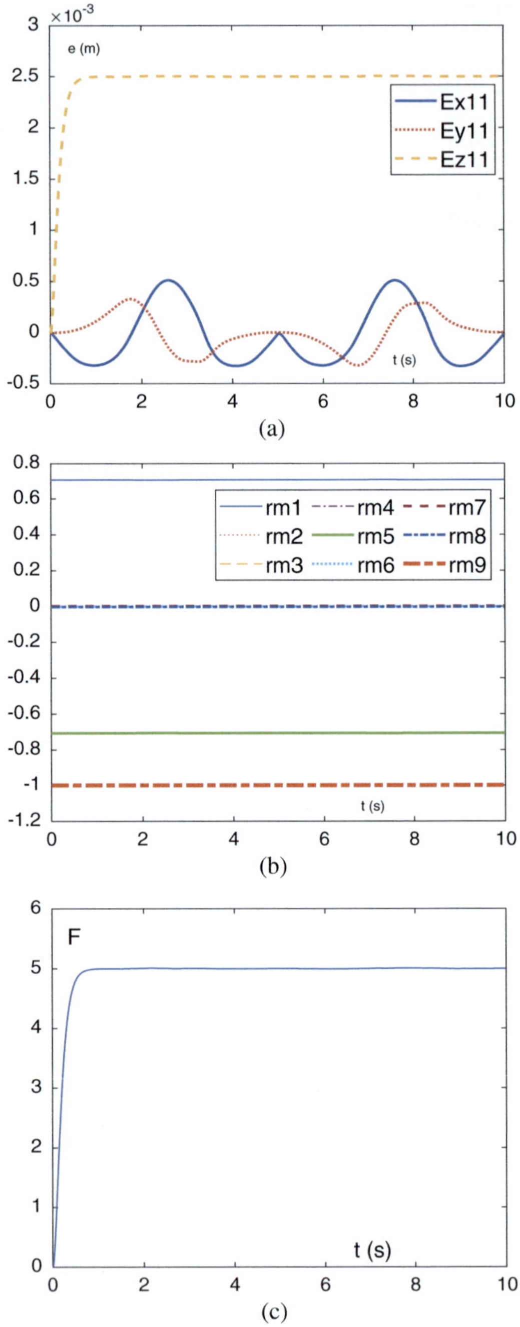

Fig. 6.4 Performance of the manipulator synthesized by the presented AMFC scheme (i.e., MFCMP strategy (6.8) aided with the improved FDNN model (6.16)) in tracking the trajectory of the number "8." **a** Position error. **b** Components of rotation matrix $\mathfrak{R}$. **c** Contact force F_c

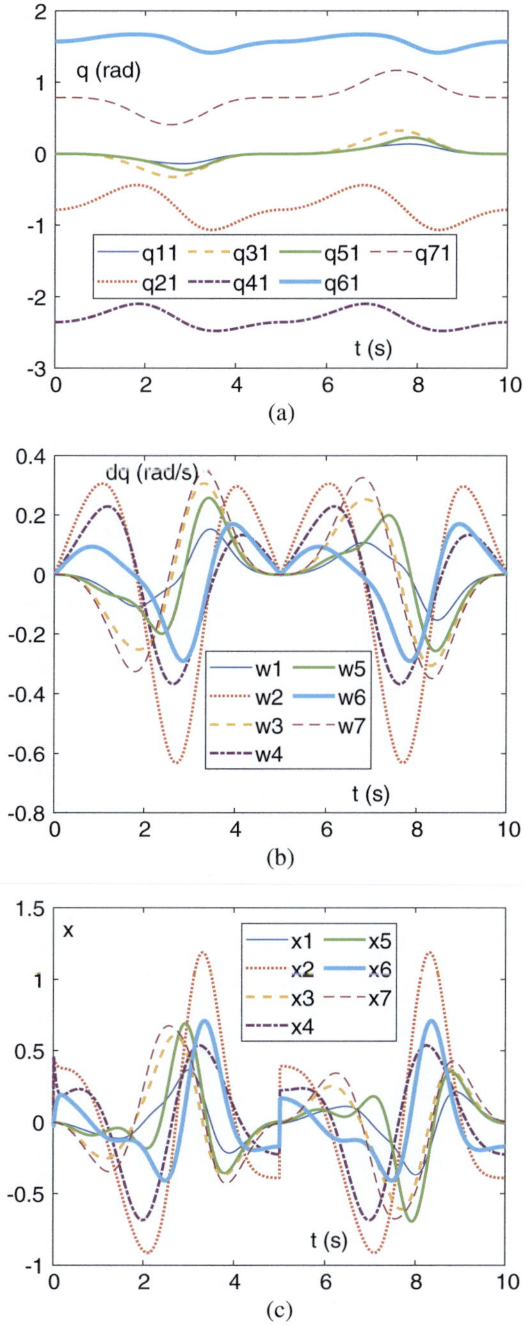

Fig. 6.5 Joint information of the manipulator synthesized by the presented AMFC scheme (i.e., MFCMP strategy (6.8) aided with the improved FDNN model (6.16)) in tracking the trajectory of the number "8." **a** Joint angle. **b** Joint velocity. **c** Joint acceleration

Fig. 6.6 Performance of the manipulator synthesized by MFCPM strategy (6.8) aided with DNN model (6.12) in tracking the trajectory of the number "8." **a** Position errors. **b** Components of rotation matrix. **c** Contact force

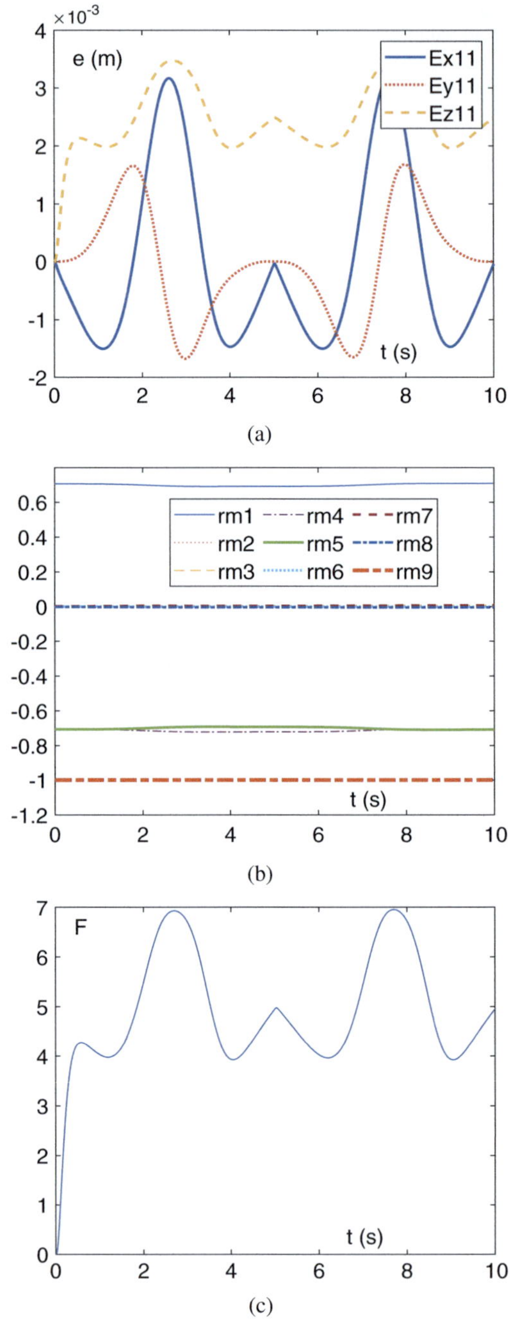

Table 6.2 Joint drifts generated by employing DNN model (6.12) and FDNN model (6.13) in noise-free case

	$\lvert\theta_1(T) - \theta_1(0)\rvert$	$\lvert\theta_2(T) - \theta_2(0)\rvert$	$\lvert\theta_3(T) - \theta_3(0)\rvert$	$\lvert\theta_4(T) - \theta_4(0)\rvert$	$\lvert\theta_5(T) - \theta_5(0)\rvert$	$\lvert\theta_6(T) - \theta_6(0)\rvert$	$\lvert\theta_7(T) - \theta_7(0)\rvert$
DNN model (6.12)	8.35×10^{-4}	9.52×10^{-4}	5.36×10^{-4}	3.14×10^{-3}	3.83×10^{-4}	2.18×10^{-3}	4.54×10^{-4}
Improved FDNN model (6.16)	1.33×10^{-5}	3.21×10^{-5}	1.09×10^{-5}	1.12×10^{-4}	7.88×10^{-6}	1.49×10^{-4}	5.86×10^{-6}

6.5.2.3 Other Time-Varying Noise Cases

In addition to polynomial noises, the improved FDNN model (6.16) can deal with noises that change slowly and can be fitted by a limited-order polynomial. Subsequently, two kinds of time-varying noises are injected to the AMFC scheme for simulations, i.e., $\mathfrak{N} = 10\cos(t) + 20\cos(2t + 1)$ and $\mathfrak{N} = 10$ sawtooth $(2t, 0.7)$ (sawtooth $(2t, 0.7)$ represents a symmetric triangle wave with a period of π and a peak of 1). The results synthesized by the AMFC scheme are shown in Fig. 6.10, from which it can be observed that position errors can converge to these in the noise-free case. Besides, according to the position error on the Z-axis, one concludes that force control performance meets the expectation.

For comparison, scheme in Zhang and Yan (2020) and MFCPM strategy (6.8) aided with DNN model (6.12) are simulated in the same scenarios, and the corresponding position errors are revealed in Fig. 6.11. It is noticeable that the tracking accuracy decreases, and the position errors are large when applying comparative schemes.

6.5.3 Experiment Verification

To further evaluate the feasibility and realizability of the AMFC scheme, experiments are performed with a Franka manipulator tracking the number "8" trajectory. The results presented in Fig. 6.12 demonstrate smooth task execution.

Besides, joint angles, joint error, position errors, and contact force of the Franka manipulator during the experiment are depicted in Fig. 6.13. As illustrated, joint error and position errors are small enough to meet physical requirements, which demonstrates that the AMFC scheme is valid and feasible.

Fig. 6.7 Position errors of the manipulator synthesized by the AMFC scheme (i.e., MFCPM strategy (6.8) aided with the improved FDNN model (6.16)) in tracking the trajectory of the number "8" in different polynomial-noise cases. **a** $\mathfrak{N} = 5$. **b** $\mathfrak{N} = 5t + 5$. **c** $\mathfrak{N} = 5t^2 + 5t + 5$

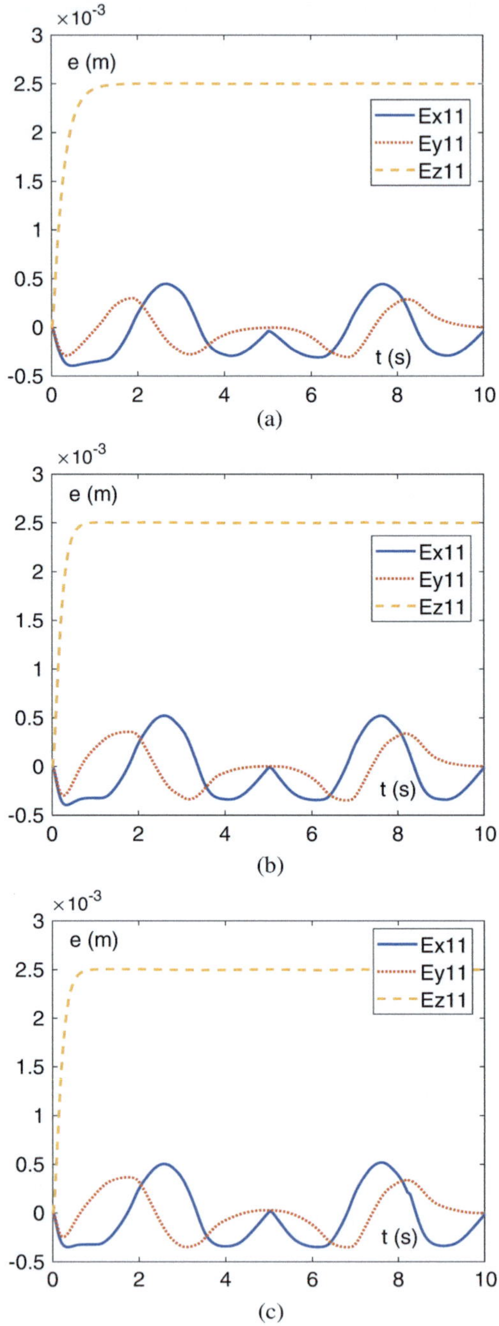

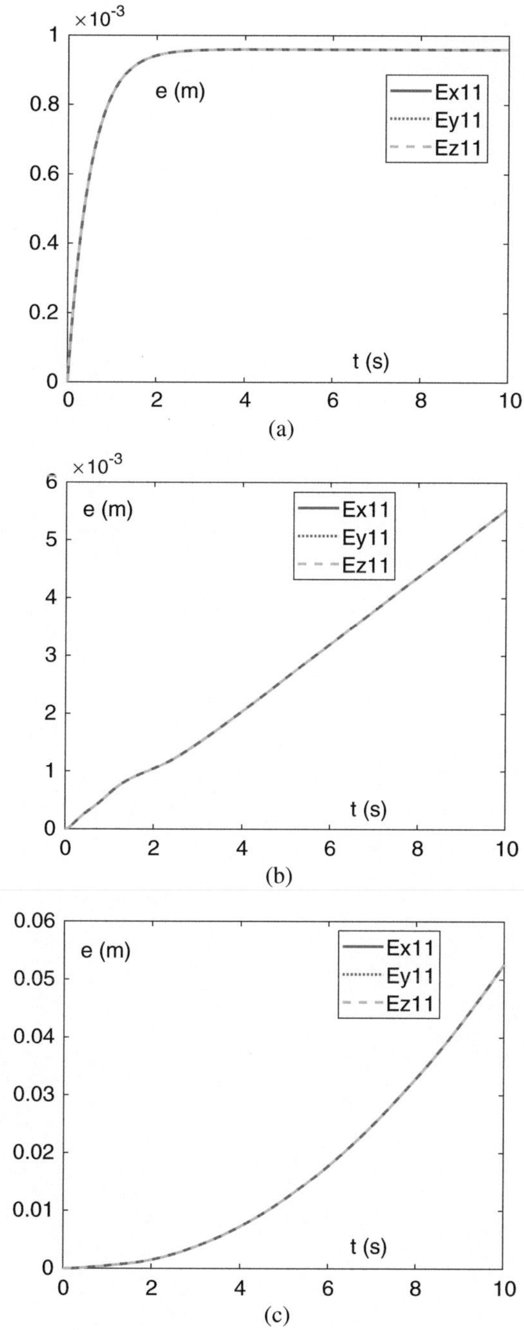

Fig. 6.8 Position errors of the manipulator synthesized by scheme in Zhang and Yan (2020) in tracking the trajectory of the number "8" in different polynomial-noise cases **a** $\mathfrak{N} = 5$. **b** $\mathfrak{N} = 5t + 5$. **c** $\mathfrak{N} = 5t^2 + 5t + 5$

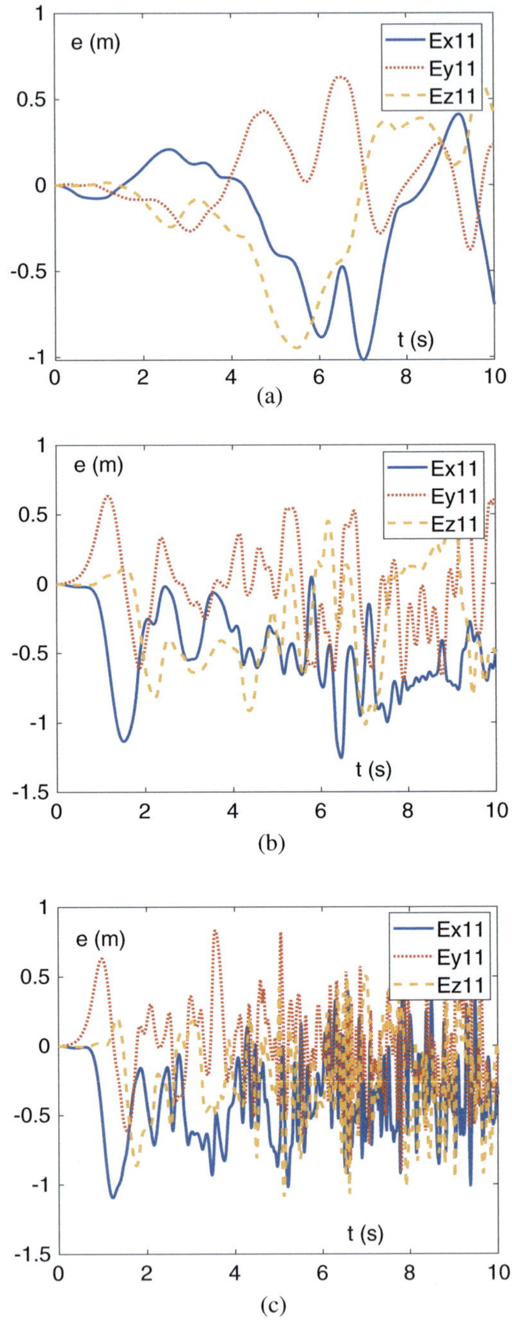

Fig. 6.9 Position errors of the manipulator synthesized by MFCPM strategy (6.8) aided with DNN model (6.12) in tracking the trajectory of the number "8" in different polynomial-noise cases. **a** $\mathfrak{N} = 5$. **b** $\mathfrak{N} = 5t + 5$. **c** $\mathfrak{N} = 5t^2 + 5t + 5$

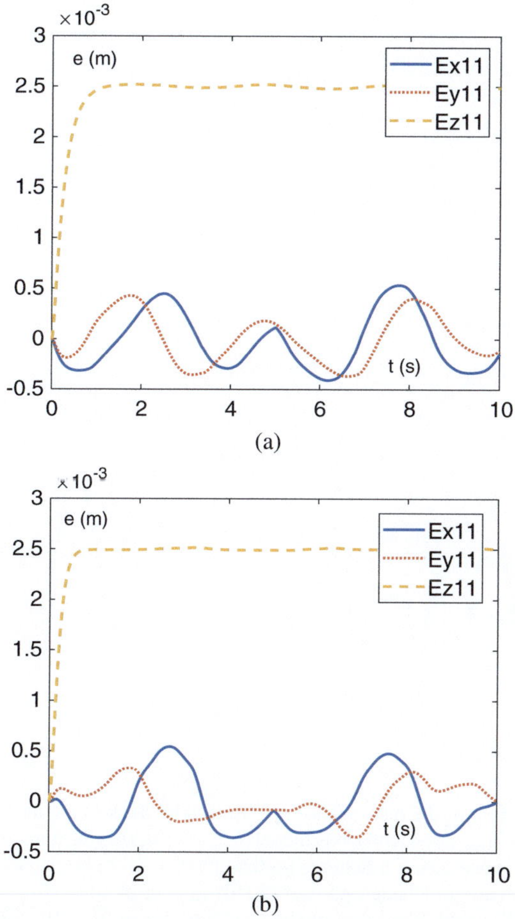

Fig. 6.10 Position errors of the manipulator synthesized by the AMFC scheme (i.e., MFCPM strategy (6.8) aided with the improved FDNN model (6.16)) tracking the trajectory of the number "8" in other time-varying noise cases. **a** $\mathfrak{N} = 10\cos(t) + 20\cos(2t + 1)$. **b** $\mathfrak{N} = 10$ sawtooth $(2t, 0.7)$

6.6 Conclusions

In this chapter, an AMFC scheme for manipulators, capable of handling polynomial noises, incorporating an MFCPM strategy and an FDNN model, has been constructed. First, the motion-force modeling has been analyzed, and the MFCPM strategy with orientation maintenance has been outlined. Following that, an FDNN model based on the conventional DNN model, which integrates a fuzzy system to adjust the convergent parameter, has been developed. Further, the FDNN model has been improved to possess the advantage of noise suppression for polynomial noises. Additionally, comparative analyses with other schemes in both noise-free and noisy cases have

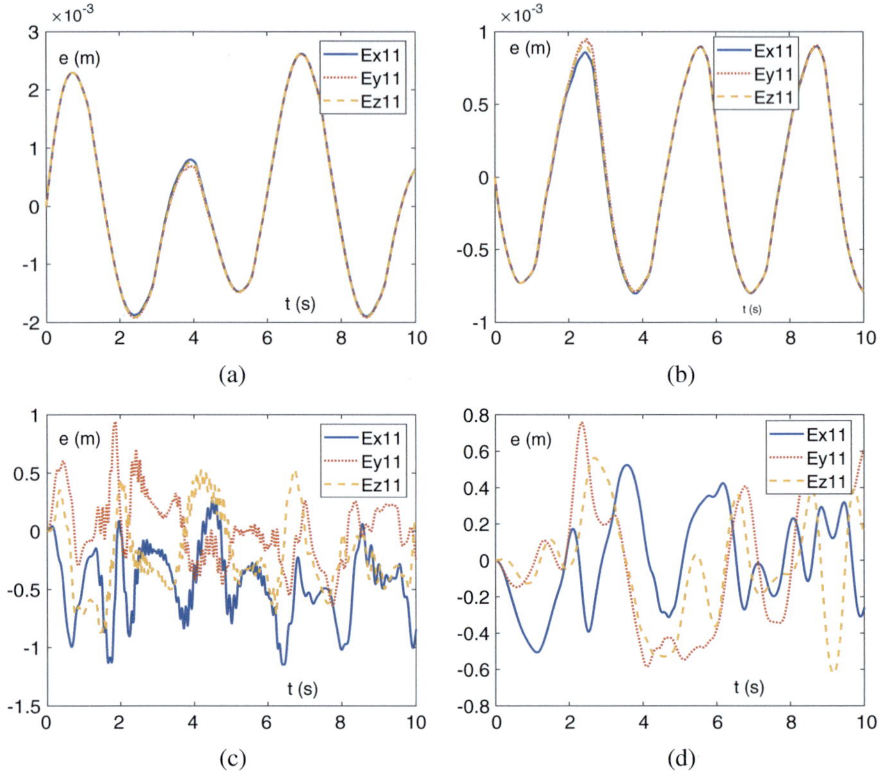

Fig. 6.11 Position errors of the manipulator synthesized by the scheme in Zhang and Yan (2020) and MFCPM strategy (6.8) aided with DNN model (6.12) in tracking the trajectory of the number "8" in other time-varying noise cases. **a** Scheme in Zhang and Yan (2020) in case of $\mathfrak{N} = 10\cos(t) + 20\cos(2t + 1)$. **b** Scheme in Zhang and Yan (2020) in case of $\mathfrak{N} = 10 \times$ sawtooth $(2t, 0.7)$. **c** MFCPM strategy (6.8) aided with DNN model (6.12) in case of $\mathfrak{N} = 10\cos(t) + 20\cos(2t + 1)$. **d** MFCPM strategy (6.8) aided with DNN model (6.12) in case of $\mathfrak{N} = 10 \times$ sawtooth $(2t, 0.7)$

demonstrated that the AMFC scheme can effectively minimize the joint drift, control motion, and force, and address problems related to end-effector orientation deviation. Finally, the experimental results have further verified the feasibility of the AMFC scheme.

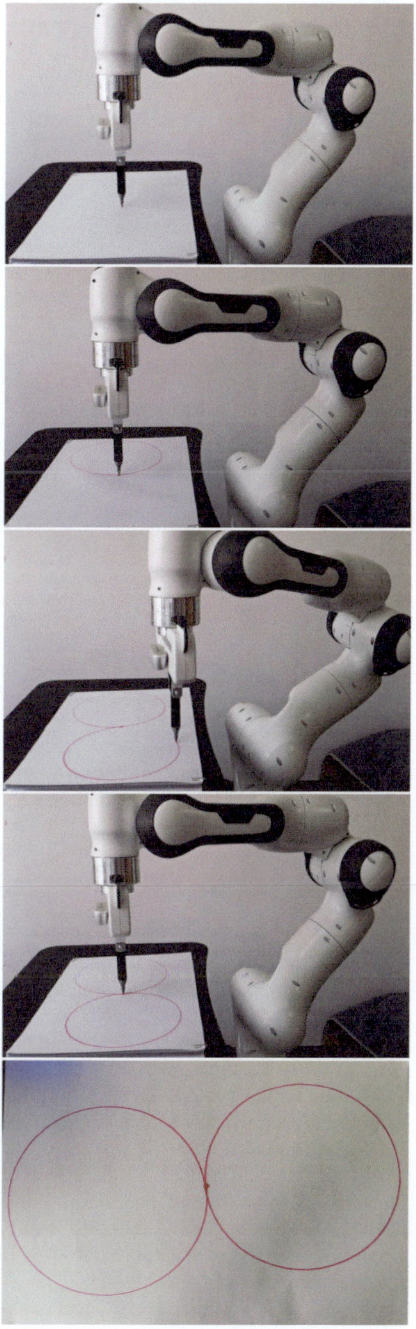

Fig. 6.12 Snapshots of the Franka manipulator synthesized by the AMFC scheme (i.e., MFCPM strategy (6.8) aided with the improved FDNN model (6.16)) in tracking the trajectory of the number "8"

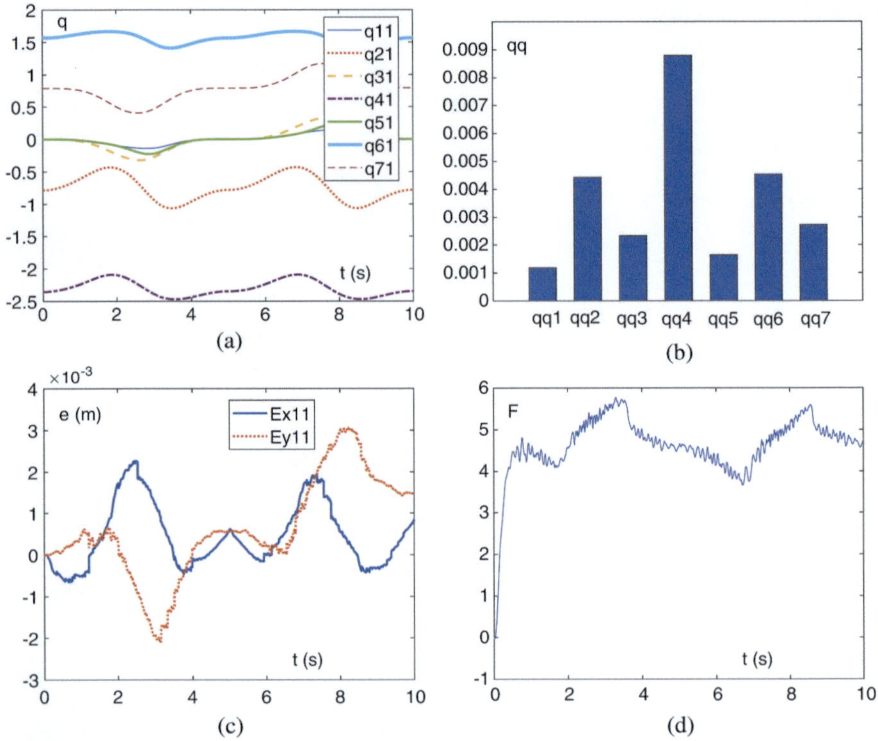

Fig. 6.13 Parameter variations during the experiment. **a** Joint angles. **b** Joint drift. **c** Position errors on the X-axis and Y-axis. **d** Contact force F_c

References

Ayusawa K, Murai A, Sagawa R, Yoshida E (2023) Fast inverse kinematics based on pseudo-forward dynamics computation: application to musculoskeletal inverse kinematics. IEEE Robot Autom Lett 8(9):5775–5782

Baek D, Nho YH, Kwon DS (2022) ViO-Com: Feed-forward compensation using vision-based optimization for high-precision surgical manipulation. IEEE Robot Autom Lett 7(1):263–270

Cao S, Sun L, Jiang J, Zuo Z (2021) Reinforcement learning-based fixed-time trajectory tracking control for uncertain robotic manipulators with input saturation. IEEE Trans Neural Netw Learn Syst 34(8):4584–4595

Chang PR, Lee CSG (1989) Residue arithmetic VLSI array architecture for manipulator pseudo-inverse Jacobian computation. IEEE Trans Robot Autom 5(5):569–582

Chen C, Modares H, Xie K, Lewis FL, Wan Y, Xie S (2019) Reinforcement learning-based adaptive optimal exponential tracking control of linear systems with unknown dynamics. IEEE Trans Autom Control 64(11):4423–4438

Dong Y, Song Y, Wei G (2021) Efficient model-predictive control for networked interval Type-2 T-S fuzzy system with stochastic communication protocol. IEEE Trans Fuzzy Syst 29(2):286–297

Eyoh I, John R, De Maere G (2018) Interval type-2 A-intuitionistic fuzzy logic for regression problems. IEEE Trans Fuzzy Syst 26(4):2396–2408

Franklin GF, Powell JD, Emami-Naeini A (2001) Feedback control of dynamic systems. Prentice Hall, New Jersey

Gao X (2003) Exponential stability of globally projected dynamic systems. IEEE Trans Neural Networks 14(2):426–431

Gold T, Völz A, Graichen K (2023) Model predictive interaction control for robotic manipulation tasks. IEEE Trans Rob 39(1):76–89

Guo M, Bürger M (2022) Geometric task networks: Learning efficient and explainable skill coordination for object manipulation. IEEE Trans Rob 38(3):1723–1734

Guo D, Li Z, Khan AH, Feng Q, Cai J (2021) Repetitive motion planning of robotic manipulators with guaranteed precision. IEEE Trans Industr Inf 17(1):356–366

Jia L, Xiao L, Dai J, Qi Z, Zhang Z, Zhang Y (2021a) Design and application of an adaptive fuzzy control strategy to zeroing neural network for solving time-variant QP problem. IEEE Trans Fuzzy Syst 29(6):1544–1555

Jia L, Xiao L, Dai J, Cao Y (2021b) A novel fuzzy-power zeroing neural network model for time-variant matrix Moore-Penrose inversion with guaranteed performance. IEEE Trans Fuzzy Syst 29(9):2603–2611

Jin L, Wei L, Li S (2023) Gradient-based differential neural-solution to time-dependent nonlinear optimization. IEEE Trans Autom Control 68(1):62–627

Jin L, Su Z, Fu D, Xiao X (2024a) Coevolutionary neural solution for nonconvex optimization with noise tolerance. IEEE Trans Neural Netw Learn Syst 35(12):17571–17581

Jin L, Zhang J, Qi Y, Li S (2024b) New distributed consensus schemes with time delays and output saturation. IEEE Trans Autom Sci Eng 21(3):2856–2867

Karami A, Sadeghian H, Keshmiri M, Oriolo G (2019) Force, orientation and position control in redundant manipulators in prioritized scheme with null space compliance. Control Eng Pract 85:23–33

Katsikis VN, Stanimirović PS, Mourtas SD, Xiao L, Karabašević D, Stanujkić D (2021) Zeroing neural network with fuzzy parameter for computing pseudoinverse of arbitrary matrix. IEEE Trans Fuzzy Syst 30(9):3426–3435

Kong Y, Jiang Y, Li X, Lei J (2021) A time-specified zeroing neural network for quadratic programming with its redundant manipulator application. IEEE Trans Industr Electron 69(5):4977–4987

Li S, He J, Li Y, Rafique MU (2017) Distributed recurrent neural networks for cooperative control of manipulators: A game-theoretic perspective. IEEE Trans Neural Netw Learn Syst 28(2):415–426

Li Z, Liao B, Xu F, Guo D (2018) A new repetitive motion planning scheme with noise suppression capability for redundant robot manipulators. IEEE Trans Syst Man Cybern Syst 50(12):5244–5254

Li W, Xie Z, Wong PK, Mei X, Zhao J (2021) Adaptive-event-trigger-based fuzzy nonlinear lateral dynamic control for autonomous electric vehicles under insecure communication networks. IEEE Trans Industr Electron 68(3):2447–2459

Li Z, Li S, Bamasag OO, Alhothali A, Luo X (2023) Diversified regularization enhanced training for effective manipulator calibration. IEEE Trans Neural Netw Learn Syst 34(11):8778–8790

Liao B, Liu W (2015) Pseudoinverse-type bi-criteria minimization scheme for redundancy resolution of robot manipulators. Robotica 33(10):2100–2113

Liu M, He L, Shang M (2023) Dynamic neural network for bicriteria weighted control of robot manipulators. IEEE Trans Neural Netw Learn Syst 34(8):4570–4583

Liu M, Zhang F, He L, Shang M (2024) Dynamic neural network for motion/force control of manipulators with polynomial noises. IEEE Trans Industr Electron 71(10):12559–12569

Lu H, Jin L, Zhang J, Sun Z, Li S, Zhang Z (2021) New joint-drift-free scheme aided with projected ZNN for motion generation of redundant robot manipulators perturbed by disturbances. IEEE Trans Syst Man Cybern Syst 51(9):5639–5651

Mai T, Wang Y (2014) Adaptive force/motion control system based on recurrent fuzzy wavelet CMAC neural networks for condenser cleaning crawler-type mobile manipulator robot. IEEE Trans Control Syst Technol 22(5):1973–1982

Qiu B, Li X-D, Guo J, Tan N (2022) New jerk-level configuration adjustment schemes applied to constrained redundant robots. IEEE Trans Industr Inf 18(4):2528–2538

Ramos MC, Koivo AJ (2002) Fuzzy logic-based optimization for redundant manipulators. IEEE Trans Fuzzy Syst 10(4):498–509

Santoso F, Garrattn MA, Anavatti SG (2021) A robust hybrid of a feedback linearization technique and an interval Type-2 fuzzy control system for the flapping angle dynamics of a biomimetic aircraft. IEEE Trans Syst Man Cybern Syst 51(9):5664–5673

Tan N, Zhong Z, Yu P, Li Z, Ni F (2022) A discrete model-free scheme for fault-tolerant tracking control of redundant manipulators. IEEE Trans Industr Inf 18(12):8595–8606

Tan J, Shang M, Jin L (2024) Metaheuristic-based RNN for manipulability optimization of redundant manipulators. IEEE Trans Industr Inf 20(4):6489–6498

Van M, Ge SS (2021) Adaptive fuzzy integral sliding-mode control for robust fault-tolerant control of robot manipulators with disturbance observer. IEEE Trans Fuzzy Syst 29(5):1284–1296

Wang J, Wang J, Han QL (2020) Neurodynamics-based model predictive control of continuous-time under-actuated mechatronic systems. IEEE/ASME Trans Mechatron 26(1):311–322

Wang Y, Dehio N, Kheddar A (2022) On inverse inertia matrix and contact-force model for robotic manipulators at normal impacts. IEEE Robot Autom Lett 7(2):3648–3655

Xiao L, Jia L, Wang Y, Dai J, Liao Q, Zhu Q (2022) Performance analysis and applications of finite-time ZNN models with constant/fuzzy parameters for TVQPEI. IEEE Trans Neural Netw Learn Syst 33(11):6665–6676

Xiao L, Li L, Huang W, Li X, Jia L (2024) A new predefined time zeroing neural network with drop conservatism for matrix flows inversion and its application. IEEE Trans Cybern 54(2):752–761

Xie Z, Jin L, Luo X, Li S, Xiao X (2021) A data-driven cyclic-motion generation scheme for kinematic control of redundant manipulators. IEEE Trans Control Syst Technol 29(1):53–63

Xie Z, Jin L, Luo X, Sun Z, Liu M (2022) RNN for repetitive motion generation of redundant robot manipulators: an orthogonal projection-based scheme. IEEE Trans Neural Netw Learn Syst 33(2):615–628

Xu Z, Li S, Zhou X, Zhou S, Cheng T, Guan Y (2021) Dynamic neural networks for motion-force control of redundant manipulators: An optimization perspective. IEEE Trans Industr Electron 68(2):1525–1536

Yan J, Jin L, Yuan Z, Liu Z (2022) RNN for receding horizon control of redundant robot manipulators. IEEE Trans Industr Electron 69(2):1608–1619

Yan J, Jin L, Hu B (2024a) Data-driven model predictive control for redundant manipulators with unknown model. IEEE Trans Cybern 54(10):5901–5911

Yan J, Jin L, Luo X, Li S (2024b) Modified RNN for solving comprehensive Sylvester equation with TDOA application. IEEE Trans Neural Netw Learn Syst 35(9):12553–12563

Yilmaz BM, Tatlicioglu E, Savran A, Alci M (2022a) Self-adjusting fuzzy logic based control of robot manipulators in task space. IEEE Trans Industr Electron 69(2):1620–1629

Yilmaz BM, Tatlicioglu E, Savran A, Alci M (2022b) Self-adjusting fuzzy logic based control of robot manipulators in task space. IEEE Trans Industr Electron 69(2):1620–1629

Zhang F, Chen Y (2021) Fuzzy adaptive containment control for nonlinear nonaffine pure-feedback multiagent systems. IEEE Trans Fuzzy Syst 29(10):2878–2889

Zhang Z, Yan Z (2020) An adaptive fuzzy recurrent neural network for solving the nonrepetitive motion problem of redundant robot manipulators. IEEE Trans Fuzzy Syst 28(4):684–691

Zhang Y, Zhang Z (2013) Repetitive motion planning and control of redundant robot manipulators. Springer, Berlin Heidelberg

Zhang Z, Chen S, Zhu X, Yan Z (2020) Two hybrid end-effector posture-maintaining and obstacle-limits avoidance schemes for redundant robot manipulators. IEEE Trans Industr Inf 16(2):754–763

Zhang Z, Chen S, Liang J (2022a) Discrete-time circadian rhythms neural network for perturbed redundant robot manipulators tracking problem with periodic noises. IEEE Trans Industr Inf 18(1):242–251

Zhang J, Jin L, Yang C (2022b) Distributed cooperative kinematic control of multiple robotic manipulators with an improved communication efficiency. IEEE/ASME Trans Mechatron 27(1):149–158

Chapter 7
Orientation-Tracking-Incorporated Control with DNN

7.1 Introduction

Redundant manipulators, which possess more degrees of freedom (DOFs) than are strictly necessary, have the ability to execute complex and demanding tasks when equipped with appropriate control strategies and computational resources Yang et al. (2018), Zhang et al. (2019). Thus, they have been investigated and equipped in industrial and engineering fields to promote production efficiency or complete extremely sophisticated tasks Liu and Shang (2024), Caccavale et al. (2008), Czimmermann et al. (2022). Furthermore, to advance the development of redundant manipulators, several subtasks are considered, such as obstacle avoidance Rubagotti et al. (2019), manipulability optimization Zhu et al. (2022), Jin et al. (2023), Li et al. (2024), and disturbance suppression Xu et al. (2021).

A critical and essential challenge in the redundancy resolution of redundant manipulators is the online solution to inverse kinematics, which has prompted significant research and development of related approaches Safeea et al. (2021), Yan et al. (2025), Kim et al. (2021). Typically, robotic motion control involves various physical constraints at multiple levels Lyu et al. (2023), such as joint angle and velocity limits, in addition to the orientation tracking of the end-effector. By converting these physical and kinematic constraints into inequality and equality constraints, quadratic programming (QP) is employed to model and solve these problems Van Steen et al. (2024), Yang et al. (2024), Li and Cui (2024). Subsequently, significant efforts have been made to advance this approach, with various schemes presented at the joint velocity level to accomplish different tasks Xie and Jin (2024), Hamad et al. (2023), Zhang et al. (2024). Moreover, joint-acceleration-based schemes have been introduced, taking into account various loss functions Jin and Zhang (2015). In many cases, minimum kinetic energy (MKE) schemes Liu and Sun (2013), derived from the joint velocity perspective, are particularly important in scenarios that demand low power consumption and continuous solutions. Additionally, the minimum-velocity-infinite-norm (MVIN) scheme Gravagne and Walker (2000) can regulate individual

M. Liu et al., *Dynamic Neural Networks for Motion Control of Redundant Manipulators*, Intelligent Control and Learning Systems 21,
https://doi.org/10.1007/978-981-96-9144-9_7

joint velocities. However, it faces issues with discontinuity in the generated minimum infinity-norm solutions. Cyclic motion has garnered significant attention in the control and planning of redundant manipulators. Cyclic motion control (CMC) schemes Li (2020), Lu et al. (2021) enable redundant manipulators to perform cyclic motions, where each joint returns to its initial state, effectively addressing the joint drift issue that could impact the precision of subsequent motions. However, in contrast to existing single-purpose schemes, multi-criteria schemes provide a more versatile solution for robotic motion control by addressing multiple tasks simultaneously. Building on advantages of the aforementioned schemes, this chapter introduces a novel multi-criteria approach that integrates the MKE scheme, MVIN scheme, and CMC scheme for robotic motion control. It is important to emphasize that the presented scheme leverages the strengths of the combined approaches to overcome the limitations of individual performance optimizations, thereby ensuring the efficient and accurate manipulator control.

Neural networks have been found to have successful applications across diverse fields, including robotics, control systems, and electronics Li and Li (2021), Li et al. (2021). Common neural network approaches used to control robot manipulators include feedforward neural networks and recurrent neural networks (RNNs) Hao et al. (2023), Godoy et al. (2022), Morita et al. (2024). In Li (2020), an RNN is introduced for achieving the CMC of robots, and subsequently adapted for the online control of redundant robots. Additionally, a zeroing neural dynamic algorithm, a specific form of RNNs, is developed to enable model-free control of redundant manipulators in Yang et al. (2020). To address residual errors in time-dependent systems, a complex-valued RNN is presented in Qi et al. (2020), which is used for the online control of robots, offering enhanced accuracy. A primal-dual RNN is employed in Xia et al. (2005) for redundant manipulators, and is demonstrated to exhibit global convergence. The dynamic neural network (DNN), an emerging variant of the RNN Shi et al. (2024), Jin et al. (2024), Yang et al. (2023), is capable of addressing complex issues by considering the dynamic properties of parameters Li et al. (2022), Nurbayeva et al. (2023), Chen et al. (2025), Zhang et al. (2024). In Li et al. (2017), a modified DNN is introduced by incorporating velocity compensation, which addresses the accumulation of position errors caused by noise pollution and achieves rapid tracking with improved accuracy. Additionally, in Yan et al. (2025), a simultaneous learning and control scheme for redundant manipulator control is presented, formulated as a QP problem and solved using a specially designed DNN. Given the high efficiency and practicality of DNNs, this chapter presents a DNN model with low computational complexity to perform robotic motion control, taking into account multiple performance indexes, including both the orientation-tracking constraint and physical constraints.

The rest of this chapter is organized into four sections. Section 7.2 outlines the development and transformation of the presented multi-criteria control scheme for redundant manipulators with physical constraints. In Sect. 7.3, a DNN model is developed to solve the presented multi-criteria control scheme, accompanied by theoretical analyses. Section 7.4 presents simulation and experimental results based on a 7-DOF Franka Emika Panda manipulator (referred to as the Franka manipulator), along with a comparison of various performance indices for robotic redundancy solutions. Section 7.5 concludes this chapter. The key contributions are summarized as follows:

- A novel multi-criteria control scheme is introduced, which incorporates multiple performance indices and considers orientation tracking. This scheme allows for the simultaneous completion of various tasks in robotic motion control. Moreover, it demonstrates the superior performance compared to existing methods in controlling k-DOF redundant manipulators with $k \geq 7$, owing to the integration of orientation tracking.
- A training-free DNN model is constructed to globally and exponentially converge to the theoretical solution of the presented multi-criteria control scheme. This model operates with low computational complexity, specifically in an $O(n^2)$ manner, where n represents the dimension of the variables.
- Simulations and physical experiments are performed to validate the effectiveness and feasibility of the multi-criteria control scheme and the corresponding DNN model.

7.2 Problem Establishment and Transformation

This section introduces a novel multi-criteria control scheme that simultaneously minimizes kinetic energy and the infinite norm of velocity, while generating cyclic motion and considering orientation tracking. Subsequently, the formulated scheme is converted into a QP problem with multiple constraints.

7.2.1 Problem Construction

For a redundant manipulator with k-DOF, its forward kinematic equation, which maps the joint space $\boldsymbol{\theta}(t) \in \mathbb{R}^k$ to the end-effector workspace $\boldsymbol{r}(t) \in \mathbb{R}^b$, is analogous to (4.1) and expressed as

$$\boldsymbol{r}(t) = \Phi\big(\boldsymbol{\theta}(t)\big), \tag{7.1}$$

where $\Phi(\cdot)$ carries the physical structure and parameter information of the given redundant manipulator, and b represents the dimension of the end-effector workspace. To develop an affine system for further implementation, taking the time derivative of both sides of (7.1) yields

$$\dot{\boldsymbol{r}}(t) = J\dot{\boldsymbol{\theta}}(t), \tag{7.2}$$

where $\dot{\boldsymbol{r}}(t)$ is the velocity of the end-effector; $J = \partial\Phi\big(\boldsymbol{\theta}\big)/\partial\boldsymbol{\theta} \in \mathbb{R}^{b\times k}$ is the Jacobian matrix; and $\dot{\boldsymbol{\theta}}(t)$ denotes the joint velocity. In contrast to Eqs. (7.1), (7.2) provides a simpler velocity-level affine form to address manipulator tracking problems.

To analyze the motion control of manipulators, it is essential to determine not only the position of the end-effector, but also its orientation, typically represented by the rotation matrix. For a redundant manipulator with $k > 6$ operating in a three-dimensional workspace, the orientation of its end-effector may vary during task execution, and such changes should be constrained to ensure high reliability. Defining rotation matrix $R = [x_1, x_2, x_3; y_1, y_2, y_3; z_1, z_2, z_3]^{\mathrm{T}}$, where ^{T} represents the transposition of a vector or matrix. For an orthogonal matrix of order ι, where the row (or column) vectors form a set of unit orthogonal vectors, the elements of a row (or column) vector are determined once any $\iota - 1$ elements of that row (or column) are specified. Based on the fact that the rotation matrix R is an orthogonal matrix with a determinant of 1, any 2×2 sub-matrix can be selected to determine the entire matrix. This can then be reconstructed as a direction vector $\boldsymbol{g} = [x_1, y_1, x_2, y_2]^{\mathrm{T}}$. During the task execution, it is required that $\boldsymbol{g}(t) = \boldsymbol{g}(0)$ at each time instant. Computing the time derivation of $\boldsymbol{g}(t)$ leads to $\dot{\boldsymbol{g}}(t) = A\dot{\boldsymbol{\theta}}$, where $A = \partial \boldsymbol{g}/\partial \boldsymbol{\theta} \in \mathbb{R}^{4\times k}$ is the Jacobian matrix. Letting $\boldsymbol{g}_{\mathrm{d}}(t)$ denote the desired direction vector and using the neural dynamic design formula to $\boldsymbol{g}(t) - \boldsymbol{g}_{\mathrm{d}}(t)$ leads to

$$\dot{\boldsymbol{g}}(t) - \dot{\boldsymbol{g}}_{\mathrm{d}}(t) = -\kappa(\boldsymbol{g}(t) - \boldsymbol{g}_{\mathrm{d}}(t)), \tag{7.3}$$

where $\kappa > 0$. To ensure the stability of the manipulator's end-effector orientation, the rotation matrix must satisfy the condition $R(t) = R(0)$ at all time instants. When $\boldsymbol{g}(t) - \boldsymbol{g}\mathrm{d}(t) = 0$ with $\boldsymbol{g}\mathrm{d}(t) = \boldsymbol{g}(0)$, it follows that $\dot{\boldsymbol{g}}(t) = \dot{\boldsymbol{g}}(0) = A\dot{\boldsymbol{\theta}} = \boldsymbol{0}$. In this case, the orientation of the end-effector remains fixed at its initial state throughout the task execution. Otherwise, $\boldsymbol{g}(t) - \boldsymbol{g}_{\mathrm{d}}(t) \neq 0$ leads to the variation of the orientation vector $\boldsymbol{g}(t)$ for a redundant robot. For simplicity, the time variable t is omitted in the subsequent content.

Due to the underdetermination of (7.2), the redundancy solution for manipulators may not have a unique solution. To simultaneously satisfy the kinematic constraints and track the orientation of the end-effector, the following multi-criteria control scheme is presented for the redundant manipulator:

$$\begin{aligned} \min_{\dot{\boldsymbol{\theta}}}\ & \frac{1}{2}(\alpha\dot{\boldsymbol{\theta}}^{\mathrm{T}}P(\boldsymbol{\theta})\dot{\boldsymbol{\theta}} + \beta\|\dot{\boldsymbol{\theta}}\|_\infty + \gamma\|\dot{\boldsymbol{\theta}} + \delta(\boldsymbol{\theta} - \boldsymbol{\theta}_0)\|_2^2), \\ \text{s.t.}\ & J\dot{\boldsymbol{\theta}} = \dot{\boldsymbol{r}}_{\mathrm{d}} + \sigma(\boldsymbol{r}_{\mathrm{d}} - \boldsymbol{r}), \\ & A\dot{\boldsymbol{\theta}} = \dot{\boldsymbol{g}}_{\mathrm{d}} - \kappa(\boldsymbol{g} - \boldsymbol{g}_{\mathrm{d}}), \\ & \xi_i^- \le \dot{\theta}_i \le \xi_i^+, i = 1, 2, \ldots, k, \end{aligned} \tag{7.4}$$

where $\|\cdot\|_\infty$ and $\|\cdot\|_2$ express the infinite norm and Euclidean norm of a vector or matrix, respectively; coefficients $\alpha,\ \beta,\ \gamma \in [0, 1]$ with $\alpha + \beta + \gamma = 1$ are introduced to adjust the index; the adjusting factor $\delta > 0$; $\boldsymbol{\theta}_0$ is the initial joint angle of the redundant manipulator; $P(\boldsymbol{\theta})$ indicates an inertial matrix which is symmetrically positive definite; $\boldsymbol{r}_{\mathrm{d}}$ represents the desired trajectory; $\sigma > 0$ represents the position error feedback coefficient; $\xi_i^+ = \min\{\dot{\theta}_i^+, a_v(\theta_i^+ - \theta_i)\}$ and $\xi_i^- = \max\{\dot{\theta}_i^-, a_v(\theta_i^- - \theta_i)\}$

with $\dot{\theta}_i^+$ (θ_i^+) and $\dot{\theta}_i^-$ (θ_i^-) denoting the upper and lower limits of the ith joint velocity (joint angle), respectively; and a_v is a coefficient used to adjust the magnitude when the joint angle exceeds its limits.

Remark 7.1 For a manipulator with DOFs number of $k \leq 6$ in a three-dimensional workspace, it can be considered non-redundant if both the orientation and position of the end-effector are taken into account. On the other hand, it becomes redundant if only the position of the end-effector is considered. Based on this, various redundancy resolution schemes are applied to 6-DOF manipulators without considering the orientation information Li et al. (2017). Furthermore, even for manipulators with more than 7 DOFs Ngo and Liu (2023), Bertino et al. (2022), it is difficult to detect failures resulting from the lack of orientation tracking when a kinematic control scheme that does not account for orientation tracking is used in simulations. For instance, various 7-DOF manipulators on the CoppeliaSim platform are used in Li (2020) and Yan et al. (2021) to validate the effectiveness of the presented schemes, with the simulation results being visualized to show success with minimal tracking position errors. However, these schemes do not account for orientation tracking, which may result in failures when applied to real-world scenarios. This issue is clearly illustrated in the experimental results presented in the following section. To clearly demonstrate the effect of orientation tracking, we set $\boldsymbol{g} - \boldsymbol{g}_\text{d} = \boldsymbol{0}$, where $\boldsymbol{g}_\text{d}$ represents the initial state of the orientation in the subsequent parts.

7.2.2 Scheme Transformation

To convert the presented multi-criteria control scheme (7.4) into a standard QP problem, the infinite-norm term in the loss function needs to be transformed into an inequality constraint. Based on the definition of the infinite norm Liu et al. (2023), we have $\|\dot{\boldsymbol{\theta}}\|_\infty = \max\{|\dot{\theta}_1|, |\dot{\theta}_2|, \ldots, |\dot{\theta}_k|\} = \max_{1\leq i\leq k} |\boldsymbol{I}_i^\text{T}\dot{\boldsymbol{\theta}}|$, where $\boldsymbol{I}_i$ denotes the ith column vector of the identity matrix $\boldsymbol{I} \in \mathbb{R}^{k\times k}$. By introducing an auxiliary variable $\varrho = |\dot{\boldsymbol{\theta}}|\infty$, the inequality relation $\varrho \geq |\boldsymbol{I}i^\text{T}\dot{\boldsymbol{\theta}}|$ can then be derived, which in turn leads to $[\boldsymbol{I},\ -\boldsymbol{\ell}_1;\ \ -\boldsymbol{I},\ -\boldsymbol{\ell}_1][\dot{\boldsymbol{\theta}}^\text{T},\ \varrho]^\text{T} \leq \boldsymbol{0}$ with $\boldsymbol{\ell}_1$ denoting a vector composing of 1. Subsequently, by reformulating the matrices and vectors, the following expression is derived: $\boldsymbol{X} = [\boldsymbol{I}, -\boldsymbol{\ell}_1; -\boldsymbol{I}, -\boldsymbol{\ell}_1] \in \mathbb{R}^{(2k)\times(k+1)}$, $\Psi = [\alpha P + \gamma \boldsymbol{I}, 0; 0, 0] \in \mathbb{R}^{(k+1)\times(k+1)}$, $\Gamma = [J, \boldsymbol{0};\ A, \boldsymbol{0}] \in \mathbb{R}^{(b+4)\times(k+1)}$, $\varpi = [\dot{\boldsymbol{r}}_\text{d} + \sigma(\boldsymbol{r}_\text{d} - \boldsymbol{r}); \dot{\boldsymbol{g}}_\text{d} - \kappa(\boldsymbol{g} - \boldsymbol{g}_\text{d})] \in \mathbb{R}^{b+4}$, $\boldsymbol{\zeta} = [\gamma\delta(\boldsymbol{\theta} - \boldsymbol{\theta}_0)^\text{T}, \beta/2]^\text{T} \in \mathbb{R}^{k+1}$, $\Upsilon = [\boldsymbol{X}; \hat{\boldsymbol{I}}; -\hat{\boldsymbol{I}}] \in \mathbb{R}^{(4k+2)\times(k+1)}$, $\boldsymbol{\lambda} = [\dot{\boldsymbol{\theta}}^\text{T}, \varrho]^\text{T} \in \mathbb{R}^{k+1}$, and $\boldsymbol{\chi} = [\boldsymbol{0}; \boldsymbol{\xi}^+; \dot{\theta}^+; \boldsymbol{\xi}^-; 0] \in \mathbb{R}^{4k+2}$, where the identity matrix $\hat{\boldsymbol{I}} \in \mathbb{R}^{(k+1)\times(k+1)}$; $0 \leq \|\dot{\boldsymbol{\theta}}\|_\infty \leq \dot{\theta}^+ = \max_{1\leq i\leq k} \dot{\theta}_i^+$; $[\boldsymbol{\xi}^-; 0] \leq \boldsymbol{\lambda} \leq [\boldsymbol{\xi}^+; \Theta]$ with Θ denoting a positive infinity constant, $\boldsymbol{\xi}^-$ ($\boldsymbol{\xi}^+$) representing the lower (upper) limit of the joint velocity. Accordingly, the presented multi-criteria control scheme (7.4) is reformulated into a simplified form as follows:

$$\begin{aligned}\min_{\boldsymbol{\lambda}}\ & \frac{1}{2}\boldsymbol{\lambda}^{\mathrm{T}}\Psi\boldsymbol{\lambda}+\boldsymbol{\zeta}^{\mathrm{T}}\boldsymbol{\lambda},\\ \text{s.t.}\ & \Gamma\boldsymbol{\lambda}=\boldsymbol{\varpi},\\ & \Upsilon\boldsymbol{\lambda}\le\boldsymbol{\chi},\end{aligned}\tag{7.5}$$

which is a convex QP with decision variable $\boldsymbol{\lambda}$.

7.3 DNN Model and Its Convergence Analysis

In this section, a DNN model is constructed to address the presented multi-criteria control formulation (7.4), accompanied by a detailed analysis of its convergence behavior and computational complexity.

7.3.1 DNN Model

The global optimum of the convex QP problem (7.5) can be determined by ensuring the satisfaction of the KKT conditions. Thus, the corresponding Lagrangian function is formulated as follows:

$$\mathfrak{L}(\boldsymbol{\lambda},\boldsymbol{\nu},\boldsymbol{\mu})=\frac{1}{2}\boldsymbol{\lambda}^{\mathrm{T}}\Psi\boldsymbol{\lambda}+\boldsymbol{\zeta}^{\mathrm{T}}\boldsymbol{\lambda}+\boldsymbol{\nu}^{\mathrm{T}}(\Gamma\boldsymbol{\lambda}-\boldsymbol{\varpi})+\boldsymbol{\mu}^{\mathrm{T}}(\Upsilon\boldsymbol{\lambda}-\boldsymbol{\chi}),\tag{7.6}$$

where $\boldsymbol{\nu}\in\mathbb{R}^{b+4}$ and $\boldsymbol{\mu}\in\mathbb{R}^{4k+2}$ are KKT multipliers. In order to obtain the solution to the optimization problem, the following KKT conditions should be satisfied:

$$\begin{aligned}&\partial\mathfrak{L}/\partial\boldsymbol{\lambda}=\Psi\boldsymbol{\lambda}+\boldsymbol{\zeta}+\Gamma^{\mathrm{T}}\boldsymbol{\nu}+\Upsilon^{\mathrm{T}}\boldsymbol{\mu}=\mathbf{0},\\ &\partial\mathfrak{L}/\partial\boldsymbol{\nu}=\Gamma\boldsymbol{\lambda}-\boldsymbol{\varpi}=\mathbf{0},\\ &\partial\mathfrak{L}/\partial\boldsymbol{\mu}=\Upsilon\boldsymbol{\lambda}-\boldsymbol{\chi}=\mathbf{0},\\ &\boldsymbol{\lambda}^{-}\le\boldsymbol{\lambda}\le\boldsymbol{\lambda}^{+},\\ &\mu_i\ge 0,\ i=1,2,\ldots,4k+2,\end{aligned}\tag{7.7}$$

where μ_i denotes the ith element of $\boldsymbol{\mu}$; $\boldsymbol{\lambda}^{+}$ and $\boldsymbol{\lambda}^{-}$ are the upper and lower bounds on $\boldsymbol{\lambda}$, with the ith elements being

$$\lambda_i^{+}=\begin{cases}\min\{a_v(\theta_i^{+}-\theta_i),\dot{\theta}_i^{+}\}, & i=1,2,\ldots,k,\\ \Theta, & i=k+1,\end{cases}$$

$$\lambda_i^{-}=\begin{cases}\max\{a_v(\theta_i^{-}-\theta_i),\dot{\theta}_i^{-}\}, & i=1,2,\ldots,k,\\ 0, & i=k+1.\end{cases}$$

To efficiently address the inequality constraints in (7.7), a DNN model is constructed to solve the above system of equations in the following structure:

$$
\begin{aligned}
\tau\dot{\boldsymbol{\lambda}} &= -\boldsymbol{\lambda} + \mathrm{F}_{\lambda}\big(\boldsymbol{\lambda} - (\Psi\boldsymbol{\lambda} + \boldsymbol{\zeta} + \Gamma^{\mathrm{T}}\boldsymbol{\nu} + \Upsilon^{\mathrm{T}}\boldsymbol{\mu})\big),\\
\tau\dot{\boldsymbol{\nu}} &= -\boldsymbol{\nu} + \mathrm{F}_{\nu}(\boldsymbol{\nu} + \Gamma\boldsymbol{\lambda} - \boldsymbol{\varpi}),\\
\tau\dot{\boldsymbol{\mu}} &= -\boldsymbol{\mu} + \mathrm{F}_{\mu}(\boldsymbol{\mu} + \Upsilon\boldsymbol{\lambda} - \boldsymbol{\chi}),
\end{aligned}
\tag{7.8}
$$

where convergence coefficient $\tau > 0$. Besides, in DNN model (7.8), projection functions F_{λ}, F_{ν}, and F_{μ} exploited to impose restrictions on ranges of variables $\boldsymbol{\lambda}$, $\boldsymbol{\nu}$, and $\boldsymbol{\mu}$, and they are defined as

$$
\mathrm{F}_{\lambda}(\lambda_i) = \begin{cases} \lambda_i^+, & \lambda_i \geq \lambda_i^+,\\ \lambda_i, & \lambda_i^- < \lambda_i < \lambda_i^+,\\ \lambda_i^-, & \lambda_i \leq \lambda_i^-, \end{cases}
$$

$$
\mathrm{F}_{\nu}(\nu_i) = \nu_i, \quad \mathrm{F}_{\mu}(\mu_i) = \begin{cases} \mu_i, & \mu_i \geq 0,\\ 0, & \mu_i < 0. \end{cases}
$$

7.3.2 Convergence Analysis on DNN Model

To analyze the convergence behavior of the presented DNN model (7.8), this section introduces the following theorem along with its corresponding proof. To simplify the proof, we present the following definitions:

$$
\boldsymbol{d} = \begin{bmatrix} \boldsymbol{\lambda} \\ \boldsymbol{\nu} \\ \boldsymbol{\mu} \end{bmatrix} \in \mathbb{R}^n, \boldsymbol{d}^+ = \begin{bmatrix} \boldsymbol{\lambda}^+ \\ \Theta \\ \Theta \end{bmatrix} \in \mathbb{R}^n, \boldsymbol{d}^- = \begin{bmatrix} \boldsymbol{\lambda}^- \\ -\Theta \\ \mathbf{0} \end{bmatrix} \in \mathbb{R}^n,
$$

where $n = b + 5k + 7$, and thresholds $\boldsymbol{d}^+$ and $\boldsymbol{d}^-$ represent the upper bound and lower bound on the vector $\boldsymbol{d}$, respectively.

Theorem 7.1 *The presented DNN model (7.8) is stable and globally converges to the theoretical solution of (7.7), i.e.,* $\boldsymbol{d}^* = [\boldsymbol{\lambda}^*; \boldsymbol{\nu}^*; \boldsymbol{\mu}^*]$*, and the first k elements of* $\boldsymbol{d}^*$ *are the optimal solution to the presented scheme (7.4).*

Proof A canonical dynamic system is leveraged to reconstruct DNN model (7.8):

$$
\tau\dot{\boldsymbol{d}} = -\boldsymbol{d} + \mathrm{F}_{d}\big(\boldsymbol{d} - \mathfrak{f}(\boldsymbol{d})\big), \tag{7.9}
$$

where $\mathrm{F}_d = [\mathrm{F}_{\lambda}; \mathrm{F}_{\nu}; \mathrm{F}_{\mu}] \in \mathbb{R}^n$ is a projection function similar to F_{λ}, with $\boldsymbol{d}^+$ and $\boldsymbol{d}^-$ as the thresholds, and

$$\mathfrak{f}(\boldsymbol{d}) = \begin{bmatrix} \Psi\boldsymbol{\lambda} + \zeta + \Gamma^{\mathrm{T}}\boldsymbol{\nu} + \Upsilon^{\mathrm{T}}\boldsymbol{\mu} \\ \varpi - \Gamma\boldsymbol{\lambda} \\ \chi - \Upsilon\boldsymbol{\lambda} \end{bmatrix}.$$

Then, the gradient of $\mathfrak{f}(\boldsymbol{d})$ can be derived as

$$\nabla\mathfrak{f}(\boldsymbol{d}) = \partial\mathfrak{f}(\boldsymbol{d})/\partial\boldsymbol{d} = \begin{bmatrix} \Psi^{\mathrm{T}} & \Gamma & \Upsilon \\ , -\Gamma^{\mathrm{T}} & \mathbf{0} & \mathbf{0} \\ -\Upsilon^{\mathrm{T}} & \mathbf{0} & \mathbf{0} \end{bmatrix}.$$

Assuming that the matrix Ψ is semi-positive definite and satisfies $\Psi^{\mathrm{T}} = \Psi$, it follows that

$$\nabla\mathfrak{f}(\boldsymbol{d}) + \nabla^{\mathrm{T}}\mathfrak{f}(\boldsymbol{d}) = \begin{bmatrix} 2\Psi & \mathbf{0} & \mathbf{0} \\ \mathbf{0} & \mathbf{0} & \mathbf{0} \\ \mathbf{0} & \mathbf{0} & \mathbf{0} \end{bmatrix}.$$

Evidently, $\nabla\mathfrak{f}(\boldsymbol{d}) + \nabla^{\mathrm{T}}\mathfrak{f}(\boldsymbol{d})$ is a semi-positive definite matrix. Furthermore, utilizing the Lagrange mean value theorem, the following results can be derived: $\mathfrak{f}(\boldsymbol{d}_1) - \mathfrak{f}(\boldsymbol{d}_2) = \nabla\mathfrak{f}(\eta)(\boldsymbol{d}_1 - \boldsymbol{d}_2)$ with $\eta \in [\boldsymbol{d}_1, \boldsymbol{d}_2]$. Then, we further have

$$\begin{aligned}(\boldsymbol{d}_1 - \boldsymbol{d}_2)^{\mathrm{T}}(\mathfrak{f}(\boldsymbol{d}_1) - \mathfrak{f}(\boldsymbol{d}_2)) &= (\boldsymbol{d}_1 - \boldsymbol{d}_2)^{\mathrm{T}}\nabla\mathfrak{f}(\eta)(\boldsymbol{d}_1 - \boldsymbol{d}_2) \\ &= (\boldsymbol{d}_1 - \boldsymbol{d}_2)^{\mathrm{T}}(\nabla^{\mathrm{T}}\mathfrak{f}(\eta) + \nabla\mathfrak{f}(\eta))(\boldsymbol{d}_1 - \boldsymbol{d}_2)/2 \geq 0.\end{aligned}$$

From this, it can be concluded that $\mathfrak{f}(\boldsymbol{d})$ exhibits a monotonically decreasing behavior. Based on Lemma 4 from Gao (2003), it is generalized that the dynamic system (7.9) is Lyapunov stable, with its trajectory converging to the equilibrium point $\boldsymbol{d}^* = [\boldsymbol{\lambda}^*; \boldsymbol{\nu}^*; \boldsymbol{\mu}^*]$ with

$$(\boldsymbol{d} - \boldsymbol{d}^*)^{\mathrm{T}}\mathfrak{f}(\boldsymbol{d}^*) \geq 0. \tag{7.10}$$

Then, (7.10) is equivalent to the projection equation:

$$\mathrm{F}_{\boldsymbol{d}}\big(\boldsymbol{d} - \mathfrak{f}(\boldsymbol{d}^*)\big) = \boldsymbol{d}^*. \tag{7.11}$$

Consequently, it is proved that DNN model (7.8) is stable and converges to the exact solution $\boldsymbol{d}^*$. Ulteriorly, it can be easily found that

$$\begin{aligned}\boldsymbol{\lambda}^* &= \mathrm{F}_{\lambda}\big(\boldsymbol{\lambda} - (\Psi\boldsymbol{\lambda}^* + \zeta + \Gamma^{\mathrm{T}}\boldsymbol{\nu}^* + \Upsilon^{\mathrm{T}}\boldsymbol{\mu}^*)\big), \\ \boldsymbol{\nu}^* &= \mathrm{F}_{\nu}\big(\boldsymbol{\nu} - (\varpi - \Gamma\boldsymbol{\lambda}^*)\big), \\ \boldsymbol{\mu}^* &= \mathrm{F}_{\mu}\big(\boldsymbol{\mu} - (\chi - \Upsilon\boldsymbol{\lambda}^*)\big).\end{aligned} \tag{7.12}$$

From the above equation, it is clear that $\boldsymbol{\lambda}$ globally converges to $\boldsymbol{\lambda}^*$, where the first k elements of $\boldsymbol{\lambda}^*$ represent the optimal solution to the presented multi-criteria control scheme (7.4) based on DNN model (7.8). The proof is completed. □

To determine the convergence rate of the DNN model (7.8), the following theorem is presented.

Theorem 7.2 *The presented DNN model (7.8) exponentially converges to the theoretical solution of (7.7), i.e.,* $\boldsymbol{d}^* = [\boldsymbol{\lambda}^*; \boldsymbol{\nu}^*; \boldsymbol{\mu}^*]$, *and the first k elements of* $\boldsymbol{d}^*$ *are the optimal solution to the presented scheme (7.4).*

Proof Before analyzing the convergent rate of DNN model (7.8), we define an auxiliary diagonal matrix $\boldsymbol{B}$ with its ith diagonal element:

$$B_i = \mathrm{F}_{di}\big(\boldsymbol{d}_i - \mathfrak{f}_i(\boldsymbol{d}_i)\big)/\big(\boldsymbol{d}_i - \mathfrak{f}_i(\boldsymbol{d}_i)\big). \tag{7.13}$$

Note that, with $\Im_i = \boldsymbol{d}_i - \mathfrak{f}_i(\boldsymbol{d}_i)$, when $\Im_i = 0$, it obtains $B_i = \mathrm{F}_{di}(\Im_i)/(\Im_i) = 0/0$. L'hospital's rule Stewart (2007) states that, if $\lim_{x\to c} f(x) = \lim_{x\to c} g(x) = 0$, $g'(x) \neq 0$ for all x in I with $x \neq c$, and $\lim_{x\to c} f'(x)/g'(x)$ exists, then $\lim_{x\to c} f(x)/g(x) = \lim_{x\to c} f'(x)/g'(x)$. Therefore, $\lim_{\Im_i\to 0} B_i = \mathrm{F}_{di}(\Im_i)/(\Im_i) = \Im_i/\Im_i = 1/1 = 1$. On this basis, we define that, when $\Im_i = 0$ (i.e., $\boldsymbol{d}_i - \mathfrak{f}_i(\boldsymbol{d}_i) = 0$), $B_i = \mathrm{F}_{di}\big(\boldsymbol{d}_i - \mathfrak{f}_i(\boldsymbol{d}_i)\big)/\big(\boldsymbol{d}_i - \mathfrak{f}_i(\boldsymbol{d}_i)\big) = 1$. Then, it can be readily deduced that $B_i \in (0, 1]$. One can further have that $\mathrm{F}_{\boldsymbol{d}}\big(\boldsymbol{d} - \mathfrak{f}(\boldsymbol{d})\big) = \boldsymbol{B}\big(\boldsymbol{d} - \mathfrak{f}(\boldsymbol{d})\big) = \boldsymbol{B}\boldsymbol{d} - \boldsymbol{B}\mathfrak{f}(\boldsymbol{d})$. In addition, $\mathfrak{f}(\boldsymbol{d})$ in (7.9) is rewritten as

$$\mathfrak{f}(\boldsymbol{d}) = \boldsymbol{H}\boldsymbol{d} + \phi, \tag{7.14}$$

where $\boldsymbol{H} = [\Psi, \Gamma^{\mathrm{T}}, \Upsilon^{\mathrm{T}}; -\Gamma, \mathbf{0}, \mathbf{0}; -\Upsilon, \mathbf{0}, \mathbf{0}] \in \mathbb{R}^{n\times n}$, and $\phi = [\zeta, \varpi, \chi]^{\mathrm{T}} \in \mathbb{R}^n$. The compact expression of DNN model (7.8), i.e., (7.9), is restated as

$$\tau\dot{\boldsymbol{d}} = -\boldsymbol{d} + \boldsymbol{B}\big(\boldsymbol{d} - \boldsymbol{H}\boldsymbol{d} - \phi\big) = -\big(\boldsymbol{I} - \boldsymbol{B} + \boldsymbol{B}\boldsymbol{H}\big)\boldsymbol{d} - \boldsymbol{B}\phi, \tag{7.15}$$

which is a first-order differential equation with $\boldsymbol{d}$ being the decision variable. It can be shown that the convergence rate of DNN model (7.8) is influenced by the matrix $\boldsymbol{I} - \boldsymbol{B} + \boldsymbol{B}\boldsymbol{H}$. Considering that the diagonal elements of $\boldsymbol{I} - \boldsymbol{B}$ are all greater than or equal to zero, and the diagonal elements of $\boldsymbol{B}$ are all strictly positive, we only need to demonstrate that the eigenvalues of $\boldsymbol{H}$ are all greater than or equal to zero to prove convergence of the DNN model (7.8). According to Theorem 7.1 in Liu et al. (2023), the positive semi-definite property of matrix $\boldsymbol{H}$ can be readily achieved. Then, matrix $\boldsymbol{I} - \boldsymbol{B} + \boldsymbol{B}\boldsymbol{H}$ is positive semi-definite and its eigenvalues are greater than zero, and it further has

$$\tau\dot{\boldsymbol{d}} \leq -\mathrm{eig}_{\min}\big(\boldsymbol{I} - \boldsymbol{B} + \boldsymbol{B}\boldsymbol{H}\big)\boldsymbol{d} - \boldsymbol{B}\phi, \tag{7.16}$$

where $\mathrm{eig}_{\min}(\cdot)$ denotes the minimum eigenvalue of a square matrix. It is concluded that DNN model (7.8) exhibits exponential convergence with rate $\mathrm{eig}_{\min}(\boldsymbol{I} - \boldsymbol{B} + \boldsymbol{B}\boldsymbol{H})/\tau$ to the theoretical solution of (7.7), i.e., $\boldsymbol{d}^* = [\boldsymbol{\lambda}^*; \boldsymbol{\nu}^*; \boldsymbol{\mu}^*]$, and the first k elements of $\boldsymbol{d}^*$ are the optimal solution to the presented scheme (7.4). The convergence coefficient in the DNN model should be positive, and decreasing its value results in the fast convergence. The proof is completed. □

7.3.3 Complexity Analysis

This part analyzes the complexity of the presented DNN model (7.8) by discretizing it using the Euler forward difference formula Qi et al. (2020). To compute the complexity, the compact form of DNN model (7.8), as shown in (7.9), is used, and the projection function $\mathrm{F}_{\boldsymbol{d}}$ is assumed to be linear. Then, based on (7.9) and (7.14), we have

$$\boldsymbol{d}_{j+1} = \boldsymbol{d}_j + \frac{h}{\tau}\left[-\boldsymbol{d}_j + \mathrm{F}_{\boldsymbol{d}}\left(\boldsymbol{d}_j - (H_j\boldsymbol{d}_j + \phi_j)\right)\right], \tag{7.17}$$

where coefficient h represents the step size, $j = 1, 2, \ldots$ stands for the updating index. Before analyzing the complexity, we introduce the following definitions based on the principle of floating-point operations (flops) Boyd and Vandenberghe (2004), where m, $\mathbf{m}$, and M represent a scalar, a vector, and a matrix, respectively.

- $m \in \mathbb{R} \times \mathbf{m} \in \mathbb{R}^{s_1}$ consumes s_1 flops.
- $M \in \mathbb{R}^{s_1 \times s_2} \times \mathbf{m} \in \mathbb{R}^{s_2}$ consumes $s_1(2s_2 - 1)$ flops.
- $\mathbf{m_1} \in \mathbb{R}^{s_1} \pm \mathbf{m_2} \in \mathbb{R}^{s_1}$ consumes s_1 flops.
- $M_1 \in \mathbb{R}^{s_1 \times s_2} \pm M_2 \in \mathbb{R}^{s_1 \times s_2}$ consumes $s_1 s_2$ flops.
- Computing the inverse of a square matrix $M \in \mathbb{R}^{s_1 \times s_1}$ consumes s_1^3 flops.
- $M_1 \in \mathbb{R}^{s_1 \times s_2} \times M_2 \in \mathbb{R}^{s_2 \times s_3}$ consumes $s_1 s_3(2s_2 - 1)$ flops.

Based on the above facts, the computational complexity of calculating $\boldsymbol{d}j - (H_j\boldsymbol{d}_j + \phi_j)$ at the jth time step requires a total of $2n^2 + n$ flops. Assuming that the projection function $\mathrm{F}_{\boldsymbol{d}}$ is linear, computing $\boldsymbol{d}_{j+1}$ requires $2n^2 + 4n$ flops in the discrete-time DNN model (7.17). In other words, the computational complexity of the discrete-time DNN model (7.17) is $O(n^2)$. It should be noted that, in the existing zeroing neural network model in Lu et al. (2021) and the pseudo-inverse technique in Katsikis et al. (2022), the online inverse of a Jacobian-like matrix is always required. As highlighted in the aforementioned facts, computing the inverse of a square matrix requires n^3 flops, not to mention the pseudoinversion operations. Moreover, for some classical and conventional QP models for manipulators, such as E47 and 94LVI Zhang and Jin (2017), Qiu et al. (2014), both exhibit a computational complexity of $O(n^2)$ Qiu et al. (2014). However, to solve the same manipulator-control QP problem, the E47 and 94LVI algorithms require multiple iterations to obtain a solution within a sampling interval, while the discrete-time DNN model (7.17) obtains the solution with just one computation over the same sampling interval. In this regard, the computational burdens of E47 and 94LVI algorithms are higher than that of the discrete-time DNN model (7.17) over a sampling interval.

Thus, the presented scheme (7.4) equipped with DNN model (7.8), which has global convergence verified, offers advantages in terms of complexity, as it does not involve matrix inversion operations and obtains the solution with only one computation during a sampling interval.

7.4 Verifications and Comparisons

In this section, simulations are performed using a 7-DOF Franka manipulator to demonstrate the feasibility and validity of the presented multi-criteria control scheme (7.4) based on DNN model (7.8). Subsequently, physical experiments and comparisons are performed to validate the effectiveness of the presented scheme (7.4) with orientation tracking considered.

7.4.1 Parameters' Settings

Primarily, the relevant parameters in the simulations and physical experiments are set as $\alpha = 0.3, \beta = 0.4, \gamma = 1 - \alpha - \beta, \delta = 10, \tau = 1 \times 10^{-4}, \sigma = 10, P = I$, and the initial joint angles are given by $\theta(0) = [0, \quad 0.785472, 0, \quad 2.35425, 0, 1.57164, 0.785465]^{\mathrm{T}}$ rad, with an execution time of $T = 20$ s. Simulations based on the Franka manipulator are conducted to track three different trajectories, namely, the desired Lissajous, rose-curve, and leaf-shaped trajectories. Moreover, in the trajectory-tracking task, the orientation of the manipulator's end-effector is required to remain unchanged, with the target orientation set as $\boldsymbol{g}_{\mathrm{d}} = [0.707, -0.707, -0.707, -0.707]^{\mathrm{T}}$, which ensures that the end-effector remains vertically downward throughout the tracking process. Note that the position errors are computed as $\boldsymbol{e}_{\mathrm{p}} = [e_{\mathrm{X}}, e_{\mathrm{Y}}, e_{\mathrm{Z}}]^{\mathrm{T}} = \boldsymbol{r} - \boldsymbol{r}_{\mathrm{d}}$, while the orientation errors are calculated as $\boldsymbol{e}_{\mathrm{o}} = [e_{x_1}, e_{y_1}, e_{x_2}, e_{y_2}]^{\mathrm{T}} = \boldsymbol{g} - \boldsymbol{g}_{\mathrm{d}}$.

7.4.2 Illustrative Simulations

Evidently, the presented scheme (7.4) based on DNN model (7.8) is capable of accurately tracking the desired trajectory, as demonstrated in Figs. 7.1a, b, and 7.3a, and b. Furthermore, the maximum position errors of the end-effector for both tasks are less than 6×10^{-5} m as shown in Figs. 7.1b and 7.3b, indicating the feasibility of the presented scheme (7.4) with DNN model (7.8) for the kinematic control of redundant manipulators. As shown in Figs. 7.2a, b, and 7.4a, and b, joint angles and joint velocities vary smoothly and stay within their respective bounds while performing the end-effector trajectory-tracking tasks. In addition, due to the influence of the CMC subscheme in the presented scheme (7.4), the final states of the joint angles and velocities of the redundant manipulator remain consistent with their initial states. The results based on the MVIN subscheme and MKE subscheme, derived from the presented multi-criteria control scheme (7.4), for the robotic motion control of the Franka manipulator are shown in Figs. 7.2c and 7.4c, respectively. Additionally, the orientation errors for the manipulator's end-effector are shown in Figs. 7.1c and 7.3c, which stabilize at the order of 10^{-4} and 10^{-5}, demonstrating the high accuracy

Fig. 7.1 Simulation results for the Franka manipulator tracking the Lissajous trajectory using the presented scheme (7.4) solved by DNN model (7.8). **a** Actual trajectory and desired trajectory. **b** Position error. **c** Orientation error

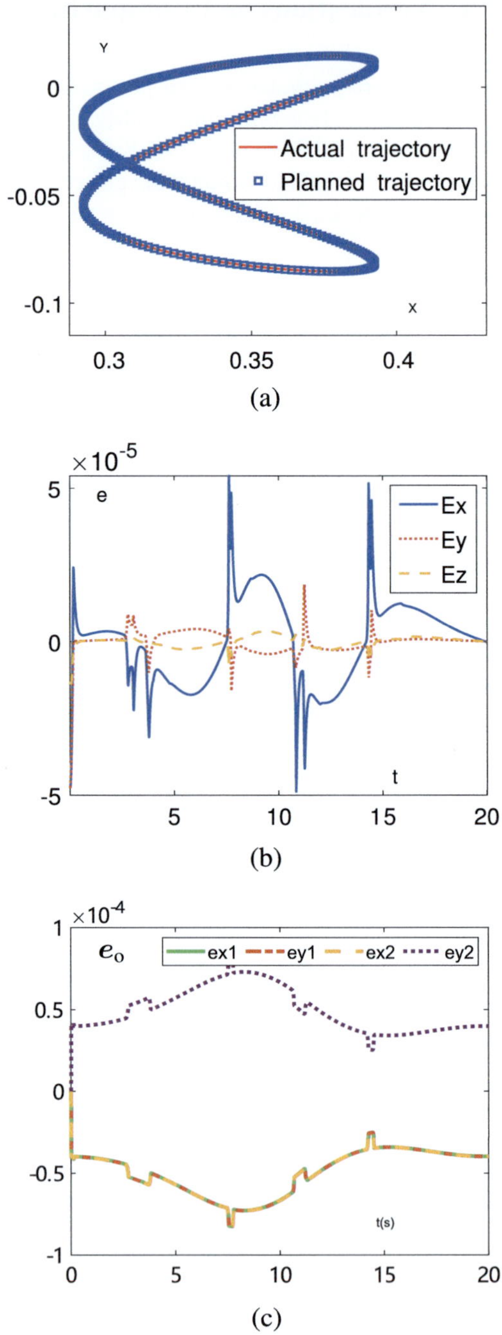

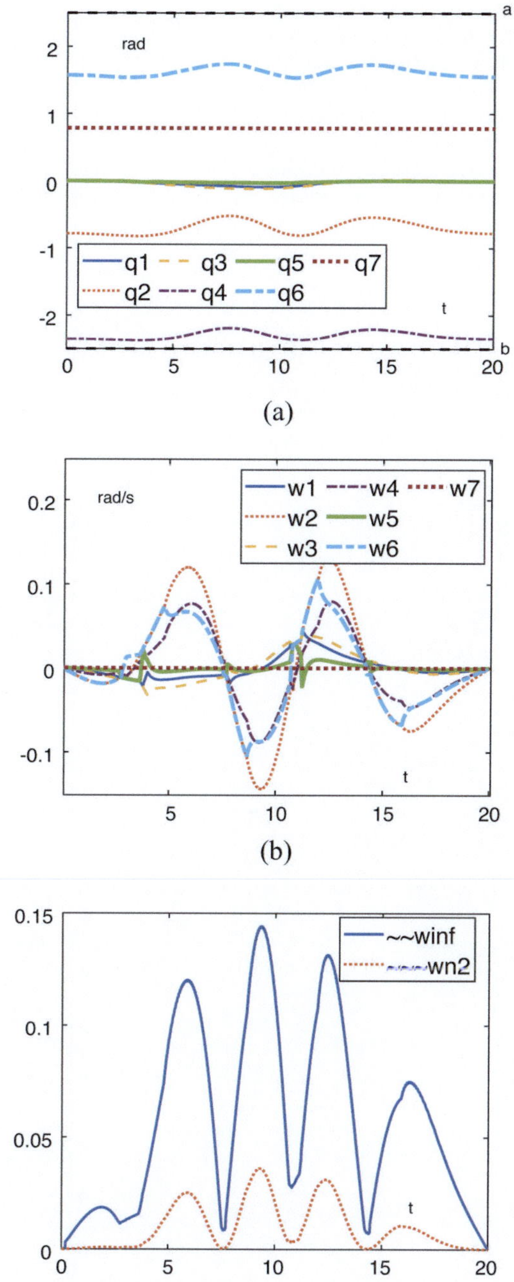

Fig. 7.2 Joint information of the Franka manipulator tracking the Lissajous trajectory using the presented scheme (7.4) solved by DNN model (7.8). **a** Joint angle. **b** Joint velocity. **c** $\|\dot{\boldsymbol{\theta}}\|_\infty$ and $\dot{\boldsymbol{\theta}}^{\mathrm{T}} P \dot{\boldsymbol{\theta}}$

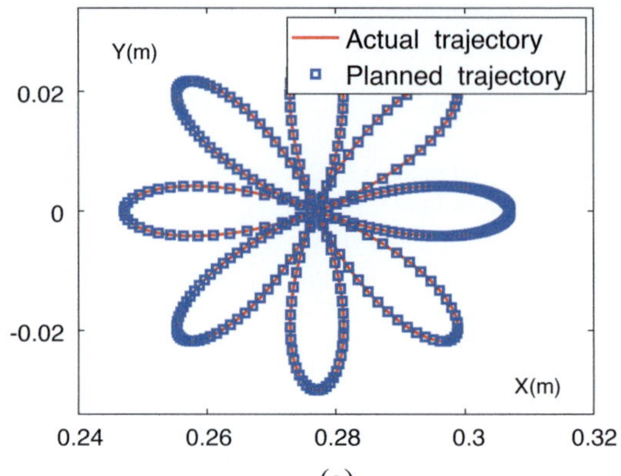

(a)

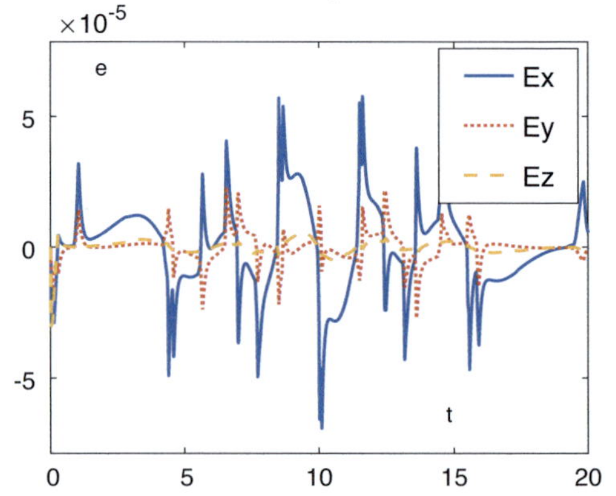

(b)

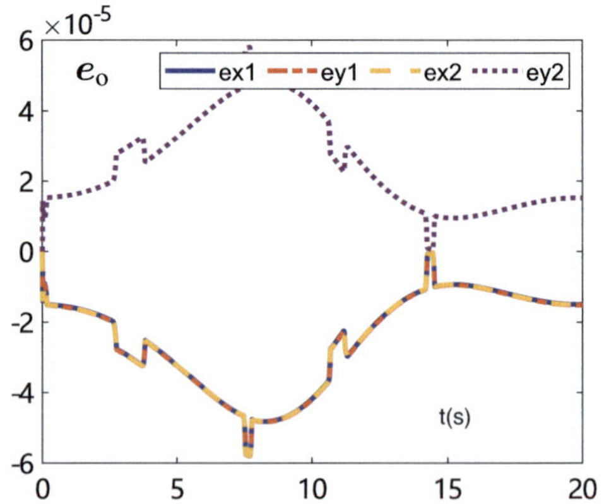

(c)

Fig. 7.3 Simulation results for the manipulator tracking the rose trajectory using the presented scheme (7.4) solved by DNN model (7.8). **a** Actual trajectory and desired trajectory. **b** Position error. **c** Orientation error

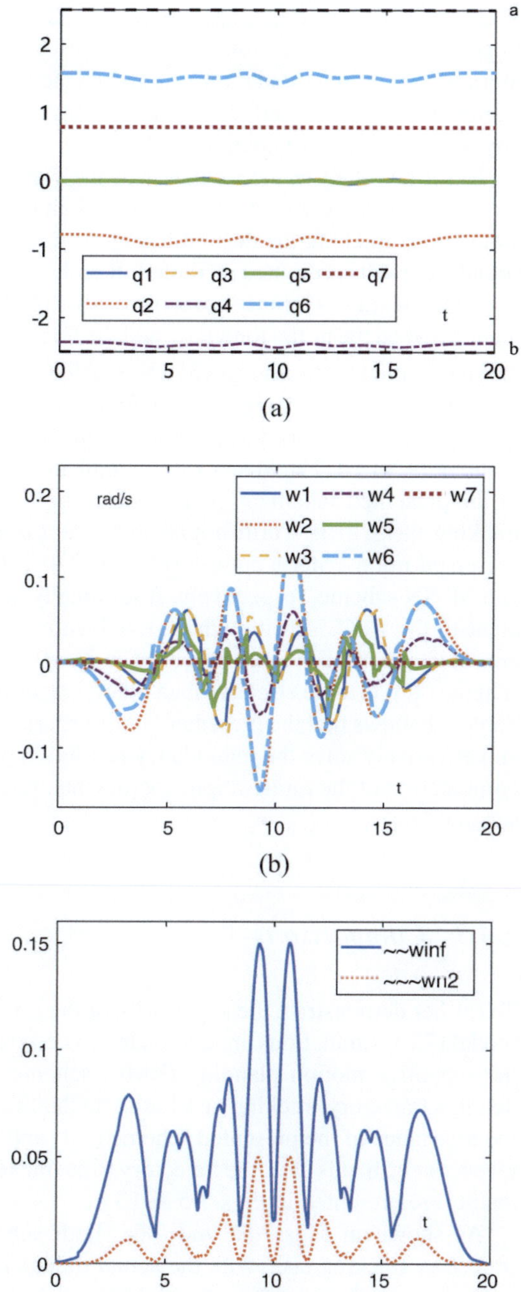

Fig. 7.4 Joint information of the Franka manipulator tracking the rose trajectory using the presented scheme (7.4) solved by DNN model (7.8). **a** Joint angle. **b** Joint velocity. **c** $\|\dot{\boldsymbol{\theta}}\|_\infty$ and $\dot{\boldsymbol{\theta}}^{\mathrm{T}} P \dot{\boldsymbol{\theta}}$

of the presented scheme (7.4) based on DNN model (7.8) for orientation tracking. In general, the presented multi-criteria control scheme (7.4) proves to be feasible for motion control in redundant manipulators, achieving multiple performance indexes simultaneously while effectively executing a given task by considering different physical limits and orientation tracking.

To further analyze the influence of the weight parameters α and β (with γ not directly considered due to $\gamma = 1 - \alpha - \beta$) in the presented scheme (7.4), a series of simulations tracking the desired Lissajous trajectory are conducted, with the corresponding results presented in Table I. It includes the maximum position error (MAX $|\boldsymbol{e}_{\mathrm{p}}|$), the maximum infinity norm of velocity (MAX $\|\dot{\boldsymbol{\theta}}\|\infty$), the maximum displacement between the initial joint state and the final joint state (MAX $|\boldsymbol{\theta}(T) - \boldsymbol{\theta}(0)|$), and the maximum kinetic energy (MAX $\dot{\boldsymbol{\theta}}^{\mathrm{T}} P\dot{\boldsymbol{\theta}}$). From Table 7.1, a clear conclusion can be drawn: When the weighting coefficients vary, the presented scheme effectively tracks the predetermined trajectories, with the maximum position errors maintained at the order of 10^{-4} m in an exact state, demonstrating the reliability and flexibility of the presented scheme (7.4) in controlling the redundant manipulator to achieve tracking tasks. It is worth noting that when $\alpha = 0, \beta = 0$ and $\alpha = 0, \beta = 1$, the presented multi-criteria control scheme (7.4) is transformed into the CMC scheme and MVIN scheme, respectively. Additionally, the presented scheme (7.4) is equivalent to the MKE scheme when $\alpha = 1$ and $\beta = 0$. Specifically, when parameters are set to $\alpha = 0.2, \beta = 0.3$ or $\alpha = 0.3, \beta = 0.2$, the motion control of the presented scheme (7.4) is more effective than other weighting coefficient values. In conclusion, Table 7.1 shows that the presented multi-criteria control scheme (7.4) can accurately and effectively solve the redundancy resolution of the manipulator, regardless of the combination of the three weight coefficients, further demonstrating the flexibility of scheme (7.4).

7.4.3 Comparisons

To further demonstrate the superiority of the presented scheme (7.4) based on DNN model (7.8), simulations are conducted to compare the presented scheme (7.4) with the repetitive motion planning (RMP) scheme presented in Xie et al. (2019). In detail, a trajectory-tracking task based on the Franka manipulator is carried out with the assistance of the presented scheme (7.4) and the RMP scheme Xie et al. (2019), respectively. In this case, the trajectory to be tracked is leaf-shaped, and the simulation results are presented in Figs. 7.5 and 7.6.

As shown in Figs. 7.5a and 7.6a, both schemes successfully accomplish the trajectory-tracking task, with the actual trajectory closely aligning with the desired trajectory, exhibiting minor position errors. It is important to note that the position errors with the presented scheme (7.4) are smaller than those with the RMP scheme Xie et al. (2019). Additionally, as shown in Fig. 7.5c and g, the manipulator driven by the presented scheme (7.4) operates at lower kinetic energy compared to the

Table 7.1 Performance evaluation under various coefficients of the presented multi-criteria control scheme (7.4)

$(\alpha,\ \beta)$	(0, 0)	(0, 0.4)	(0, 0.6)	(0.1, 0)	(0.1, 0.1)	(0.2, 0.2)	(0.2, 0.3)
MAX $\lvert \boldsymbol{e}_\mathrm{p} \rvert (10^{-4})$	0.711	0.657	0.792	0.641	0.795	0.641	0.432
MAX $\lVert \dot{\boldsymbol{\theta}} \rVert_\infty$	0.146	0.145	0.145	0.146	0.146	0.145	0.145
MAX $\lvert \theta(T) - \theta(0) \rvert$	4.258×10^{-8}	0.010	0.021	1.956×10^{-7}	0.002	0.005	0.009
MAX $\dot{\boldsymbol{\theta}}^\mathrm{T} P \dot{\boldsymbol{\theta}}$	0.036	0.036	0.036	0.036	0.036	0.036	0.036
$(\alpha,\ \beta)$	(0.4, 0.5)	(0.5, 0.4)	(0.5, 0.5)	(0.6, 0)	(0.3, 0.2)	(0.3, 0.4)	(0.4, 0.2)
MAX $\lvert \boldsymbol{e}_\mathrm{p} \rvert (10^{-4})$	0.663	0.560	0.626	0.617	0.489	0.559	0.690
MAX $\lVert \dot{\boldsymbol{\theta}} \rVert_\infty$	0.130	0.130	0.130	0.146	0.146	0.144	0.145
MAX $\lvert \theta(T) - \theta(0) \rvert$	0.028	0.038	0.008	0.174×10^{-7}	0.006	0.019	0.007
MAX $\dot{\boldsymbol{\theta}}^\mathrm{T} P \dot{\boldsymbol{\theta}}$	0.036	0.036	0.036	0.038	0.039	0.040	0.036
$(\alpha,\ \beta)$	(0.6, 0.3)	(0.7, 0.1)	(0.7, 0.3)	(0.8, 0)	(0.8, 0.1)	(0.9, 0)	
MAX $\lvert \boldsymbol{e}_\mathrm{p} \rvert (10^{-4})$	0.674	0.811	0.772	0.614	0.530	0.596	
MAX $\lVert \dot{\boldsymbol{\theta}} \rVert_\infty$	0.134	0.145	0.131	0.146	0.143	0.145	
MAX $\lvert \theta(T) - \theta(0) \rvert$	0.042	0.007	0.004	0.159×10^{-6}	0.014	0.175×10^{-5}	
MAX $\dot{\boldsymbol{\theta}}^\mathrm{T} P \dot{\boldsymbol{\theta}}$	0.038	0.036	0.039	0.036	0.036	0.036	

manipulator driven by the RMP scheme Xie et al. (2019). Regarding the orientation errors shown in Fig. 7.6d and 7.5d, the end-effector of the manipulator driven by the presented scheme (7.4) remains vertical downward during trajectory tracking with tiny orientation errors, while the one driven by the RMP scheme Xie et al. (2019) generates significant orientation errors, which could potentially lead to failure in the tracking task. Therefore, the presented scheme (7.4) better meets the requirements when the task involves orientation considerations.

In summary, the manipulator driven by the presented scheme (7.4), based on DNN model (7.8), demonstrates excellent accuracy in both position and orientation tracking at low kinetic energy. This gives it a wider range of potential applications compared to pure position-control schemes. Considering the manipulator's characteristics in simulations, the end-effector is treated as a fixed point, and its three-dimensional deviations in orientation during task execution cannot be visually observed in the actual trajectories. However, in practical scenarios, the end-effector does possess directionality, making the inclusion of orientation-tracking essential. This fact is demonstrated and confirmed through physical experiments.

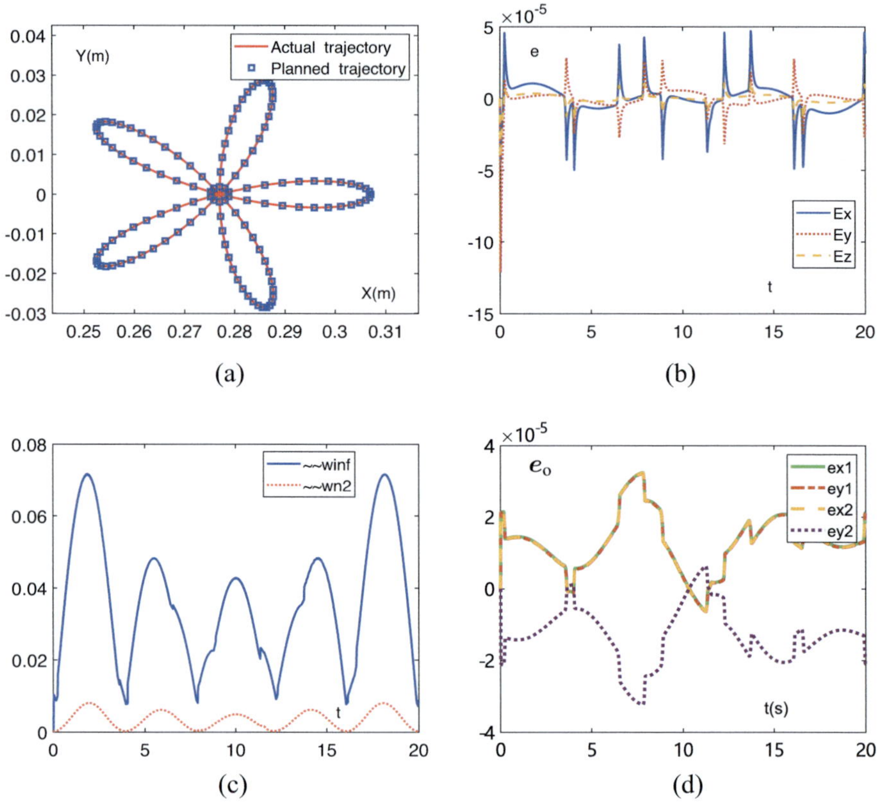

Fig. 7.5 Simulation results for the Franka manipulator tracking the leaf-shaped trajectory using the presented scheme (7.4). **a** Actual trajectory and desired trajectory. **b** Position error. **c** $\|\dot{\theta}\|_\infty$ and $\dot{\theta}^{\mathrm{T}} P \dot{\theta}$. **d** Orientation error

For further comparisons, Table 7.2 presents a comparison of different optimization schemes for solving robotic redundancy in the execution of their specific tasks. It is particularly noteworthy that the presented scheme (7.4), which incorporates three distinct performance indexes, enhances its flexibility in the optimal control of the redundant manipulator. This scheme is capable of ensuring joint constraint avoidance and achieving global convergence of tracking error in the manipulator operation optimization. Meanwhile, the presented scheme (7.4) considers the end-effector orientation-tracking index for the first time.

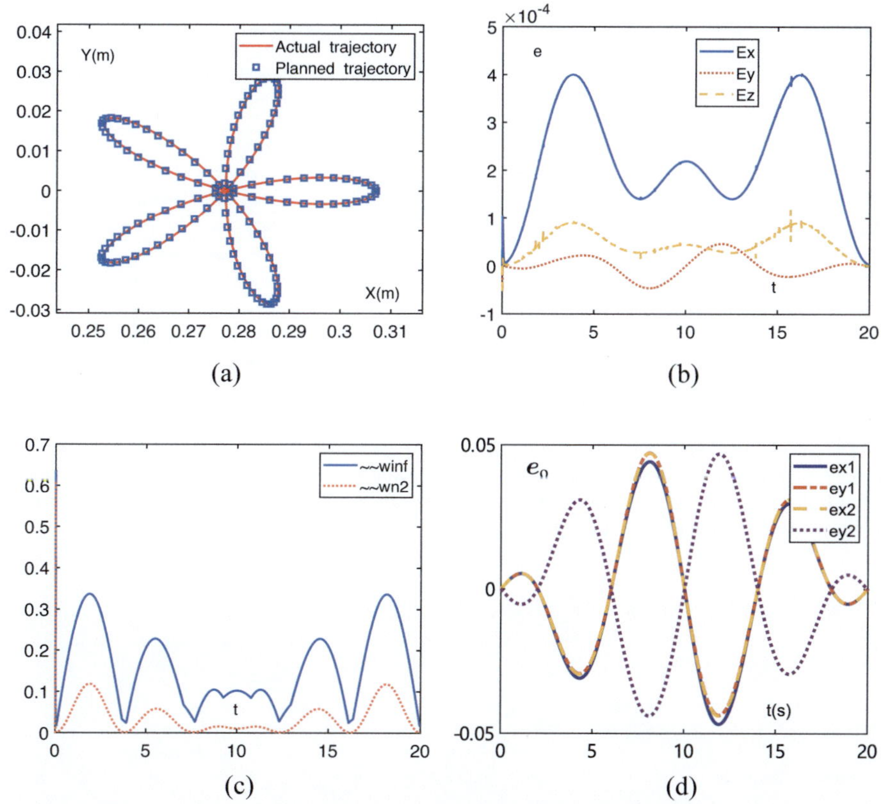

Fig. 7.6 Simulation results for the Franka manipulator tracking the leaf-shaped trajectory using the RMP scheme Xie et al. (2019). **a** Actual trajectory and desired trajectory. **b** Position error. **c** $\|\dot{\boldsymbol{\theta}}\|_\infty$ and $\dot{\boldsymbol{\theta}}^{\mathrm{T}} P\dot{\boldsymbol{\theta}}$. **d** Orientation error

7.4.4 Physical Experiments

To validate the operability and superiority of the presented scheme (7.4) integrated with DNN model (7.8), physical experiments are conducted on the experimental platform of the 7-DOF Franka manipulator, as shown in Fig. 7.7, which is equipped with the robotic operating system. It is important to highlight that the tracking trajectory of a rose curve, depicted in Figs. 7.8 and 7.9a, is executed perfectly, with the tracking errors shown in Fig. 7.9b remaining on the order of 10^{-3} m. Furthermore, the final states of the redundant manipulator align with the initial states, as recorded on the right side of Fig. 7.8. As mentioned in Remark 7.1, generally, a manipulator with its DOF number $k \leq 6$ in a three-dimensional workspace is considered non-redundant, as the orientation of the end-effector must be taken into account. Therefore, solutions that do not include end-effector orientation tracking perform adequately for robots with $k \leq 6$. However, when these existing schemes are applied to the 7-DOF Franka

Table 7.2 Comparisons of various schemes for the control of redundant manipulators

	Joint constraint avoidance	Minimum infinite velocity norm	Multiple criterion	CMC	Theoretical tracking error♯	Scheme level	Orientation tracking
Scheme in (7.4)	√	√	√	√	Zero	Velocity	√
Scheme in Jin and Zhang (2015)	√	×	×	√	None-Zero	Acceleration	×
Scheme in Li (2020)	√	×	×	√	None-Zero	Velocity	×
Scheme in Xia et al. (2005)	√	×	×	×	None-Zero	Acceleration	×
Scheme in Li et al. (2017)	√	×	×	×	None-Zero	Velocity	×
Scheme in Bertino et al. (2022)	√	×	×	√	None-Zero	Acceleration	×
Scheme in Ngo and Liu (2023)	√	×	×	×	None-Zero	Velocity	×
Scheme in Liu et al. (2023)	√	√	√	√	None-Zero	Velocity	×
Scheme in Li et al. (2019)	√	×	×	×	None-Zero	Velocity	×
Scheme in Zhang et al. (2019)	×	×	×	√	None-Zero	Velocity	×
Scheme in Tan et al. (2024)	√	×	×	×	Zero	Angle	×

♯ "Theoretical tracking error" denotes the theoretical error in both the position and orientation of a manipulator's end-effector. It should be noted that for a redundant robot with $k > 6$, a control scheme that neglects orientation tracking will produce an orientation error, and thus its "Theoretical tracking error" is considered as none-zero

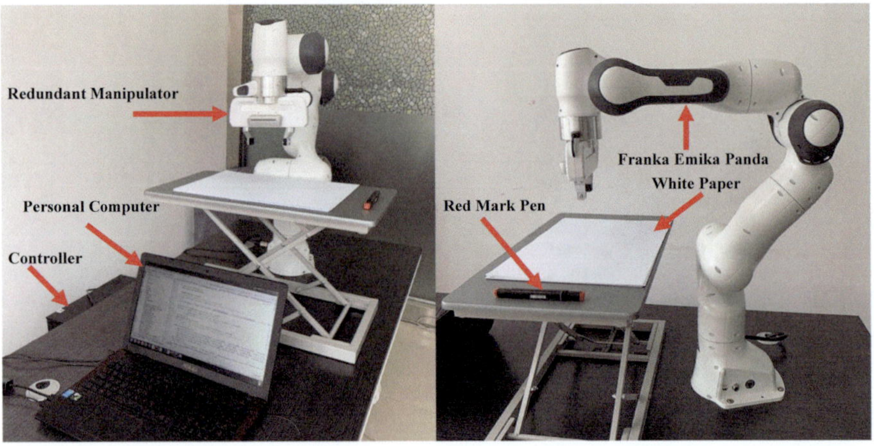

Fig. 7.7 The experimental platform of the robot

manipulator, they fail to produce satisfactory results. As shown in Fig. 7.11, aided with the scheme in Li (2020), the manipulator is unable to track the desired trajectory accurately. The trajectory depicted in Fig. 7.10 is nonclosed, and significant tracking errors are observed in Fig. 7.11a and b. In contrast, by employing the presented multi-criteria control scheme (7.4) along with DNN model (7.8), the manipulator can reliably and accurately perform the trajectory-tracking task, achieving the desired outcome.

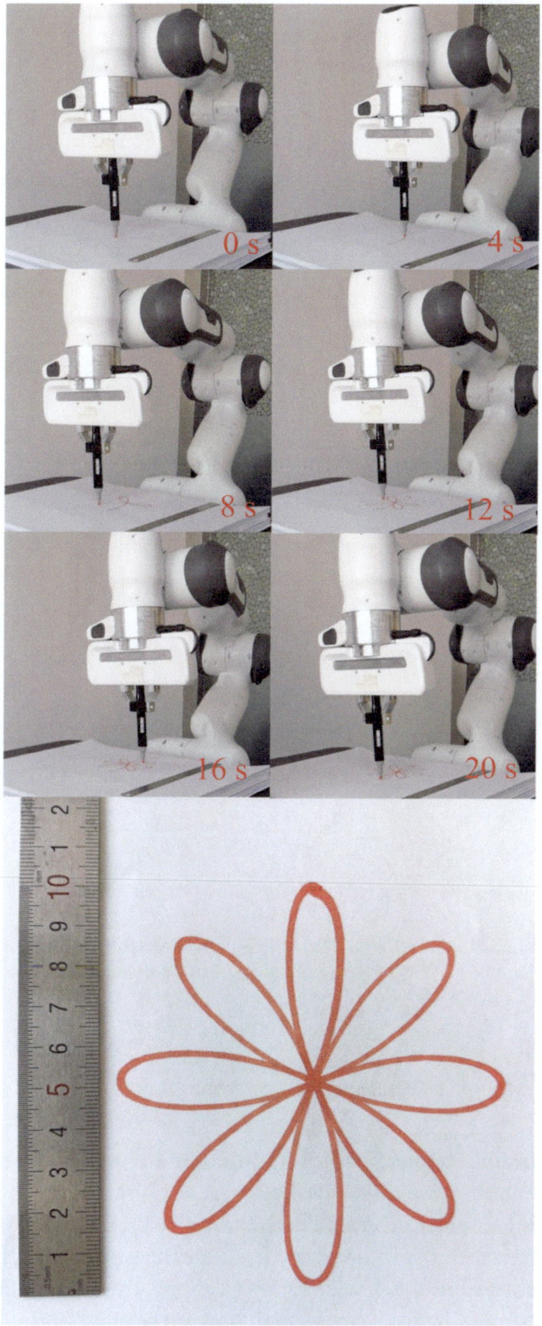

Fig. 7.8 Snapshots of experiment results for the Franka manipulator tracking a rose trajectory using the presented scheme (7.4) with the orientation tracking considered

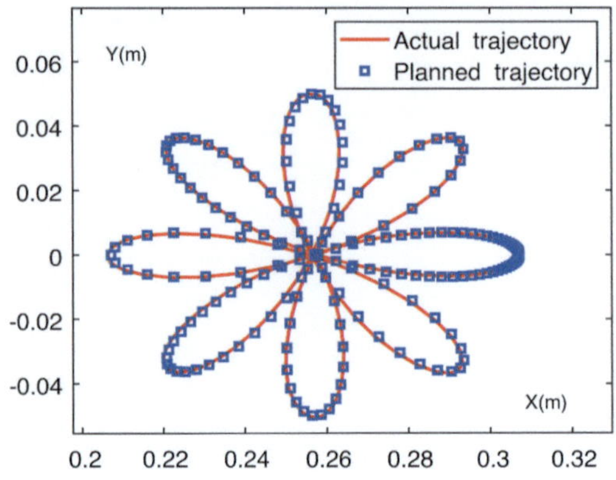

(a)

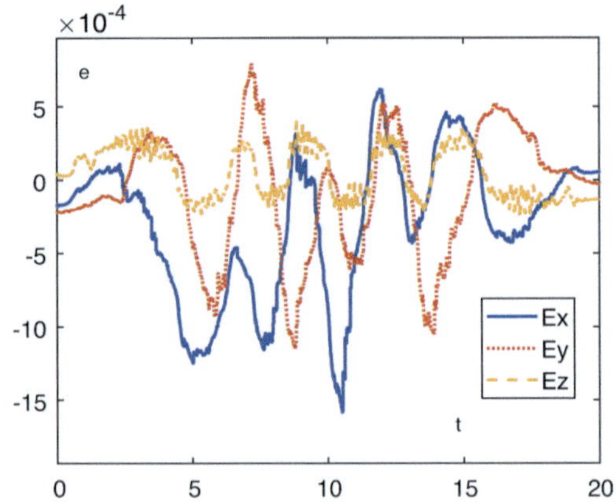

(b)

Fig. 7.9 Experiment results for the Franka manipulator tracking a rose trajectory using the presented scheme (7.4) with the orientation tracking considered. **a** Actual trajectory and desired trajectory. **b** Position error

7.5 Conclusion

In this chapter, a novel multi-criteria control scheme that integrates end-effector orientation tracking and physical constraints has been presented for the kinematic tracking control of redundant manipulators. Specifically, multiple performance indexes have been developed to formulate the presented scheme, imbuing it with remarkable adaptability to a wide range of application scenarios. Additionally, a DNN model has been utilized to investigate the regenerated optimization problem, with the theoretical analysis provided to validate the convergence of the new multi-criteria control scheme. In conclusion, illustrative simulations, performance comparisons, and physical experiments have systematically confirmed the effectiveness, characteristics, and

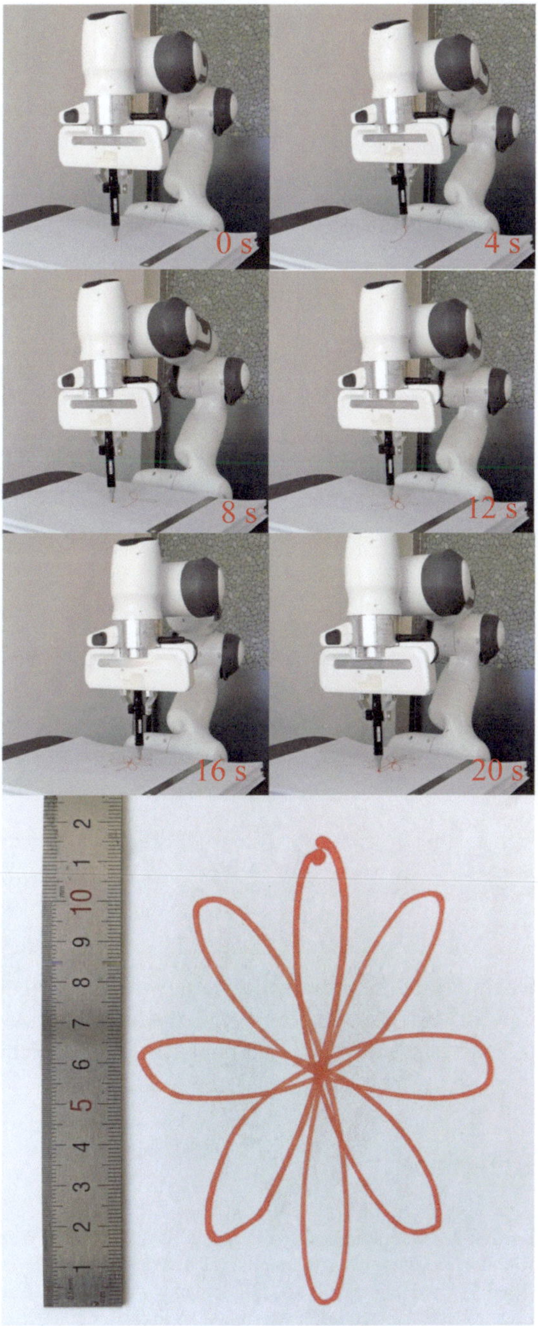

Fig. 7.10 Snapshots of experiment results on the Franka manipulator tracking a rose trajectory using the scheme in Li (2020) without the orientation tracking considered

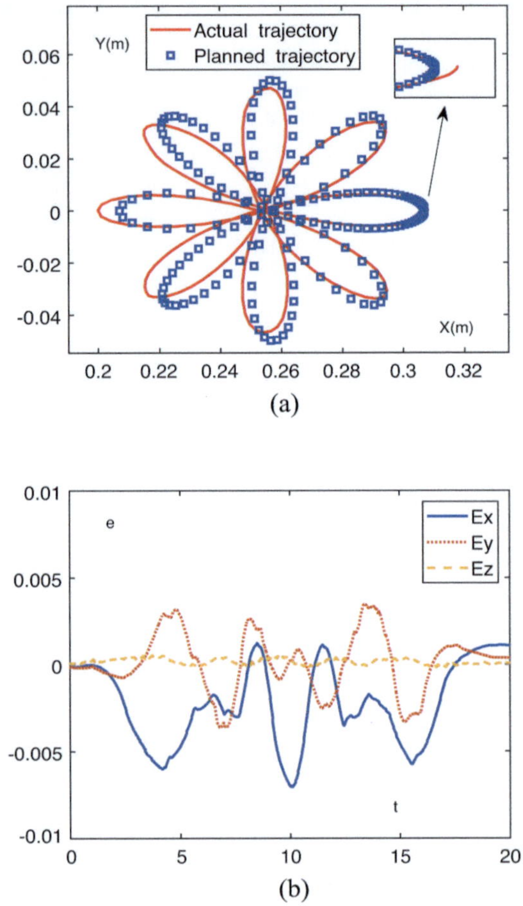

Fig. 7.11 Experiment results on the Franka manipulator tracking a rose trajectory using the scheme in Li (2020) without the orientation tracking considered. **a** Actual trajectory and desired trajectory. **b** Position error

feasibility of the presented scheme with the DNN model. Future work will incorporate data-driven technologies into the orientation-aware multi-criteria control scheme to address systems with unknown structural parameters.

References

Bertino A, Naseradinmousavi P, Krstić M (2022) Delay-adaptive control of a 7-DOF robot manipulator: design and experiments. IEEE Trans Control Syst Technol 30(6):2506–2521

Boyd S, Vandenberghe L (2004) Convex Optimization. Cambridge University Press

Caccavale F, Chiacchio P, Marino A, Villani L (2008) Six-DOF impedance control of dual-arm cooperative manipulators. IEEE/ASME Trans Mechatron 13(5):576–586

Chen D, Shao Y, Chen Z, Li S (2025) Robust neural dynamics for depth maintenance tracking control of robot manipulators with uncertainty and perturbation. IEEE Trans Autom Sci Eng. 22:7052–7063

Czimmermann T, Chiurazzi M, Milazzo M, Roccella S, Barbieri M, Dario P, Ciuti G (2022) An autonomous robotic platform for manipulation and inspection of metallic surfaces in industry 4.0. IEEE Trans Autom Sci Eng 19(3):1691–1706

Gao XB (2003) Exponential stability of globally projected dynamic systems. IEEE Trans Neural Netw 14(2):426–431

Godoy RV, Dwivedi A, Liarokapis M (2022) Electromyography based decoding of dexterous, in-hand manipulation motions with temporal multichannel vision transformers. IEEE Trans Neural Syst Rehabil Eng 30:2207–2216

Gravagne I, Walker I (2000) On the structure of minimum effort solutions with application to kinematic redundancy resolution. IEEE Trans Robot Autom 16(6):855–863

Hamad M, Kurdas A, Mansfeld N, Abdolshah S, Haddadin S (2023) Modularize-and-conquer: a generalized impact dynamics and safe precollision control framework for floating-base tree-like robots. IEEE Trans Robot 39(4):3200–3221

Hao J, Duan J, Wang K, Hu C, Shi C (2023) Inverse kinematic modeling of the tendon-actuated medical continuum manipulator based on a lightweight timing input neural network. IEEE Trans Med Robot Bionics 5(4):916–928

Jin L, Zhang Y (2015) G2-type SRMPC scheme for synchronous manipulation of two redundant robot manipulators. IEEE Trans Cybern 45(2):153–164

Jin L, Wei L, Li S (2023) Gradient-based differential neural-solution to time-dependent nonlinear optimization. IEEE Trans Autom Control 68(1):62–627

Jin L, Su Z, Fu D, Xiao X (2024) Coevolutionary neural solution for nonconvex optimization with noise tolerance. IEEE Trans Neural Netw Learn Syst 35(12):17571–17581

Katsikis VN, Stanimirovic PS, Mourtas SD, Xiao L, Karabasevic D (2022) Zeroing neural network with fuzzy parameter for computing pseudoinverse of arbitrary matrix. IEEE Trans Fuzzy Syst 30(9):3426–3435

Kim J, Jie W, Kim H, Lee MC (2021) Modified configuration control with potential field for inverse kinematic solution of redundant manipulator. IEEE/ASME Trans Mechatron 26(4):1782–1790

Li W (2020) Predefined-time convergent neural solution to cyclical motion planning of redundant robots under physical constraints. IEEE Trans Ind Electron 67(12):10732–10743

Li Z, Cui J (2024) Fault-tolerant kinematic control of redundant manipulators with joint failures under joint velocity disturbances. IEEE Trans Circuits Syst II Express Briefs 71(4):2104–2108

Li Z, Li S (2021) Neural network model-based control for manipulator: an autoencoder perspective. IEEE Trans Neural Netw Learn Syst 34(6):2854–2868

Li S, Zhang Y, Jin L (2017) Kinematic control of redundant manipulators using neural networks. IEEE Trans Neural Netw Learn Syst 28(10):2243–2254

Li Z, Li C, Li S, Cao X (2019) A fault-tolerant method for motion planning of industrial redundant manipulator. IEEE Trans Ind Inform 16(12):7469–7478

Li G, Yu J, Chen X (2021) Adaptive fuzzy neural network command filtered impedance control of constrained robotic manipulators with disturbance observer. IEEE Trans Neural Netw Learn Syst 34(8):5171–5180

Li J, Ganguly A, Figueredo LFC, Haddadin S (2024) The fingertip manipulability assessment of tendon-driven multi-fingered hands. IEEE Robot Autom Lett 9(3):2726–2733

Li G, Stalin T, y Alvarado PV, (2022) DNN-based predictive model for a batoid-inspired soft robot. IEEE Robot Autom Lett 7(2):1024–1031

Liu M, Shang M (2024) Orientation tracking incorporated multicriteria control for redundant manipulators with dynamic neural network. IEEE Trans Ind Electron 71(4):3801–3810

Liu S, Sun D (2013) Minimizing energy consumption of wheeled mobile robots via optimal motion planning. IEEE/ASME Trans Mechatron 19(2):401–411

Liu M, He L, Shang M (2023) Dynamic neural network for bicriteria weighted control of robot manipulators. IEEE Trans Neural Netw Learn Syst 34(8):4570–4583

Lu H, Jin L, Zhang J, Sun Z, Li S, Zhang Z (2021) New joint-drift-free scheme aided with projected ZNN for motion generation of redundant robot manipulators perturbed by disturbances. IEEE Trans Syst Man Cybern Syst 51(9):5639–5651

Lyu E, Liu T, Wang J, Song S, Meng MQH (2023) Motion planning of manipulator by points-guided sampling network. IEEE Trans Autom Sci Eng 20(2):821–831

Morita T, Armleder S, Zhu Y, Iino H, Aoyama T, Cheng G, Hasegawa Y (2024) RNN-based visual guidance for enhanced sense of agency in teleoperation with time-varying delays. IEEE Robot Autom Lett 9(12):11537–11544

Ngo V-T, Liu Y-C (2023) Human-robot coordination control for heterogeneous Euler-Lagrange systems under communication delays and relative position. IEEE Trans Ind Electron 70(2):1761–1771

Nurbayeva A, Shintemirov A, Rubagotti M (2023) Deep imitation learning of nonlinear model predictive control laws for a safe physical human-robot interaction. IEEE Trans Ind Inform 19(7):8384–8395

Qi Y, Jin L, Wang Y, Xiao L, Zhang J (2020) Complex-valued discrete-time neural dynamics for perturbed time-dependent complex quadratic programming with applications. IEEE Trans Neural Netw Learn Syst 31(9):3555–3569

Qiu J, Li W, Zhang Y, Fu S, Tan H (2014) Two numerical algorithms and numerical experiments for efficiently solving inequality-and-bound constrained QP. In: 2014 2nd international conference on systems and informatics, pp 39–44

Rubagotti M, Tasbolat T, Bukeikhan O, Almas S (2019) Semi-autonomous robot teleoperation with obstacle avoidance via model predictive control. IEEE Robot Autom Lett 4(3):2746–2753

Safeea M, Bearee R, Neto P (2021) A modified DLS scheme with controlled cyclic solution for inverse kinematics in redundant robots. IEEE Trans Ind Inform 17(12):8014–8023

Shi Y, Ding C, Li S, Li B, Sun X (2024) New RNN algorithms for different time-variant matrix inequalities solving under discrete-time framework. IEEE Trans Neural Netw Learn Syst

Stewart J (2007) Calculus. Cengage Learning

Tan J, Shang M, Jin L (2024) Metaheuristic-based RNN for manipulability optimization of redundant manipulators. IEEE Trans Ind Inform 20(4):6489–6498

Van Steen J, Van Der Brandt G, Van Der Wouw N, Kober J, Saccon A (2024) Quadratic programming-based reference spreading control for dual-arm robotic manipulation with planned simultaneous impacts. IEEE Trans Robot 40:3341–3355

Xia Y, Feng G, Wang J (2005) A primal-dual neural network for online resolving constrained kinematic redundancy in robot motion control. IEEE Trans Syst Man Cybern Part B Cybern 35(1):54–64

Xie Z, Jin L (2024) A fuzzy neural controller for model-free control of redundant manipulators with unknown kinematic parameters. IEEE Trans Fuzzy Syst 32(3):1589–1601

Xie Z, Du X, Xiao X, Li H (2019) On generalized RMP scheme for redundant robot manipulators aided with dynamic neural networks and nonconvex bound constraints. IEEE Trans Ind Inform 15(9):5172–5181

Xu Z, Zhou X, Wu H, Li X, Li S (2021) Motion planning of manipulators for simultaneous obstacle avoidance and target tracking: an RNN approach with guaranteed performance. IEEE Trans Ind Electron 69(4):3887–3897

Yan J, Jin L, Yuan Z, Liu Z (2021) RNN for receding horizon control of redundant robot manipulators. IEEE Trans Ind Electron 69(2):1608–1619

Yan J, Su Z, Ma X, Jin L (2025) An obstacle avoidance scheme for manipulators aided by noise-tolerant neural dynamics. IEEE Trans Ind Inform 21(1):703–712

Yang C, Jiang Y, He W, Na J, Li Z, Xu B (2018) Adaptive parameter estimation and control design for robot manipulators with finite-time convergence. IEEE Trans Ind Electron 65(10):8112–8123

Yang M, Zhang Y, Zhang Z, Hu H (2020) Adaptive discrete ZND models for tracking control of redundant manipulator. IEEE Trans Ind Inform 16(12):7360–7368

Yang W, Li S, Li Z, Luo X (2023) Highly accurate manipulator calibration via extended Kalman filter-incorporated residual neural network. IEEE Trans Ind Inform 19(11):10831–10841

Yang X, Ma B, Zhao Z, Zhao J, Liu H (2024) A kinematics control scheme of redundant manipulators under unknown loads or external forces. IEEE Trans Syst Man Cybern Syst 54(8):5048–5060
Zhang Y, Jin L (2017) Robot manipulator redundancy resolution. Wiley
Zhang Y, Li S, Zou J, Khan AH (2019) A passivity-based approach for kinematic control of manipulators with constraints. IEEE Trans Ind Inform 16(5):3029–3038
Zhang Y, Li S, Kardry S, Liao B (2019) Recurrent neural network for kinematic control of redundant manipulators with periodic input disturbance and physical constraints. IEEE Trans Cybern 49(12):4194–4205
Zhang F, Zhang J, Cheng M, Ding R, Xu B (2024) Flow manipulability polytope-a novel velocity performance evaluation tool for hydraulic robotic manipulators. IEEE Trans Ind Inform 20(5):7698–7707
Zhang Z, He H, Deng X (2024) An FPGA-implemented antinoise fuzzy recurrent neural network for motion planning of redundant robot manipulators. IEEE Trans Neural Netw Learn Syst 35(9):12263–12275
Zhu C, Jiang Y, Yang C (2022) Fixed-time neural control of robot manipulator with global stability and guaranteed transient performance. IEEE Trans Ind Electron 70(1):803–812

Chapter 8
Position and Orientation-Tracking MPC with Finite-Time DNN

8.1 Introduction

Over recent decades, robotic manipulators have been widely deployed across numerous industrial and scientific domains Zhou et al. (2020), Jin et al. (2023), Du et al. (2018). Considering economic efficiency, three-, four-, and six-DOF manipulators dominate industrial applications, particularly 6-DOF collaborative robots (e.g., universal robots series) that are essential for automotive assembly, packaging systems, and supply chain operations Maric et al. (2020), Consolini et al. (2024), Khan et al. (2020). As industrial tasks grow increasingly sophisticated, demand has risen for manipulators with the enhanced flexibility and intelligence, driving the adoption of higher-DOF configurations Xu et al. (2016), Kim et al. (2022). While 6-DOF manipulators achieve complete workspace pose accessibility, the inverse kinematics yields singular solutions for specified position-orientation pairs. Consequently, 6-DOF manipulators lack optimization capacity for joint limit avoidance or kinematic/dynamic enhancement, limiting their applicability in complex and high-flexibility scenarios Zhang et al. (2023), Huang et al. (2020), Zhang et al. (2022). Redundant manipulators possess additional degrees of freedom beyond task requirements, enabling concurrent execution of secondary objectives alongside primary tasks Yang et al. (2018), Jin et al. (2023), Peng et al. (2021).

The operational flexibility requirements of redundant manipulators necessitate precise concurrent end-effector position and orientation tracking. Numerous control approaches are developed for redundant manipulator trajectory tracking Yang et al. (2018), Dietrich and Ott (2020), Liu et al. (2024). In Jin and Zhang (2015), an acceleration-level cyclic motion control framework is presented to integrate trajectory planning and execution, which effectively compensates joint-angle drift during coordinated tracking by dual redundant manipulators through position-velocity error feedback mechanisms. In Faulwasser et al. (2017), a sampled-data nonlinear model predictive control (MPC) framework is introduced for constrained path-following tasks, demonstrating the superior control performance under both input and state

M. Liu et al., *Dynamic Neural Networks for Motion Control of Redundant Manipulators*, Intelligent Control and Learning Systems 21,
https://doi.org/10.1007/978-981-96-9144-9_8

limitations. A velocity-compensated cyclic motion planning framework with lagging error suppression is developed for redundant manipulator trajectory tracking Xie et al. (2019). These approaches focus exclusively on positional tracking, significantly constraining the practical deployment of redundant manipulators. Limited control approaches incorporate end-effector orientation considerations for redundant manipulators Zhang et al. (2020), Li et al. (2022). In Zhang et al. (2020), two hybrid end-effector posture-maintaining and obstacle-limits avoidance schemes are presented for simultaneous path tracking and end-effector posture regulation in cluttered environments, integrating obstacle-limit avoidance capabilities. In Li et al. (2022), a data-driven motion-force control scheme with end-effector orientation maintenance is developed to address the motion-force control problem under unknown manipulator dynamics. However, these approaches are restricted to maintaining fixed end-effector orientations during operation, consequently constraining the application scope of redundant manipulators.

Beyond concurrent position-orientation tracking, redundant manipulators must maintain operational safety during flexible task execution. Consequently, the control framework must incorporate constraints on joint angles, velocities, and accelerations. Most existing trajectory-tracking approaches for redundant manipulators, including widely adopted minimum-velocity-norm (MVN) Xie et al. (2019), Zhang et al. (2020), Li et al. (2022), and minimum-acceleration-norm (MAN) methods Jin and Zhang (2015), exhibit limitations in managing multi-level joint constraints. Through conversion methodologies Li et al. (2022), Jin and Zhang (2015), MVN and MAN approaches demonstrate distinct joint limit handling capacities: velocity-level implementations accommodate only angle and velocity constraints, whereas acceleration-level formulations simultaneously enforce angle, velocity, and acceleration limits. Nonetheless, the feasible region of decision variables is reduced by these conversion techniques Zhang et al. (2021), Carron et al. (2019), which may prevent redundant manipulators from attaining the optimal solution and thus diminish the flexibility of redundant manipulators.

To address issues related to trajectory tracking and multiple constraints, MPC serves as a powerful approach. Derived from industrial experience, MPC is an advanced control technique that excels in managing multiple-input and multiple-output systems. It involves three primary steps: (1) forecasting the future state of the system using the model; (2) resolving the optimization problem; (3) implementing the initial element of the optimal solution to the system. The majority of trajectory-tracking schemes based on MPC utilize dynamic models Faulwasser et al. (2017), Yan et al. (2024), Cisneros et al. (2018), Kleff et al. (2021). In Yan et al. (2024), a control system for unmanned satellite-manipulator systems is introduced, comprising a trajectory planning module and a model predictive controller. This system demonstrates benefits such as high accuracy in end-effector trajectory tracking and robustness against disturbances. A practical framework for nonlinear MPC of a manipulator with nonlinear state constraints, as presented in Cisneros et al. (2018), demonstrates efficiency and robustness to unmodelled dynamics. A hardware realization of closed-loop nonlinear MPC for 7-DOF torque-driven robotic trajectory tracking is demonstrated in Kleff et al. (2021), achieving the high-performance dynamic cyclic task execution. These model-based approaches face challenges in enforcing joint constraints, including angular displacement boundaries, rotational speed thresholds, and acceleration ceilings. Moreover, the orientation control of the manipulator's end-effector is not considered in these cases. Therefore, this chapter introduces a model predictive tracking control of position and orientation (MPTCPO) scheme built upon MPC, aiming to compensate for the limitations of most kinematics-based trajectory-tracking approaches for redundant manipulators. Specifically, the MPTCPO scheme can simultaneously track the end-effector's position and orientation for redundant manipulators, minimize tracking error along with the norms of joint velocity and acceleration, and directly manage three tiers of joint constraints without shrinking the feasible region, as no transformation is necessary. In addition, the MPTCPO scheme employs the rotation vector to represent orientation with only three parameters, which imposes a lower computational load compared to the commonly used rotation matrix that involves nine parameters.

The receding horizon optimization inherent to MPC necessitates the efficient solution of robotic optimization problems Shi et al. (2022), Xiao and Zhang (2016), Shi et al. (2021). Neural networks have been extensively investigated as a solution to these challenges Liufu et al. (2024), Saito et al. (2022), Tan et al. (2022), with both theoretical insights and practical implementations being thoroughly explored Cao et al. (2022), Katsikis et al. (2021), Jeon et al. (2024). Prior research shows that neural networks are capable of efficiently addressing optimization problems Liu et al. (2021), Schnelle and Eberhard (2017), Jin et al. (2024), providing significant benefits in terms of computational efficiency and flexibility when compared to conventional approaches Incremona et al. (2017), Yan et al. (2024), Peng et al. (2022), Ke et al. (2018). Liao et al. (2016) develop a Taylor-series-based discretization framework for continuous neural networks to address robotic optimization challenges. Jin et al. (2021) introduce a saturation-enabled neural network capable of solving optimization problems for manipulators in the presence of various additive noises. Furthermore, Xu et al. (2021) present a motion-force control strategy for redundant manipulators within the projection recurrent neural network (RNN) framework, extending a projection RNN originally designed for position control to one that incorporates position-force control. Moreover, the selection of activation functions significantly impacts the performance of neural network models Godoy et al. (2022). Building upon the successful work in Xiao et al. (2021), a special activation function is incorporated into the neural network model for the quadratic programming (QP) problem derived from the MPTCPO scheme to enable finite-time control.

The rest of this chapter is arranged as four sections as follows. Section 8.2 develops an MPTCPO scheme and simplifies it into a convex QP problem. Section 8.3 introduces a finite-time dynamic neural network (FTDNN) model, along with the corresponding convergence analysis. Section 8.4 presents computer simulations and physical experiments using a 7-DOF redundant manipulator. Section 8.5 concludes the chapter. The contributions of this chapter are outlined as follows.

1. In addition to controlling the position of the end-effector of redundant manipulators, the presented MPTCPO scheme manages the orientation, addressing the lim-

itation of most trajectory-tracking schemes that focus solely on the end-effector's position. Furthermore, the orientation is characterized using a three-parameter rotation vector, which reduces computational complexity compared to the commonly used nine-parameter rotation matrix.
2. The presented MPTCPO scheme simultaneously minimizes tracking error, joint-velocity norm, and joint-acceleration norm, while directly addressing three levels of joint limits without reducing the feasible region, as no conversion is needed. This overcomes the limitations of most trajectory-tracking schemes for redundant manipulators.
3. An FTDNN model is developed for the presented MPTCPO scheme, and it is theoretically demonstrated to ensure the finite-time convergence.

8.2 Scheme Design

This section presents an MPTCPO scheme for the trajectory tracking of redundant manipulators, capable of tracking both the position and orientation of the end-effector. Additionally, definitions of variables and operators are provided in Table 8.1.

To begin with, the orientation of the end-effector is represented as follows:

$$\boldsymbol{\varphi}(t) = \theta(t)\boldsymbol{e}(t) = (\varphi_{\mathrm{x}}, \varphi_{\mathrm{y}}, \varphi_{\mathrm{z}})^{\mathrm{T}}, \tag{8.1}$$

where $\boldsymbol{\varphi}(t) \in \mathbb{R}^3$ denotes the rotation vector, $\boldsymbol{e}(t) \in \mathbb{R}^3$ represents the rotation axis, with its direction determined by the right-hand rule, $\theta(t) \in \mathbb{R}$ denotes the counterclockwise rotation angle around the rotation axis, and the superscript ^{T} denotes the transposition operation. Additionally, $\theta(t) = \sqrt{\|\boldsymbol{\varphi}(t)\|^2}$ and $\boldsymbol{e}(t) = \boldsymbol{\varphi}(t)/\theta(t)$. Furthermore, the relationship between the end-effector and the joints of redundant manipulators is analogous to (7.1) and expressed as

Table 8.1 Definitions of variables and operators

Symbol	Definition
m	Number of joints in the manipulator
$\otimes$	Kronecker product
Γ	Predictive horizon
Γ_u	Control horizon ($0 < \Gamma_u \leq \Gamma$)
$\mathrm{vec}(\cdot)$	Reshaping operation into a vector
δ	Sampling interval
$\xi \in \mathbb{R}^+$	Index related to convergence rate
$\mathscr{R} \in \mathbb{R}^{6\times 6}$	Weight matrix of the tracking error
$\mathscr{Q}_1 \in \mathbb{R}^{m\times m}$	Weight matrix of $\dot{\boldsymbol{\theta}}(t)$
$\mathscr{Q}_2 \in \mathbb{R}^{m\times m}$	Weight matrix of $\Delta\dot{\boldsymbol{\theta}}(t)$

$$\boldsymbol{r}(t) = f_p(\boldsymbol{\theta}(t)), \tag{8.2}$$

$$\boldsymbol{\varphi}(t) = f_\varphi(\boldsymbol{\theta}(t)), \tag{8.3}$$

$$\dot{\boldsymbol{r}}(t) = J_p(\boldsymbol{\theta}(t))\dot{\boldsymbol{\theta}}(t), \tag{8.4}$$

$$\dot{\boldsymbol{\varphi}}(t) = J_\varphi(\boldsymbol{\theta}(t))\dot{\boldsymbol{\theta}}(t). \tag{8.5}$$

For convenience, let

$$\boldsymbol{\omega}(t) = \begin{bmatrix} \boldsymbol{r}(t) \\ \boldsymbol{\varphi}(t) \end{bmatrix}, f(\cdot) = \begin{bmatrix} f_p(\cdot) \\ f_\varphi(\cdot) \end{bmatrix}, J(\cdot) = \begin{bmatrix} J_p(\cdot) \\ J_\varphi(\cdot) \end{bmatrix},$$

and (8.2)–(8.5) can be rewritten as

$$\boldsymbol{\omega}(t) = f(\boldsymbol{\theta}(t)), \tag{8.6}$$

$$\dot{\boldsymbol{\omega}}(t) = \dot{f}(\boldsymbol{\theta}(t)) = J(\boldsymbol{\theta}(t))\dot{\boldsymbol{\theta}}(t). \tag{8.7}$$

The discrete form of joint velocity $\dot{\boldsymbol{\theta}}(k)$ is expressed as follows:

$$\dot{\boldsymbol{\theta}}(k) = \frac{\boldsymbol{\theta}(k+1) - \boldsymbol{\theta}(k)}{\delta}. \tag{8.8}$$

By applying the backward Euler method, the joint acceleration $\ddot{\boldsymbol{\theta}}(k)$ can be written as

$$\ddot{\boldsymbol{\theta}}(k) \approx \frac{\dot{\boldsymbol{\theta}}(k) - \dot{\boldsymbol{\theta}}(k-1)}{\delta} = \frac{\Delta\dot{\boldsymbol{\theta}}(k)}{\delta}. \tag{8.9}$$

Thus, one has

$$\begin{aligned} \dot{\boldsymbol{\theta}}(k+j) &= \Delta\dot{\boldsymbol{\theta}}(k+j) + \Delta\dot{\boldsymbol{\theta}}(k+j-1) + \cdots + \Delta\dot{\boldsymbol{\theta}}(k) + \dot{\boldsymbol{\theta}}(k-1), \\ & j = 0, 1, \ldots, \Gamma_u - 1. \end{aligned} \tag{8.10}$$

Furthermore, by relating (8.8) and (8.10), the joint angle $\boldsymbol{\theta}$ is expressed as

$$\begin{aligned} \boldsymbol{\theta}(k+j) &= \boldsymbol{\theta}(k) + j\delta\Delta\dot{\boldsymbol{\theta}}(k) + (j-1)\delta\Delta\dot{\boldsymbol{\theta}}(k+1) + \cdots + \delta\Delta\dot{\boldsymbol{\theta}}(k+j-1) \\ &\quad + j\delta\dot{\boldsymbol{\theta}}(k-1), \ j = 1, 2, \ldots, \Gamma, \end{aligned} \tag{8.11}$$

where $\Delta\dot{\boldsymbol{\theta}}(k+j) = 0$ for $j > \Gamma_u$. Next, by applying the first-order Taylor expansion to (8.7), the following expression is obtained:

$$f(\boldsymbol{\theta}(k+j)) = f(\boldsymbol{\theta}(k)) + J(\boldsymbol{\theta}(k))[\boldsymbol{\theta}(k+j) - \boldsymbol{\theta}(k)], \ j = 1, 2, \ldots, \Gamma. \tag{8.12}$$

Then, the MPTCPO scheme is constructed as

$$\min_{\boldsymbol{\theta}(k+j),\dot{\boldsymbol{\theta}}(k+j),} \sum_{j=1}^{\Gamma} \|f(\boldsymbol{\theta}(k+j)) - \boldsymbol{\omega}_{\mathrm{d}}(k+j)\|_{\mathscr{R}}^2 + \sum_{j=0}^{\Gamma_u-1} \|\dot{\boldsymbol{\theta}}(k+j)\|_{\mathscr{Q}_1}^2 + \sum_{j=0}^{\Gamma_u-1} \|\Delta\dot{\boldsymbol{\theta}}(k+j)\|_{\mathscr{Q}_2}^2,$$

$$\text{s.t. } \boldsymbol{\theta}^- \le \boldsymbol{\theta}(k+j) \le \boldsymbol{\theta}^+,\ j = 1, 2, \ldots, \Gamma, \tag{8.13a}$$

$$\dot{\boldsymbol{\theta}}^- \le \dot{\boldsymbol{\theta}}(k+j) \le \dot{\boldsymbol{\theta}}^+,\ j = 0, 1, \ldots, \Gamma_u - 1, \tag{8.13b}$$

$$\ddot{\boldsymbol{\theta}}^- \le \frac{\Delta\dot{\boldsymbol{\theta}}(k+j)}{\delta} \le \ddot{\boldsymbol{\theta}}^+,\ j = 0, 1, \ldots, \Gamma_u - 1, \tag{8.13c}$$

where $\|\cdot\|_{\mathscr{R}} = \sqrt{(\cdot)^{\mathrm{T}}\mathscr{R}(\cdot)}$ and the same definition applies to $\|\cdot\|_{\mathscr{Q}_1}$ and $\|\cdot\|_{\mathscr{Q}_2}$. For simplicity, the following matrices are defined:

$$V = \begin{bmatrix} \delta & 0 & \ldots & 0 \\ 2\delta & \delta & \ldots & 0 \\ \vdots & \vdots & \ddots & \vdots \\ \Gamma_u\delta & (\Gamma_u - 1)\delta & \ldots & \delta \\ \vdots & \vdots & \ddots & \vdots \\ \Gamma\delta & (\Gamma - 1)\delta & \ldots & (\Gamma - \Gamma_u + 1)\delta \end{bmatrix} \in \mathbb{R}^{\Gamma\times\Gamma_u},\ \widetilde{I} = \begin{bmatrix} 1 & 0 & \ldots & 0 \\ 1 & 1 & \ldots & 0 \\ \vdots & \vdots & \ddots & \vdots \\ 1 & 1 & \ldots & 1 \end{bmatrix} \in \mathbb{R}^{\Gamma_u\times\Gamma_u},$$

$$\begin{aligned}
J &= J(\boldsymbol{\theta}(k)),\ \Omega_k = [\boldsymbol{\omega}_{\mathrm{d}}(k+1), \ldots, \boldsymbol{\omega}_{\mathrm{d}}(k+\Gamma)]^{\mathrm{T}} \in \mathbb{R}^{\Gamma\times 6},\\
\Theta_k &= [\boldsymbol{\theta}(k+1), \ldots, \boldsymbol{\theta}(k+\Gamma)]^{\mathrm{T}} \in \mathbb{R}^{\Gamma\times m},\\
U_k &= [\dot{\boldsymbol{\theta}}(k), \ldots, \dot{\boldsymbol{\theta}}(k+\Gamma_u-1)]^{\mathrm{T}} \in \mathbb{R}^{\Gamma_u\times m},\\
\Delta U_k &= [\Delta\dot{\boldsymbol{\theta}}(k), \ldots, \Delta\dot{\boldsymbol{\theta}}(k+\Gamma_u-1)]^{\mathrm{T}} \in \mathbb{R}^{\Gamma_u\times m},\\
E_\theta &= [\boldsymbol{\theta}(k), \boldsymbol{\theta}(k), \ldots, \boldsymbol{\theta}(k)]^{\mathrm{T}} \in \mathbb{R}^{\Gamma\times m},\\
E_u &= [\dot{\boldsymbol{\theta}}(k-1), \dot{\boldsymbol{\theta}}(k-1), \ldots, \dot{\boldsymbol{\theta}}(k-1)]^{\mathrm{T}} \in \mathbb{R}^{\Gamma_u\times m},\\
E_f &= [f(\boldsymbol{\theta}(k)), f(\boldsymbol{\theta}(k)), \ldots, f(\boldsymbol{\theta}(k))]^{\mathrm{T}} \in \mathbb{R}^{\Gamma\times 6},\\
f_k(\boldsymbol{\theta}) &= [f(\boldsymbol{\theta}(k+1)), \ldots, f(\boldsymbol{\theta}(k+\Gamma))]^{\mathrm{T}} \in \mathbb{R}^{\Gamma\times 6},\\
A &= [\delta\dot{\boldsymbol{\theta}}(k-1), \ldots, \Gamma_u\delta\dot{\boldsymbol{\theta}}(k-1), \ldots, \Gamma\delta\dot{\boldsymbol{\theta}}(k-1)]^{\mathrm{T}} \in \mathbb{R}^{\Gamma\times m},\\
\Theta^+ &= [\boldsymbol{\theta}^+, \ldots, \boldsymbol{\theta}^+]^{\mathrm{T}} \in \mathbb{R}^{\Gamma\times m},\ \Theta^- = [\boldsymbol{\theta}^-, \ldots, \boldsymbol{\theta}^-]^{\mathrm{T}} \in \mathbb{R}^{\Gamma\times m},
\end{aligned}$$

$$\begin{aligned}
\dot{\Theta}^+ &= [\dot{\boldsymbol{\theta}}^+, \ldots, \dot{\boldsymbol{\theta}}^+]^{\mathrm{T}} \in \mathbb{R}^{\Gamma_u\times m},\ \dot{\Theta}^- = [\dot{\boldsymbol{\theta}}^-, \ldots, \dot{\boldsymbol{\theta}}^-]^{\mathrm{T}} \in \mathbb{R}^{\Gamma_u\times m},\\
\ddot{\Theta}^+ &= [\ddot{\boldsymbol{\theta}}^+, \ldots, \ddot{\boldsymbol{\theta}}^+]^{\mathrm{T}} \in \mathbb{R}^{\Gamma_u\times m},\ \ddot{\Theta}^- = [\ddot{\boldsymbol{\theta}}^-, \ldots, \ddot{\boldsymbol{\theta}}^-]^{\mathrm{T}} \in \mathbb{R}^{\Gamma_u\times m}.
\end{aligned}$$

Hence, MPTCPO scheme (8.13) is rewritten as

$$
\begin{aligned}
\min_{\Delta U_k}\quad & \|\mathrm{vec}(E_f + V\Delta U_k J^{\mathrm{T}} + AJ^{\mathrm{T}} - \Omega_k)\|_R^2 \\
& + \|\mathrm{vec}(E_u + \widetilde{I}\Delta U_k)\|_{Q_1}^2 + \|\mathrm{vec}(\Delta U_k)\|_{Q_2}^2, && \text{(8.14a)}\\
\text{s.t.}\quad & \Theta^- \le \Theta_k \le \Theta^+, && \text{(8.14b)}\\
& \dot{\Theta}^- \le U_k \le \dot{\Theta}^+, && \text{(8.14c)}\\
& \ddot{\Theta}^- \le \frac{\Delta U_k}{\delta} \le \ddot{\Theta}^+, && \text{(8.14d)}
\end{aligned}
$$

where $R = I \otimes \mathscr{R}$, $Q_1 = I \otimes \mathscr{Q}_1$, and $Q_2 = I \otimes \mathscr{Q}_2$. Let $H = 2[(J \otimes V)^{\mathrm{T}} R (J \otimes V) + (I \otimes \widetilde{I})^{\mathrm{T}} Q_1 (I \otimes \widetilde{I}) + Q_2]$ and $\boldsymbol{g} = 2(J \otimes V)^{\mathrm{T}} R[\mathrm{vec}(E_f) + \mathrm{vec}(AJ^{\mathrm{T}}) - \mathrm{vec}(\Omega_k)] + 2(I \otimes \widetilde{I})^{\mathrm{T}} Q_1 \mathrm{vec}(E_u)$. Therefore, by omitting the constant terms in (8.14a), which are independent of ΔU_k, the cost function (8.14a) can be expressed in a compact form:

$$
\min_{\Delta U_k}\quad \frac{1}{2}\mathrm{vec}(\Delta U_k)^{\mathrm{T}} H \mathrm{vec}(\Delta U_k) + \boldsymbol{g}^{\mathrm{T}} \mathrm{vec}(\Delta U_k).
$$

Besides, (8.14b)–(8.14d) can be reformulated as

$$
W \mathrm{vec}(\Delta U_k) \le \boldsymbol{h},
$$

where $W = I \otimes W' \in \mathbb{R}^{m(2\Gamma + 4\Gamma_u) \times m\Gamma_u}$ and $\boldsymbol{h} = \mathrm{vec}(h') \in \mathbb{R}^{m(2\Gamma+4\Gamma_u)}$ with

$$
W' = \begin{bmatrix} V \\ -V \\ \widetilde{I} \\ -\widetilde{I} \\ I \\ -I \end{bmatrix}, \quad h' = \begin{bmatrix} \Theta^+ - E_\theta - A \\ -\Theta^- + E_\theta + A \\ \dot{\Theta}^+ - E_u \\ -\dot{\Theta}^- + E_u \\ \delta\ddot{\Theta}^+ \\ -\delta\ddot{\Theta}^- \end{bmatrix}.
$$

Next, (8.14) is transformed into the following QP problem form:

$$
\begin{aligned}
\min_{\boldsymbol{y}(t)}\quad & \frac{1}{2}\boldsymbol{y}(t)^{\mathrm{T}} H \boldsymbol{y}(t) + \boldsymbol{g}^{\mathrm{T}} \boldsymbol{y}(t), && \text{(8.15a)}\\
\text{s.t.}\quad & W\boldsymbol{y}(t) \le \boldsymbol{h}, && \text{(8.15b)}
\end{aligned}
$$

where $\boldsymbol{y}(t) = \mathrm{vec}(\Delta U_k) \in \mathbb{R}^{m\Gamma_u}$. The solution (i.e., ΔU_k) to the QP problem (8.15) can be obtained at time instant k, with its first element ($\Delta\dot{\boldsymbol{\theta}}(k)$) being used to calculate $\dot{\boldsymbol{\theta}}(k)$, which is then applied to the redundant manipulator.

8.3 Model Construction and Theoretical Analysis

This section develops an FTDNN model to solve QP problem (8.15), which is derived from the MPTCPO scheme (8.13). Additionally, theoretical proofs are provided to demonstrate the accuracy and finite-time characteristics of the MPTCPO scheme (8.13) when solved by the FTDNN model.

8.3.1 FTDNN Model Construction

With the slack variable vector $\boldsymbol{\varrho}(t) \in \mathbb{R}^{m(2\Gamma+4\Gamma_u)}$ ($\boldsymbol{\varrho}(t) \geq \mathbf{0}$), QP problem (8.15) is transformed as

$$\min_{\boldsymbol{y}(t)} \quad \frac{1}{2}\boldsymbol{y}(t)^{\mathrm{T}} H \boldsymbol{y}(t) + \boldsymbol{g}^{\mathrm{T}} \boldsymbol{y}(t), \tag{8.16a}$$

$$\text{s.t.} \quad W \boldsymbol{y}(t) - \boldsymbol{h} + \boldsymbol{\varrho}(t) = \mathbf{0}. \tag{8.16b}$$

The constrained optimization problem is then transformed into an unconstrained one by introducing a penalty term:

$$L(\boldsymbol{y}(t), \boldsymbol{\varrho}(t)) = \frac{1}{2}\boldsymbol{y}(t)^{\mathrm{T}} H \boldsymbol{y}(t) + \boldsymbol{g}^{\mathrm{T}} \boldsymbol{y}(t) + \frac{1}{2}\mu \| W \boldsymbol{y}(t) - \boldsymbol{h} + \boldsymbol{\varrho}(t) \|^2, \tag{8.17}$$

where $\mu \in \mathbb{R}$ represents the penalty factor. Moreover, since the QP problem (8.15) is a convex optimization problem, a global optimal solution exists, which satisfies the following conditions:

$$\begin{cases} \nabla_{\boldsymbol{y}} L = H \boldsymbol{y}^* + \boldsymbol{g} + \mu W^{\mathrm{T}}(W \boldsymbol{y}^* - \boldsymbol{h} + \boldsymbol{\varrho}^*) = \mathbf{0}, & (8.18a) \\ \nabla_{\boldsymbol{\varrho}} L = \mu(W \boldsymbol{y}^* - \boldsymbol{h} + \boldsymbol{\varrho}^*) = \mathbf{0}, & (8.18b) \end{cases}$$

where $\boldsymbol{y}^*$ and ϱ^* denote the optimal solutions. Therefore, (8.15) is equivalent to the following equation:

$$G \boldsymbol{x}(t) + \boldsymbol{b} = \mathbf{0}, \tag{8.19}$$

with

$$G = \begin{bmatrix} H + \mu W^{\mathrm{T}} W & \mu W^{\mathrm{T}} \\ \mu W & \mu I \end{bmatrix}, \boldsymbol{x}(t) = \begin{bmatrix} \boldsymbol{y}(t) \\ \boldsymbol{\varrho}(t) \end{bmatrix}, \boldsymbol{b} = \begin{bmatrix} \boldsymbol{g} - \mu W^{\mathrm{T}} \boldsymbol{h} \\ -\mu \boldsymbol{h} \end{bmatrix}.$$

An FTDNN model is then developed to address problem (8.19). Initially, an error function is introduced as $\boldsymbol{\varepsilon}(t) = G\boldsymbol{x}(t) + \boldsymbol{b}$, and the solution to problem (8.19) can be obtained when $\boldsymbol{\varepsilon}(t)$ approaches zero. Building on this fact, a design formula Godoy et al. (2022) with an activation function is introduced to drive the error function toward zero, as given by $\dot{\boldsymbol{\varepsilon}}(t) = -\xi \Phi(\boldsymbol{\varepsilon}(t))$, where $\Phi(\cdot)$ represents an activation function composed of $\phi(\cdot)$, which is odd and monotonically increasing. At the same time, $\phi(\cdot)$ is defined as

$$\phi(\varepsilon_i) = \frac{1}{2}\alpha(\mathrm{Lip}^{\tau}(\varepsilon_i) + \mathrm{Lip}^{1/\tau}(\varepsilon_i)) + \frac{1}{2}\beta\varepsilon_i,$$

where $\tau \in (0, 1)$ is a design parameter; ε_i means the ith element of $\boldsymbol{\varepsilon}(t)$; α and β denote positive parameters; and

$$\mathrm{Lip}^{\tau}(\varepsilon_i) = \begin{cases} |\varepsilon_i|^{\tau}, & \varepsilon_i > 0, \\ 0, & \varepsilon_i = 0, \\ -|\varepsilon_i|^{\tau}, & \varepsilon_i < 0. \end{cases}$$

Therefore, the FTDNN model solving (8.19) is as follows:

$$G\dot{\boldsymbol{x}}(t) = -\xi\Phi(G\boldsymbol{x}(t) + \boldsymbol{b}). \tag{8.20}$$

8.3.2 Theoretical Analyses

Theorem 8.1 *Concerning the MPTCPO scheme (8.13), starting from any initial value* $\boldsymbol{x}(0) = [\boldsymbol{y}^{\mathrm{T}}(0), \boldsymbol{\varrho}^{\mathrm{T}}(0)]^{\mathrm{T}}$, *the variable* $\boldsymbol{x}(t) = [\boldsymbol{y}^{\mathrm{T}}(t), \boldsymbol{\varrho}^{\mathrm{T}}(t)]^{\mathrm{T}}$ *of FTDNN model (8.20) globally converges to the theoretical solution* $\boldsymbol{x}^*(t) = [\boldsymbol{y}^{*\mathrm{T}}(t), \boldsymbol{\varrho}^{*\mathrm{T}}(t)]^{\mathrm{T}}$, *where* $\boldsymbol{y}^*(t)$ *represents the joint-velocity increment during the sampling period* δ *for redundant manipulators.*

Proof A Lyapunov function is defined as

$$\mathscr{X}(t) = \boldsymbol{\varepsilon}^{\mathrm{T}}(t)\boldsymbol{\varepsilon}(t). \tag{8.21}$$

It is clear that $\mathscr{X}(t)$ is positive semi-definite. Differentiating (8.21) with respect to t yields

$$\dot{\mathscr{X}}(t) = \boldsymbol{\varepsilon}^{\mathrm{T}}(t)\dot{\boldsymbol{\varepsilon}}(t) = -\xi\boldsymbol{\varepsilon}^{\mathrm{T}}(t)\Phi(\boldsymbol{\varepsilon}(t)). \tag{8.22}$$

Given that $\Phi(\cdot)$ is an odd and monotonically increasing function and $\xi > 0$, it follows that $\dot{\mathscr{X}}(t) \leq \mathbf{0}$. According to Lyapunov stability theory, it can be inferred that $\boldsymbol{\varepsilon}(t)$ converges to zero. Therefore, starting from any initial value $\boldsymbol{x}(0) = [\boldsymbol{y}^{\mathrm{T}}(0), \boldsymbol{\varrho}^{\mathrm{T}}(0)]^{\mathrm{T}}$, the variable $\boldsymbol{x}(t) = [\boldsymbol{y}^{\mathrm{T}}(t), \boldsymbol{\varrho}^{\mathrm{T}}(t)]^{\mathrm{T}}$ of FTDNN model (8.20) is globally convergent to the theoretical solution $\boldsymbol{x}^*(t) = [\boldsymbol{y}^{*\mathrm{T}}(t), \boldsymbol{\varrho}^{*\mathrm{T}}(t)]^{\mathrm{T}}$, where $\boldsymbol{y}^*(t)$ denotes the joint-velocity increment over the sampling period δ for redundant manipulators. The proof is completed. □

Theorem 8.2 *Considering the MPTCPO scheme (8.13), the joint-velocity increment $\mathbf{y}^*(t)$ over the sampling period δ for redundant manipulators is derived to achieve finite-time control within a finite time t_c, where t_c is given by*

$$t_c \leq \frac{2|\varepsilon_{\max}(0)|^{1-\tau}}{\xi(1-\tau)},$$

of which $\varepsilon_{\max}(0)$ is the element with the largest absolute value from the initial error function $\boldsymbol{\varepsilon}(0)$, defined as $\boldsymbol{\varepsilon}(t) = G\boldsymbol{x}(t) + \boldsymbol{b}$.

Proof Define $\varepsilon_{\max}(t)$ to be the element in $\boldsymbol{\varepsilon}(t)$ with $\varepsilon_{\max}(0)$. From $\dot{\boldsymbol{\varepsilon}}(t) = -\xi\Phi(\boldsymbol{\varepsilon}(t))$, the following equation can be obtained:

$$\dot{\varepsilon}_i(t) = -\xi\phi(\varepsilon_i(t)),$$

of which $\dot{\varepsilon}_i(t)$ denotes the time derivative of $\varepsilon_i(t)$ (i.e., the ith element of $\boldsymbol{\varepsilon}(t)$). According to the sign of $\varepsilon_{\max}(0)$, the analysis can be divided into the following three cases.

- The case of $\varepsilon_{\max}(0) > 0$. Depending on $\varepsilon_{\max}(0) \geq \varepsilon_i(0)$ for all possible i as well as the comparison theorem Godoy et al. (2022), $\varepsilon_{\max}(t) \geq \varepsilon_i(t)$ with $t \geq 0$ for all possible i. Similarly, $-\varepsilon_{\max}(t) \leq \varepsilon_i(t)$ with $t \geq 0$ for all possible i. Hence, $-\varepsilon_{\max}(t) \leq \varepsilon_i(t) \leq \varepsilon_{\max}(t)$ for all possible i as time t goes on, which indicates that $\varepsilon_i(t)$ converges to zero when $\varepsilon_{\max}(t)$ reaches zero. In other words, the convergence time of FTDNN model (8.20) is bounded by t_{cmax}, with t_{cmax} being the convergence time of the dynamics of $\varepsilon_{\max}(t)$. That is, $t_c \leq t_{\text{cmax}}$. As for t_{cmax},

$$\dot{\varepsilon}_{\max}(t) = -\xi\phi(\varepsilon_{\max}(t)).$$

Then, define a Lyapunov function candidate as $L(t) = |\varepsilon_{\max}(t)|^2$, whose time derivation is as follows:

$$\begin{aligned}\dot{L}(t) &= -2\xi\varepsilon_{\max}(t)\phi(\varepsilon_{\max}(t))\\ &= -\xi\left(\alpha|\varepsilon_{\max}(t)|^{\tau+1} + \alpha|\varepsilon_{\max}(t)|^{(1/\tau)+1} + \beta\varepsilon_{\max}^2(t)\right)\\ &\leq -\xi\alpha|\varepsilon_{\max}(t)|^{\tau+1} = -\xi\alpha L^{(\tau+1)/2}(t).\end{aligned}$$

Solving the above inequality with $L(0) = |\varepsilon_{\max}(0)|^2$ yields

$$L^{(1-\tau)/2}\begin{cases}\leq |\varepsilon_{\max}(0)|^{1-\tau} - \frac{\xi\alpha t(1-\tau)}{2}, & t \leq \frac{2|\varepsilon_{\max}(0)|^{1-\tau}}{\xi\alpha(1-\tau)},\\ = 0, & t > \frac{2|\varepsilon_{\max}(0)|^{1-\tau}}{\xi\alpha(1-\tau)}.\end{cases}$$

From the above analysis, $L(t)$ converges to zero after at most a time duration. That is, $\varepsilon_{\max}(t)$ becomes zero when $t > 2|\varepsilon_{\max}(0)|^{1-\tau}/\xi\alpha(1-\tau)$. Therefore, we have

$$t_{\text{cmax}} < \frac{2|\varepsilon_{\max}(0)|^{1-\tau}}{\xi(1-\tau)}.$$

- The case of $\varepsilon_{\max}(0) < 0$. Similarly, the following result can be available:

$$t_{\text{cmax}} < \frac{2|\varepsilon_{\max}(0)|^{1-\tau}}{\xi(1-\tau)}.$$

- The case of $\varepsilon_{\max}(0) = 0$. One has

$$t_{\text{cmax}} = 0 = \frac{2|\varepsilon_{\max}(0)|^{1-\tau}}{\xi(1-\tau)}.$$

Summarizing the above analysis, t_{c} is as follows:

$$t_{\text{c}} \leq t_{\text{cmax}} \leq \frac{2|\varepsilon_{\max}(0)|^{1-\tau}}{\xi(1-\tau)},$$

which implies that the variable $\boldsymbol{x}(t) = [\boldsymbol{y}^{\text{T}}(t), \boldsymbol{\varrho}^{\text{T}}(t)]^{\text{T}}$ in the FTDNN model (8.20) globally converges to the theoretical solution $\boldsymbol{x}^*(t) = [\boldsymbol{y}^{*\text{T}}(t), \boldsymbol{\varrho}^{*\text{T}}(t)]^{\text{T}}$ within the finite time t_{c}. In other words, the joint-velocity increment $\boldsymbol{y}^*(t)$ during the sampling period δ for redundant manipulators can be obtained from $\boldsymbol{x}^*(t)$ within finite time t_{c}. Therefore, the MPTCPO scheme (8.13), solved by the FTDNN model (8.20), is capable of achieving the finite-time control. The proof is completed. □

8.4 Simulations and Experiments

This section presents simulations and experiments to demonstrate the effectiveness and high accuracy of the MPTCPO scheme (8.13), solved by the FTDNN model (8.20), in fulfilling trajectory-tracking tasks.

The redundant manipulator used in simulations is a 7-DOF Franka Emika Panda (abbreviated as Franka) manipulator, and the trajectory-tracking task involves tracking a sheep-shaped trajectory. Based on official specifications of the manipulator, relevant parameters are set as follows: joint number $m = 7$, joint limits $\boldsymbol{\theta}^+ = [2.89, 1.76, 2.89, -0.07, 2.89, 3.75, 2.89]^{\text{T}}$ rad, $\boldsymbol{\theta}^- = [-2.89, -1.76, -2.89, -3.07, -2.89, -0.01, -2.89]^{\text{T}}$ rad, $\dot{\boldsymbol{\theta}}^+ = -\dot{\boldsymbol{\theta}}^- = [2.18, 2.18, 2.18, 2.18, 2.61, 2.61, 2.61]^{\text{T}}$ rad/s, $\ddot{\boldsymbol{\theta}}^+ = -\ddot{\boldsymbol{\theta}}^- = [15, 7.5, 10, 12.5, 15, 20, 20]^{\text{T}}$ rad/s^2. Additionally, other parameters are: $R = 500000I$, $Q_1 = I$, $Q_2 = 20I$, $\Gamma = \Gamma_u = 5$, $\xi = 10$, $\alpha = 1$, $\beta = 30$, $\tau = 0.2$, $\mu = 5$, the sampling period $\delta = 0.01$ s, the task execution time $T = 10$ s, and the initial joint angle $\boldsymbol{\theta}_0 = [-0.30, -0.78, 0, -2.35, 0, 0.87, 0.78]^{\text{T}}$ rad.

The results obtained from the simulation using the MPTCPO scheme (8.13), solved by the FTDNN model (8.20), are presented in Figs. 8.1 and 8.2. Figure 8.1a

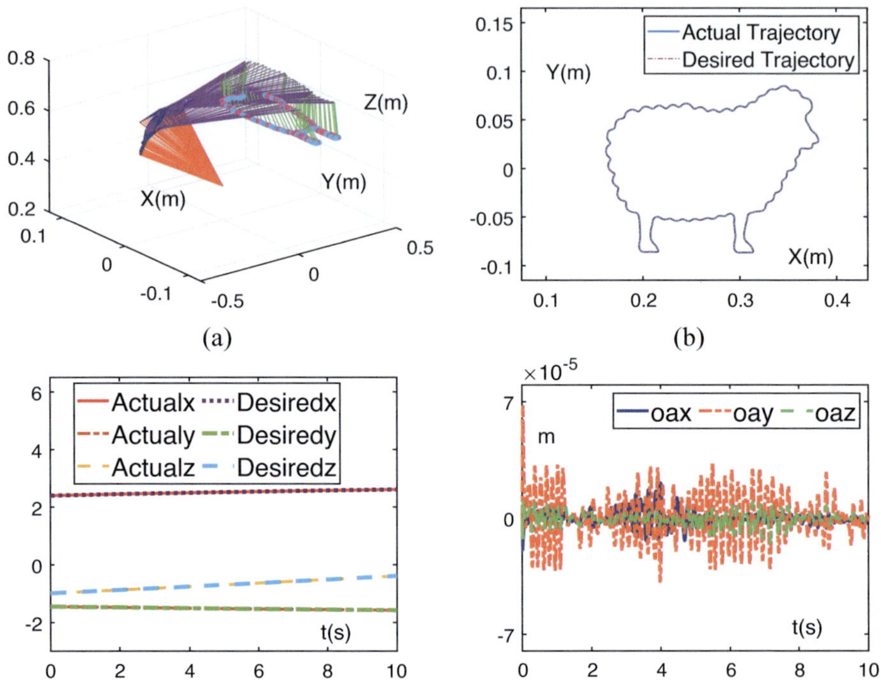

Fig. 8.1 Simulation results for tracking a sheep-shaped trajectory synthesized by MPTCPO scheme (8.13) solved by FTDNN model (8.20). **a** Motion process. **b** Actual trajectory and desired trajectory. **c** Actual and desired orientation with rotation vector. **d** Orientation error with rotation vector

illustrates the motion process of the redundant manipulator. Figure 8.1b shows the tracking trajectory, while the tracking errors for the position in Cartesian coordinates remain at 5×10^{-4} m, as shown in Fig. 8.2a. Additionally, all joint angles, joint velocities, and joint accelerations stay within the joint limits, as evidenced in Fig. 8.2b–d. Figure 8.1c represents both the desired and actual orientation of the redundant manipulator's end-effector during the motion process. As observed in Fig. 8.1d, orientation tracking errors stabilize at 7×10^{-5}. Therefore, the MPTCPO scheme (8.13) solved by the FTDNN model (8.20) proves to be both effective and highly accurate. To further demonstrate the feasibility and advantages of the MPTCPO scheme (8.13), solved by the FTDNN model (8.20), experiments are conducted using the 7-DOF Franka manipulator. Specifically, experiments include a task where the redundant manipulator tracks a sheep-shaped trajectory, as well as comparative experiments where the manipulator tracks a semicircular trajectory in the Y-Z plane, simulating an arc-welding task, as shown in Fig. 8.3.

In detail, the comparative schemes include the MPC scheme Li et al. (2016) and the MAN scheme Jin and Zhang (2015). Some parameters are updated as $R = 200000I$, $Q_2 = 180I$, the sampling period $\delta = 0.001$ s, and the initial joint

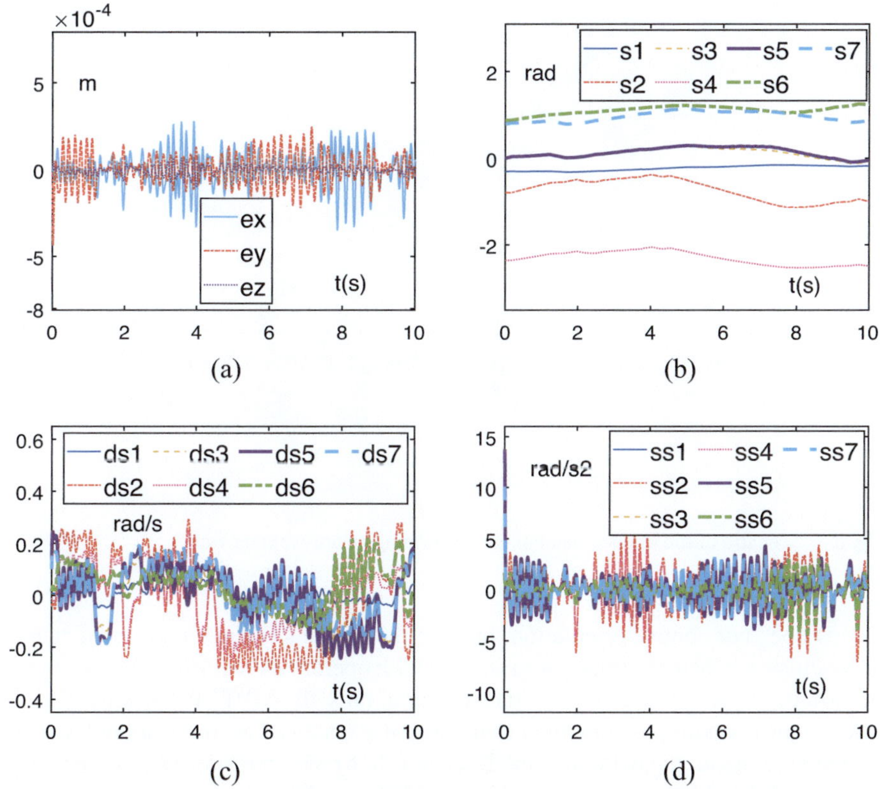

Fig. 8.2 Position error and joint information with MPTCPO scheme (8.13) solved by FTDNN model (8.20) tracking a sheep-shaped trajectory. **a** Position error. **b** Joint angle. **c** Joint velocity. **d** Joint acceleration

angle $\boldsymbol{\theta}_0 = [0, -0.78, 0, -2.35, 0, 1.57, 0.78]^{\mathrm{T}}$ rad, while other parameters remain the same as those used in the simulations. The experimental results are shown in Figs. 8.4 and 8.5. Note that the trajectories in Figs. 8.4 and 8.5 are not real trajectories that are provided to help visualize the motion of the redundant manipulator's end-effector.

Unlike simple position tracking experiments, this chapter considers orientation tracking. Figure 8.4 demonstrates that the trajectory-tracking task is successfully completed, and the variation in the end-effector's orientation aligns with expectations, thus validating the effectiveness of the presented MPTCPO scheme (8.13). As seen in Fig. 8.5, the position tracking is successfully achieved by all three schemes. In Fig. 8.5a, as anticipated, the direction of the redundant manipulator's flange continuously changes, maintaining alignment with the normal direction of the semicircular trajectory during the motion. Figure 8.5b shows that the MPC scheme Li et al. (2016) with posture-maintaining technology keeps the flange orientation constant throughout operation. Meanwhile, Fig. 8.5c reveals that the MAN scheme Jin and Zhang

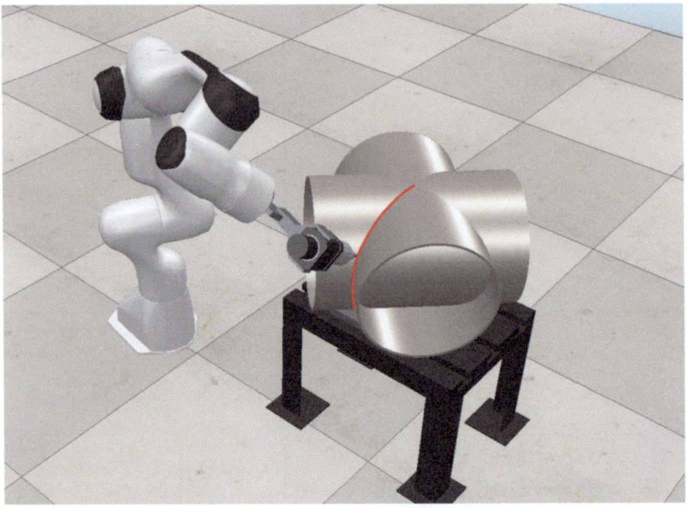

Fig. 8.3 Scenario of the Franka manipulator performing an arc-welding work

(2015) does not control orientation, making the flange's orientation unpredictable. In conclusion, when a manipulator is tasked with operations requiring changes in the end-effector's orientation, such as arc-welding, only the MPTCPO scheme (8.13), which controls both position and orientation, is capable of accomplishing the task. On the other hand, the MPC scheme Li et al. (2016) with posture-maintaining technology and the MAN scheme Jin and Zhang (2015) with pure position control cannot complete such tasks. This highlights the practicality and feasibility of the MPTCPO scheme (8.13) in real-world applications.

Finally, a comparison among various trajectory-tracking schemes for redundant manipulators is presented in Table 8.2. It is important to note that all schemes are based on the same manipulator, and parameters are selected according to the corresponding references. From the perspective of joint constraints, the MPTCPO scheme (8.13), MPC scheme Yang et al. (2018), MAN scheme Jin and Zhang (2015), and MPC scheme Li et al. (2016) all account for three levels of joint constraints, while other schemes only consider two levels. However, the MPTCPO scheme (8.13) uniquely includes control of the orientation of the end-effector. In terms of the position tracking, the maximal position tracking error norm (MPTEN) of the MPTCPO scheme (8.13) is the smallest compared to the other three schemes. Regarding orientation control, the MPTCPO scheme (8.13), MVN scheme Zhang et al. (2020), and MVN scheme Li et al. (2022) all control the orientation of the end-effector. However, only the MPTCPO scheme (8.13) enables the end-effector to achieve any orientation, while the other two schemes only allow for maintaining a fixed orientation. In terms of feasible region reduction, none of the MPC-based schemes reduce the feasible region, while other schemes like the MVN and MAN schemes make the feasible region smaller. Finally, in terms of MPTEN, the MVN scheme Chen et al.

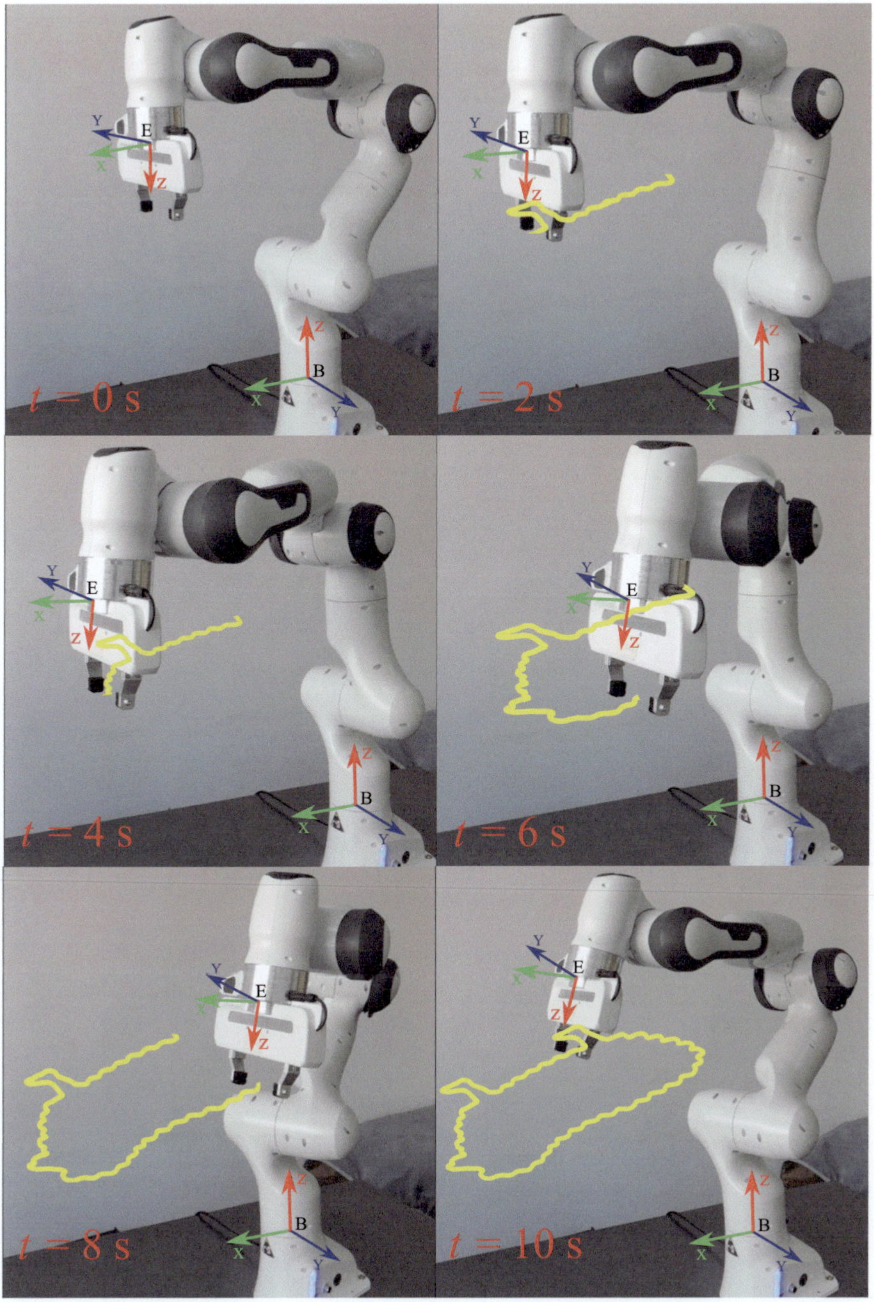

Fig. 8.4 Experimental results of the Franka manipulator tracking a sheep-shaped trajectory

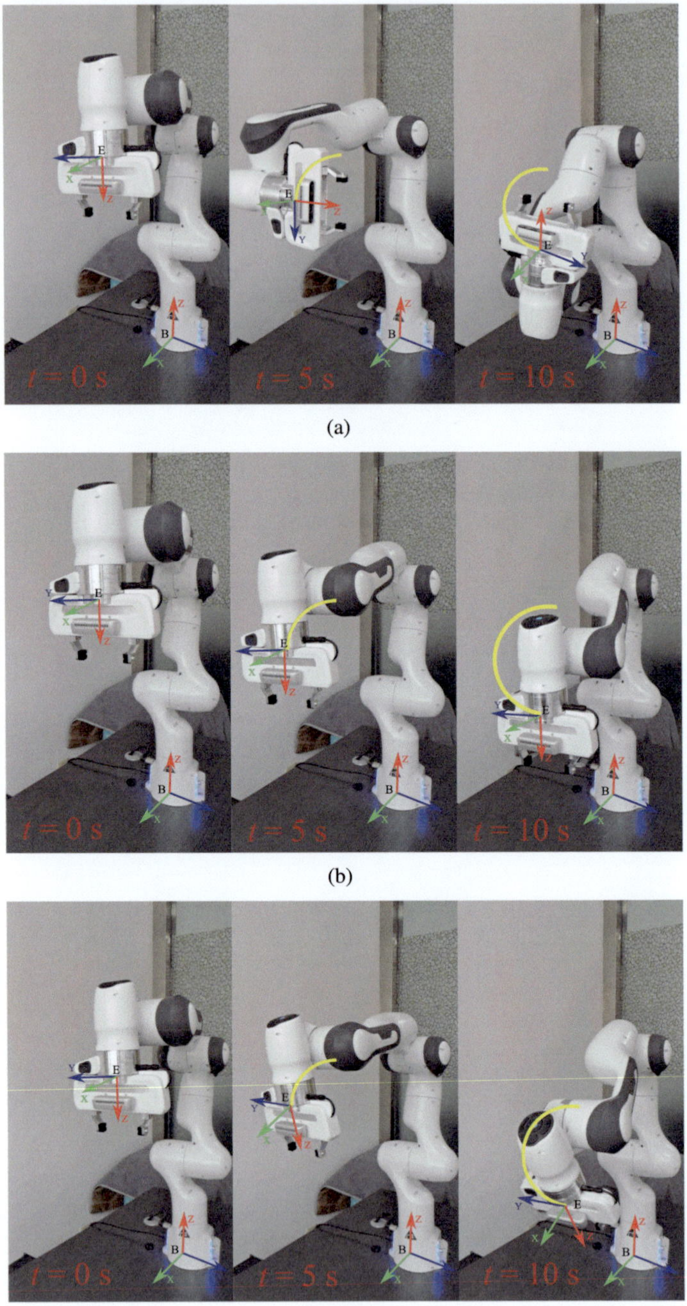

(c)

Fig. 8.5 Experimental results of the Franka manipulator tracking a semicircle trajectory. **a** Results with MPTCPO scheme (8.13). **b** Results with MPC scheme Li et al. (2016). **c** Results with MAN scheme Jin and Zhang (2015)

Table 8.2 Comparisons of various control schemes for redundant manipulators

	Optimization level	Joint constraints	Orientation control	Reduced feasible region	MPTEN ♯
This chapter	Acceleration	$\boldsymbol{\theta}, \dot{\boldsymbol{\theta}}, \ddot{\boldsymbol{\theta}}$	Any‡	×	4.471×10^{-4} m
Paper Yang et al. (2018)	Acceleration	$\boldsymbol{\theta}, \dot{\boldsymbol{\theta}}, \ddot{\boldsymbol{\theta}}$	N/A	×	1.892×10^{-3} m
Paper Jin and Zhang (2015)	Acceleration	$\boldsymbol{\theta}, \dot{\boldsymbol{\theta}}, \ddot{\boldsymbol{\theta}}$	N/A	√	3.515×10^{-3} m
Paper Zhang et al. (2020)	Velocity	$\boldsymbol{\theta}, \dot{\boldsymbol{\theta}}$	Single§	√	2.763×10^{-3} m
Paper Li et al. (2022)	Velocity	$\boldsymbol{\theta}, \dot{\boldsymbol{\theta}}$	Single§	√	1.096×10^{-3} m
Paper Li et al. (2016)	Acceleration	$\boldsymbol{\theta}, \dot{\boldsymbol{\theta}}, \ddot{\boldsymbol{\theta}}$	N/A	×	2.217×10^{-4} m
Paper Chen et al. (2020)	Velocity	$\boldsymbol{\theta}, \dot{\boldsymbol{\theta}}$	N/A	√	1.155×10^{-4} m

♯ "MPTEN" denotes the maximal position tracking error norm.
‡ "Any" implies that the scheme enables the end-effector to achieve any desired orientation.
§ "Single" implies that the scheme only enables the end-effector to maintain one specific orientation

(2020), which has a slightly smaller tracking error than the MPTCPO scheme (8.13), achieves the smallest error among all the compared schemes. However, the MVN scheme Chen et al. (2020), which considers only two levels of joint constraints, has a reduced feasible regio and does not control the orientation of the end-effector. Overall, the MPTCPO scheme (8.13) demonstrates effectiveness and enhanced safety in scenarios where the simultaneous control of both position and orientation of the end-effector is essential.

8.5 Conclusions

To address the limitation of most trajectory-tracking schemes that focus primarily on the position of the end-effector while neglecting its orientation, and to avoid reducing the feasible region when incorporating multiple levels of joint limits, an MPTCPO scheme has been developed for redundant manipulators in this chapter. The presented MPTCPO scheme not only enables the control of both the position and orientation of the end-effector, but also minimizes the tracking error, joint-velocity norm, and joint-acceleration norm. Additionally, the scheme handles three levels of joint limits—joint angle, joint velocity, and joint acceleration—without reducing the feasible region, as no conversion is necessary. Furthermore, an FTDNN model has been created to solve the MPTCPO scheme and achieve finite-time control. The effectiveness and precision of the presented MPTCPO scheme, when solved by the FTDNN model, have been validated through computer simulations and physical experiments.

References

Cao HQ, Nguyen HX, Tran TNC, Tran HN, Jeon JW (2022) A robot calibration method using a neural network based on a butterfly and flower pollination algorithm. IEEE Trans Ind Electron 69(4):3865–3875

Carron A, Arcari E, Wermelinger M, Hewing L, Hutter M, Zeilinger MN (2019) Data-driven model predictive control for trajectory tracking with a robotic arm. IEEE Robot Autom Lett 4(4):3758–3765

Chen D, Li S, Wu Q, Luo X (2020) New disturbance rejection constraint for redundant robot manipulators: an optimization perspective. IEEE Trans Ind Inform 16(4):2221–2232

Cisneros PSG, Sridharan A, Werner H (2018) Constrained predictive control of a robotic manipulator using quasi-LPV representations. IFAC-PapersOnLine 51(26):118–123

Consolini L, Laurini M, Locatelli M (2024) A dynamic programming approach for cooperative pallet-loading manipulators. IEEE Trans Autom Sci Eng 21(4):5280–5296

Dietrich A, Ott C (2020) Hierarchical impedance-based tracking control of kinematically redundant robots. IEEE Trans Robot 36(1):204–221

Du G, Chen M, Liu C, Zhang B, Zhang P (2018) Online robot teaching with natural human-robot interaction. IEEE Trans Ind Electron 65(12):9571–9581

Faulwasser T, Weber T, Zometa P, Findeisen R (2017) Implementation of nonlinear model predictive path-following control for an industrial robot. IEEE Trans Control Syst Technol 25(4):1505–1511

Godoy RV, Dwivedi A, Liarokapis M (2022) Electromyography based decoding of dexterous, in-hand manipulation motions with temporal multichannel vision transformers. IEEE Trans Neural Syst Rehabil Eng 30:2207–2216

Huang H, Zhang T, Yang C, Chen CLP (2020) Motor learning and generalization using broad learning adaptive neural control. IEEE Trans Ind Electron 67(10):8608–8617

Incremona GP, Ferrara A, Magni L (2017) MPC for robot manipulators with integral sliding modes generation. IEEE/ASME Trans Mechatron 22(3):1299–1307

Jeon S, Jung M, Choi S, Kim B, Hwangbo J (2024) Learning whole-body manipulation for quadrupedal robot. IEEE Robot Autom Lett 9(1):699–706

Jin L, Zhang Y (2015) G2-type SRMPC scheme for synchronous manipulation of two redundant robot arms. IEEE Trans Cybern 45(2):153–164

Jin L, Liufu Y, Lu H, Zhang Z (2021) Saturation-allowed neural dynamics applied to perturbed time-dependent system of linear equations and robots. IEEE Trans Ind Electron 68(10):9844–9854

Jin L, Wei L, Li S (2023) Gradient-based differential neural-solution to time-dependent nonlinear optimization. IEEE Trans Autom Control 68(1):62–627

Jin L, Zhang F, Liu M, Xu SSD (2023) Finite-time model predictive tracking control of position and orientation for redundant manipulators. IEEE Trans Ind Electron 70(6):6017–6026

Jin L, Su Z, Fu D, Xiao X (2024) Coevolutionary neural solution for nonconvex optimization with noise tolerance. IEEE Trans Neural Netw Learn Syst 35(12):17571–17581

Katsikis VN, Stanimirović PS, Mourtas SD, Xiao L, Karabasevic D, Stanujkic D (2021) Zeroing neural network with fuzzy parameter for computing pseudoinverse of arbitrary matrix. IEEE Trans Fuzzy Syst 30(9):3426–3435

Ke F, Li Z, Yang C (2018) Robust tube-based predictive control for visual servoing of constrained differential-drive mobile robots. IEEE Trans Ind Electron 65(4):3437–3446

Khan AH, Li S, Luo X (2020) Obstacle avoidance and tracking control of redundant robotic manipulator: an RNN-based metaheuristic approach. IEEE Trans Ind Inform 16(7):4670–4680

Kim U, Maeng CY, Bak G, Kim YJ (2022) High-stiffness torque sensor with a strain amplification mechanism for cooperative industrial manipulators. IEEE Trans Ind Electron 69(3):3131–3141

Kleff S, Meduri A, Budhiraja R, Mansard N, Righetti L (2021) High-frequency nonlinear model predictive control of a manipulator. In: IEEE international conference on robotics and automation, pp 7330–7336

Li Z, Deng J, Lu R, Xu Y, Bai J, Su C (2016) Trajectory-tracking control of mobile robot systems incorporating neural-dynamic optimized model predictive approach. IEEE Trans Syst Man Cybern Syst 46(6):740–749

Li W, Chiu PWY, Li Z (2022) A novel neural approach to infinity-norm joint-velocity minimization of kinematically redundant robots under joint limits. IEEE Trans Neural Netw Learn Syst 34(1):409–420

Liao B, Zhang Y, Jin L (2016) Taylor $O(h^3)$ discretization of ZNN models for dynamic equality-constrained quadratic programming with application to manipulators. IEEE Trans Neural Netw Learn Syst 27(2):225–237

Liu M, Chen L, Du X, Jin L, Shang M (2021) Activated gradients for deep neural networks. IEEE Trans Neural Netw Learn Syst 34(4):2156–2168

Liu M, Li Y, Chen Y, Qi Y, Jin L (2024) A distributed competitive and collaborative coordination for multirobot systems. IEEE Trans Mob Comput 23(2):11436–11448

Liufu Y, Jin L, Shang M, Wang X, Wang FY (2024) ACP-incorporated perturbation-resistant neural dynamics controller for autonomous vehicles. IEEE Trans Intell Veh 9(4):4675–4686

Maric B, Mutka A, Orsag M (2020) Collaborative human-robot framework for delicate sanding of complex shape surfaces. IEEE Robot Autom Lett 5(2):2848–2855

Peng G, Chen CLP, He W, Yang C (2021) Neural-learning-based force sensorless admittance control for robots with input deadzone. IEEE Trans Ind Electron 68(6):5184–5196

Peng G, Chen CP, Yang C (2022) Neural networks enhanced optimal admittance control of robot-environment interaction using reinforcement learning. IEEE Trans Neural Netw Learn Syst 33(9):4551–4561

Saito N, Shimizu T, Ogata T, Sugano S (2022) Utilization of image/force/tactile sensor data for object-shape-oriented manipulation: wiping objects with turning back motions and occlusion. IEEE Robot Autom Lett 7(2):968–975

Schnelle F, Eberhard P (2017) Adaptive nonlinear model predictive control design of a flexible-link manipulator with uncertain parameters. Acta Mech Sin 33(3):529–542

Shi Y, Zhao W, Li S, Li B, Sun X (2021) Novel discrete-time recurrent neural network for robot manipulator: a direct discretization technical route. IEEE Trans Neural Netw Learn Syst 34(6):2781–2790

Shi Y, Jin L, Li S, Li J, Qiang J, Gerontitis DK (2022) Novel discrete-time recurrent neural networks handling discrete-form time-variant multi-augmented Sylvester matrix problems and manipulator application. IEEE Trans Neural Netw Learn Syst 33(2):587–599

Tan Z, Xiao L, Chen S, Lv X (2022) Gradient-based differential kWTA network with application to competitive coordination of multiple robots. IEEE/CAA J Autom Sin 9(8):1452–1463

Xiao L, Zhang Y (2016) Dynamic design, numerical solution, and effective verification of acceleration-level obstacle-avoidance scheme for robot manipulators. Int J Syst Sci 47(4):932–945

Xiao L, Cao Y, Dai J, Jia L, Tan H (2021) Finite-time and predefined-time convergence design for zeroing neural network: theorem, method, and verification. IEEE Trans Ind Inform 17(7):4724–4732

Xie Z, Jin L, Du X, Xiao X, Li H, Li S (2019) On generalized RMP scheme for redundant robot manipulators aided with dynamic neural networks and nonconvex bound constraints. IEEE Trans Ind Inform 15(9):5172–5181

Xu W, Zhang J, Liang B, Li B (2016) Singularity analysis and avoidance for robot manipulators with nonspherical wrists. IEEE Trans Ind Electron 63(1):277–290

Xu Z, Li S, Zhou X, Zhou S, Cheng T, Guan Y (2021) Dynamic neural networks for motion-force control of redundant manipulators: an optimization perspective. IEEE Trans Ind Electron 68(2):1525–1536

Yan J, Jin L, Hu B (2024) Data-driven model predictive control for redundant manipulators with unknown model. IEEE Trans Cybern 54(10):5901–5911

Yan J, Jin L, Luo X, Li S (2024) Modified RNN for solving comprehensive Sylvester equation with TDOA application. IEEE Trans Neural Netw Learn Syst 35(9):12553–12563

Yang C, Jiang Y, He W, Na J, Li Z, Xu B (2018) Adaptive parameter estimation and control design for robot manipulators with finite-time convergence. IEEE Trans Ind Electron 65(10):8112–8123

Yang C, Luo J, Pan Y, Liu Z, Su CY (2018) Personalized variable gain control with tremor attenuation for robot teleoperation. IEEE Trans Syst Man Cybern Syst 48(10):1759–1770

Zhang Z, Chen S, Zhu X, Yan Z (2020) Two hybrid end-effector posture-maintaining and obstacle-limits avoidance schemes for redundant robot manipulators. IEEE Trans Ind Inform 16(2):754–763

Zhang Z, Chen S, Xie J, Yang S (2021) Two hybrid multiobjective motion planning schemes synthesized by recurrent neural networks for wheeled mobile robot manipulators. IEEE Trans Syst Man Cybern Syst 51(5):3270–3281

Zhang J, Jin L, Yang C (2022) Distributed cooperative kinematic control of multiple robotic manipulators with an improved communication efficiency. IEEE/ASME Trans Mechatron 27(1):149–158

Zhang M, Tong W, Li P, Hou Y, Xu X, Zhu L (2023) Robust neural dynamics method for redundant robot manipulator control with physical constraints. IEEE Trans Ind Inform 19(12):11721–11729

Zhou M, Yu Q, Huang K, Mahov S, Eslami A, Maier M, Lohmann CP, Navab N, Zapp D, Knoll A, Nasseri MA (2020) Towards robotic-assisted subretinal injection: a hybrid parallel-serial robot system design and preliminary evaluation. IEEE Trans Ind Electron 67(8):6617–6628

Part III
Different End-Effectors Aided Robot Control with DNN

Chapter 9
Robust Grasping Force Control With Noise-Tolerant DNN

9.1 Introduction

Multi-fingered dexterous gripping is one of the important research directions in robotics Yang et al. (2022). Compared with a traditional parallel gripper, a multi-fingered dexterous hand with a higher degree of freedom has more diversified grasping options Liu et al. (2023b), while it also enables precise control of the applied force on a target. The advantages of multi-fingered dexterous hands are expected to be more effective than traditional parallel grippers in grasping tasks with increasingly diverse targets and scenarios.

One of the key challenges in gripping control is the gripping force optimization (GFO) problem Zheng et al. (2012). When the target in a grasping task is fragile or prone to sliding, the grasping control requires that the force exerted by the gripper should avoid crushing the target or grasping it in an insecure manner. It requires that the force exerted by the fingers of the multi-fingered dexterous hand on the target during the execution of the grasping task should be as small as possible while keeping the two from sliding relative to each other. In systems involving multiple contacts between the robot and the environment, similar problems arise frequently Klein and Kittivatcharapong (1990).

A number of gripping force control schemes are used to solve the GFO problem Liu et al. (2004). Proportional integral differential controller, as a mature and widely used kind of control method, is widely used for the control of grasping force. A proportional-integral-differential (PID) based method for controlling the gripping force when grasping a target with an artificial elastic finger is presented in Maeno et al. (2000). An improved PID controller is constructed in Nanayakkara et al. (2016) by using the failure probability expectation as a feedback signal. It reduces the difficulty of parameter adjustment and ultimately improves the probability of gripping success with respect to the conventional PID control. In addition, sliding mode control is used for gripping force control. An improved sliding mode control method is presented in Heydarabad et al. (2017) for optimal force grasping of targets with unknown shape

M. Liu et al., *Dynamic Neural Networks for Motion Control of Redundant Manipulators*, Intelligent Control and Learning Systems 21,
https://doi.org/10.1007/978-981-96-9144-9_9

and mass distribution. Although PID and sliding mode control models are improved in various ways, they do not perform well in the face of nonlinear systems and dynamic changes due to their non-mathematical-model-based nature Liu (2023).

To cope with the above issues, the GFO problem is mathematically modeled. In Lobo et al. (1998), the GFO problem is modeled as a second-order cone programming problem, where the force exerted by each finger of a multi-fingered dexterous hand on the target should satisfy a second-order cone constraint. Based on this, various control methods are investigated. In Buss et al. (1996), the GFO problem is analyzed as a linear matrix inequality problem and solved using the interior point method. In Gazeau et al. (2005), the GFO problem is transformed into a minimum distance problem and solved using a gradient method, and other traditional control methods. However, the above methods have some limitations in directly solving constrained optimization problems. Compared to traditional methods, dynamic neural networks (DNN) are introduced to deal with the GFO problem due to their excellent performance in solving constrained optimization problems Jin et al. (2023) and (2024).

In particular, reinforcement learning is widely used in dealing with the GFO problem due to its strong adaptability to various grasping tasks. A deep reinforcement learning framework is presented in Liu et al. (2023a) to improve the grasping efficiency of soft multimodal grippers. In Hu et al. (2023), to reduce the damage to the organism while grasping it, a reinforcement learning-based algorithm is presented to train the grasping strategy so that the multi-fingered dexterous hand can grasp the organism with minimum force. However, considering that reinforcement learning requires a lot of time and computing power to train in advance before applications and cannot cope with untrained emergencies and that there are modeling errors during simulation training, its performance may deviate from the simulation training results when applied to the real world Wang et al. (2024), Xi et al. (2021, 2004) and Xie and Jin (2022).

DNN methods are also applied to solve the GFO problem due to their property of not requiring prior training once the parameters are selected Xie and Jin (2022). A DNN method for solving a convex second-order cone programming problem is constructed in Nazemi (2019) and applied to control a multi-fingered dexterous hand to achieve optimal force grasping of a target. However, considering the diverse application scenarios of multi-fingered dexterous hands, such as household robots Hasegawa et al. (2022) and disaster relief robots Matsuura et al. (2022), there are different disturbances in different grasping scenarios. It can be seen that the existing works do not take into account the effect of diverse grasping scenarios on grasping tasks, which can be referred to as noise affecting the force exerted on targets by multi-fingered dexterous hands.

To overcome the drawbacks of the above-mentioned control schemes and considering diverse grasping scenarios of multi-fingered dexterous hands, this chapter proposes a noise-tolerant dynamic neural network (NTDNN) and constructs a noise-tolerant grasping force optimization (NTGFO) scheme to address the challenge of optimizing the grasping force in noisy scenarios without the need for prior training.

This chapter consists of five sections, of which Sect. 9.2 introduces the kinematic basics of grasping and constructs the optimal gripping force model. The NTDNN is presented in Sect. 9.3 for solving the optimal gripping force in the presence of noisy disturbances. In addition, theoretical analyses on the convergence performance of the proposed NTGFO scheme under different noises are also presented. In Sect. 9.4, simulations and physical experiments are carried out to illustrate the feasibility of the proposed NTGFO scheme. Comparisons are also carried out, the results show the superiority of the proposed NTGFO scheme. Finally, a conclusion is given in Sect. 9.5. The main contributions of this chapter are summarized below.

- A multi-fingered dexterous hand optimal gripping force model based on a projective theory is proposed, enabling efficient solving of the optimal multi-fingered dexterous hand grasping force.
- A noise-tolerant dynamic neural network is constructed for the optimal gripping force model. It can deal with various noises or their superposition that may exist during the execution of the task and finally get the optimal solution.

9.2 Preliminaries and Model Construction

In this section, the kinematic basis of grasping is given. An optimal gripping force model based on a projective theory is also constructed, which forms the basis for subsequent works.

9.2.1 Kinematic Basics of Grasping

To ensure the successful execution of a task by a multi-fingered dexterous hand, the following two constraints need to be met. **(i)** There is no relative displacement between the fingers of the multi-fingered dexterous hand and the target. **(ii)** The dexterous hand can keep the target stable by adjusting the force at the point of contact, regardless of the external force on the target being grasped. Besides, definitions of mathematical symbols of this chapter are given in Table 9.1.

By splitting the force $\boldsymbol{f}_i(t)$ exerted on the target by the i-th finger of the multi-fingered dexterous hand into $\boldsymbol{f}_{i,\mathrm{x}}(t)$, $\boldsymbol{f}_{i,\mathrm{y}}(t)$, and $\boldsymbol{f}_{i,\mathrm{z}}(t)$, the first constraint can be expressed as

$$\sqrt{\boldsymbol{f}_{i,\mathrm{x}}^2(t) + \boldsymbol{f}_{i,\mathrm{y}}^2(t)} \leq \mu_i \boldsymbol{f}_{i,\mathrm{z}}(t). \tag{9.1}$$

The physical meaning of the first constraint is to keep the force exerted by the fingers on the target within the friction cone to ensure no relative displacement between the fingers and the target Chen et al. (2021). Accordingly, the second constraint is expressed as

Table 9.1 Definition of symbols

O_o	Coordinate system of grasping target
$O_{\text{c}i}$	Coordinate system of contact between the i-th finger of the dexterous hand and the target
n	Number of fingers of multi-fingered dexterous hand
$\boldsymbol{f}_\text{e}(t) \in \mathbb{R}^6$	External force on a target and the force required to maintain the motion of a target at time t
$\boldsymbol{f}_i(t) \in \mathbb{R}^3$	Force vector exerted by the i-th finger of the dexterous hand on the target at time t
$\boldsymbol{f}(t) \in \mathbb{R}^{3n}$	Vector formed by merging of $\boldsymbol{f}_i$ at time t
$\boldsymbol{f}_{i,(\cdot)}(t)$	Components of $\boldsymbol{f}_i$ on x, y, or z axes at time t
μ_i	Friction coefficient of the multi-fingered dexterous hand against the target
G_i	Grasp transformation matrix for converting forces from $O_{\text{c}i}$ to O_o coordinates
$\boldsymbol{G}$	Matrix formed by merging of G_i
$\boldsymbol{R}_{\text{c}_i} \in \mathbb{R}^{3\times3}$	Rotation matrix of coordinate system $O_{\text{c}i}$ to O_o
$\boldsymbol{p}_{\text{c}_i} \in \mathbb{R}^3$	Translation vector of coordinate system $O_{\text{c}i}$ to O_o
$\hat{\boldsymbol{p}}_{\text{c}_i} \in \mathbb{R}^{3\times3}$	Antisymmetric matrix of $\boldsymbol{p}_{\text{c}_i}$
$p_{i_1}, p_{i_2}, p_{i_3}$	Three components of vector $\boldsymbol{p}_{\text{c}_i}$
$I_{(\cdot)}$	Unit matrix of dimension $(\cdot)$
$\boldsymbol{K} \in \mathbb{R}^{3n\times3n}$	Symmetric and positive definite matrix
$\boldsymbol{\omega} \in \mathbb{R}^{3n}$	Parameter vector
$\boldsymbol{\lambda}_1(t) \in \mathbb{R}^n$, $\boldsymbol{\lambda}_2(t) \in \mathbb{R}^6$	Lagrange multipliers
$\eta(t) \in \mathbb{R}^{4n+6}$	Noises vector
γ, κ	Design parameters
$\boldsymbol{r}_i \in \mathbb{R}^3$	Coordinates of the contact point of the i-th dexterous hand finger with the target in the target coordinate system
e	Positive number infinitely close to 0
$\nabla(\cdot)$	Gradient symbol
$(\cdot)^{-1}$	Inverse of a matrix
$(\cdot)^\text{T}$	Transposition of matrices or vectors
$\\|(\cdot)\\|_2$	Euclidean norm of $(\cdot)$

$$\boldsymbol{G}\boldsymbol{f}(t) + \boldsymbol{f}_\text{e}(t) = 0, \tag{9.2}$$

where G is obtained by combining the transformation matrices G_i of the three fingers, and G_i for each finger is obtained as follows:

$$G_i = \begin{bmatrix} \boldsymbol{R}_{\text{c}_i} \\ \hat{\boldsymbol{p}}_{\text{c}_i}\boldsymbol{R}_{\text{c}_i} \end{bmatrix}. \tag{9.3}$$

The physical meaning of the second constraint is that the force exerted on the target by the multi-fingered dexterous hand can balance the external force on the target and maintain its motion. This chapter assumes that the multi-fingered dexterous hand is inelastic at the contact point with the target. Therefore, a hard finger model is chosen. Under this model, the value of $\hat{\boldsymbol{p}}_{c_i}$ is taken as follows:

$$\hat{\boldsymbol{p}}_{c_i} = \begin{bmatrix} 0 & -p_{i_3} & p_{i_2} \\ p_{i_3} & 0 & -p_{i_1} \\ -p_{i_2} & p_{i_1} & 0 \end{bmatrix}. \tag{9.4}$$

By determining the use of the hard finger model, the dimensions of each matrix can be determined as $G_i \in \mathbb{R}^{6\times 3}$ and $\boldsymbol{G} \in \mathbb{R}^{6\times 3m}$.

9.2.2 *Optimal Gripping Force Model*

In grasping tasks, the texture of the target varies. This makes it necessary to adjust the gripping force of the dexterous hand to the texture of the target so that no damage is caused to the target while ensuring the successful execution of the task. It means that the contact force exerted by the dexterous fingers on the target is as small as possible while meeting task requirements. Taking into account the discussions above, as well as the form-closure constraint and the force-closure constraint, an optimal gripping force model is proposed as

$$\begin{aligned} \min_{\boldsymbol{f}} \quad & \frac{1}{2}\boldsymbol{f}^{\mathrm{T}}(t)\boldsymbol{K}\boldsymbol{f}(t) + \boldsymbol{\omega}^{\mathrm{T}}\boldsymbol{f}(t), \\ \text{s.t.} \quad & \boldsymbol{G}\boldsymbol{f}(t) + \boldsymbol{f}_{\mathrm{e}}(t) = 0, \\ & \boldsymbol{\sigma}(\boldsymbol{f}(t)) \leq 0, \end{aligned} \tag{9.5}$$

where $\boldsymbol{\sigma}(\boldsymbol{f}(t)) = \left[\sigma(\boldsymbol{f}_1(t));\ \sigma(\boldsymbol{f}_2(t));\ \cdots;\ \sigma(\boldsymbol{f}_m(t))\right]$ with $\sigma(\boldsymbol{f}_i(t)) = \sqrt{\boldsymbol{f}_{i,\mathrm{x}}^2(t) + \boldsymbol{f}_{i,\mathrm{y}}^2(t)} - \mu_i \boldsymbol{f}_{i,\mathrm{z}}(t)$; $\boldsymbol{f}_{\mathrm{e}}(t)$ is calculated by Newton-Euler method Marghitu (2009) based on the desired motion trajectory of the target. Using the Lagrange multiplier method, the Lagrangian function of model (9.5) is obtained as follows:

$$\begin{aligned} \mathsf{L}(\boldsymbol{f}(t),\ \boldsymbol{\lambda}_1(t),\ \boldsymbol{\lambda}_2(t)) = & \frac{1}{2}\boldsymbol{f}^{\mathrm{T}}(t)\boldsymbol{K}\boldsymbol{f}(t) + \boldsymbol{\omega}^{\mathrm{T}}\boldsymbol{f}(t) \\ & + \boldsymbol{\lambda}_1^{\mathrm{T}}(t)\boldsymbol{\sigma}(\boldsymbol{f}(t)) \\ & - (\boldsymbol{G}\boldsymbol{f}(t) + \boldsymbol{f}_{\mathrm{e}}(t))^{\mathrm{T}}\boldsymbol{\lambda}_2(t). \end{aligned} \tag{9.6}$$

To obtain the optimal solution of this model, the Karush–Kuhn–Tucker (KKT) conditions Bi et al. (2023), Yan et al. (2025) for the above Lagrangian function are given as

$$
\begin{cases}
\boldsymbol{K}\boldsymbol{f}(t)+\boldsymbol{\omega}+\nabla\boldsymbol{\sigma}(\boldsymbol{f}(t))\boldsymbol{\lambda}_1(t)-G^{\mathrm{T}}\boldsymbol{\lambda}_2(t)=0,\\
\boldsymbol{\sigma}(\boldsymbol{f}(t))=0,\\
\boldsymbol{G}\boldsymbol{f}(t)+\boldsymbol{f}_{\mathrm{e}}(t)=0,\\
\boldsymbol{\lambda}_1(t)\geq 0,\\
\boldsymbol{\lambda}_1^{\mathrm{T}}(t)\boldsymbol{\sigma}(\boldsymbol{f}(t))=0,
\end{cases}
\tag{9.7}
$$

where $\nabla\boldsymbol{\sigma}(t)\in\mathbb{R}^{3n\times n}$ is the gradient of $\boldsymbol{\sigma}(t)$. Now assuming that $(\boldsymbol{f}^*(t),\boldsymbol{\lambda}_1^*(t),\boldsymbol{\lambda}_2^*(t))$ is the optimal solution of (9.7), the following system of equations is obtained:

$$
\begin{cases}
\boldsymbol{K}\boldsymbol{f}^*(t)+\boldsymbol{\omega}+\nabla\boldsymbol{\sigma}(\boldsymbol{f}^*(t))\boldsymbol{\lambda}_1^*(t)-\boldsymbol{G}^{\mathrm{T}}\boldsymbol{\lambda}_2^*(t)=0,\\
(\boldsymbol{\lambda}_1(t)-\boldsymbol{\lambda}_1^*(t))^{\mathrm{T}}(-\boldsymbol{\sigma}(\boldsymbol{f}^*(t)))\geq 0\quad\forall\boldsymbol{\lambda}_1(t)\geq 0,\\
\boldsymbol{G}\boldsymbol{f}^*(t)+\boldsymbol{f}_{\mathrm{e}}(t)=0.
\end{cases}
\tag{9.8}
$$

From the projection theorem Kinderlehrer and Stampacchia (2000), $\boldsymbol{\lambda}_1^*(t)$ is a solution of $(\boldsymbol{\lambda}_1(t)-\boldsymbol{\lambda}_1^*(t))^{\mathrm{T}}(-\boldsymbol{\sigma}(\boldsymbol{f}^*(t)))\geq 0$, $\boldsymbol{\lambda}_1(t)\geq 0$ only if $\Gamma(\boldsymbol{\lambda}_1^*(t)-\alpha(-\boldsymbol{\sigma}(\boldsymbol{f}^*(t))))=\boldsymbol{\lambda}_1^*(t)$, where α is positive defined. Therefore, equation (9.8) is transformed into the following form:

$$
\begin{cases}
\boldsymbol{K}\boldsymbol{f}(t)+\boldsymbol{\omega}+\nabla\boldsymbol{\sigma}(\boldsymbol{f}(t))\boldsymbol{\lambda}_1(t)-\boldsymbol{G}^{\mathrm{T}}\boldsymbol{\lambda}_2(t)=0,\\
\Gamma\big(\boldsymbol{\lambda}_1(t)+\boldsymbol{\sigma}(\boldsymbol{f}(t))\big)-\boldsymbol{\lambda}_1(t)=0,\\
\boldsymbol{G}\boldsymbol{f}(t)+\boldsymbol{f}_{\mathrm{e}}(t)=0,
\end{cases}
\tag{9.9}
$$

where $\Gamma(x)=\max(0,\ x)$. To ensure that $\boldsymbol{\sigma}(\boldsymbol{f}(t))$ converges to values less than zero rather than greater than zero, i.e., to avoid relative sliding between the fingers of the multi-fingered dexterous hand and the target, $\boldsymbol{\sigma}(\boldsymbol{f}_i(t))$ is redefined as $\boldsymbol{\sigma}(\boldsymbol{f}_i(t))=\sqrt{\boldsymbol{f}_{i,\mathrm{x}}^2(t)+\boldsymbol{f}_{i,\mathrm{y}}^2(t)}-\mu_i\boldsymbol{f}_{i,\mathrm{z}}(t)+e$, where e is a small positive number.

9.3 Scheme Design and Theoretical Analyses

In this section, a noise-tolerant dynamic neural network (NTDNN) for solving the proposed optimal gripping force model in noise-influenced scenarios is given. The convergence of the proposed solver under different noise conditions is also proven.

9.3.1 *Noise-Tolerant Dynamic Neural Network*

To simplify the expression, (9.9) is rewritten as

$$
\boldsymbol{M}\boldsymbol{x}(t)-\boldsymbol{b}(t)=0,
\tag{9.10}
$$

where

$$M = \begin{bmatrix} K & 0_{3n\times 3} & -G^{\mathrm{T}} \\ 0_{3\times 3n} & -I_{3\times 3} & 0_{3\times 6} \\ G & 0_{6\times 3} & 0_{6\times 6} \end{bmatrix}, x(t) = \begin{bmatrix} f(t) \\ \lambda_1(t) \\ \lambda_2(t) \end{bmatrix},$$

$$b(t) = \begin{bmatrix} -\omega - \nabla\sigma(f(t))\lambda_1(t) \\ -\Gamma(\lambda_1(t) + \sigma(f(t))) \\ -f_{\mathrm{e}}(t) \end{bmatrix}.$$

In matrix M, K is positive definite and G is a grasping matrix, both of which are full rank. Therefore, equation (9.10) is full rank and must have a unique solution. The error function of (9.10) is then constructed as

$$\delta(t) = Mx(t) - b(t). \tag{9.11}$$

With the defined error function $\delta(t)$, model (9.5) effectively becomes solving for the minimum value of the error function $\delta(t)$. Combined with the dynamic neural network (DNN), the derivative of the error function $\delta(t)$ is expressed as Jin et al. (2023)

$$\dot{\delta}(t) = -\gamma\delta(t). \tag{9.12}$$

The utilization of optimization methods to solve GFO problems necessitates the input of precise kinematic parameters of the target, including the target's mass, the center of mass position, and the contact point coordinates of the multi-fingered dexterous hand with the target. Nevertheless, there are inherent errors in measuring these input parameters, along with fluctuations in the motor's output torque. These aforementioned errors can be referred to as noises, leading to the inability to achieve an optimal solution to the GFO problem. Considering that noises are inevitable in the execution of a task, an integral term containing historical information is added to eliminate the effects of noises as follows:

$$\dot{\delta}(t) = -\gamma\delta(t) - \kappa\int_0^t \delta(t)\mathrm{d}t + \eta(t), \tag{9.13}$$

where $\eta(t)$ can be constant noises, linear noises, random noises, or a superposition of any of them. The definition of the error function $\delta(t)$ is then substituted into (9.13), the noise-tolerant dynamic neural network is finally expressed as

$$\begin{aligned} \dot{x}(t) =& (M - \frac{\partial b(t)}{\partial x(t)})^{-1}\big(-\gamma(Mx(t) - b(t)) \\ & - \kappa\int_0^t (Mx(t) - b(t)) + \eta(t)\big). \end{aligned} \tag{9.14}$$

Remark: To intuitively measure the deviation between the obtained grasping force and the theoretically optimal grasping force, an energy function based on (9.9) is introduced as follows:

$$
\begin{aligned}
E(\boldsymbol{f}(t),\ \boldsymbol{\lambda}_1(t),\ \boldsymbol{\lambda}_2(t)) &= \frac{1}{2}\big(\|\boldsymbol{G}\boldsymbol{f}(t) + \boldsymbol{f}_{\mathrm{e}(t)}\|_2^2 \\
&+ \|\varGamma(\boldsymbol{\lambda}_1(t) + \boldsymbol{\sigma}(\boldsymbol{f}(t))) - \boldsymbol{\lambda}_1(t)\|_2^2 \\
&+ \|K\boldsymbol{f}(t) + \boldsymbol{\omega} + \nabla\boldsymbol{\sigma}(\boldsymbol{f}(t))\boldsymbol{\lambda}_1(t) - \boldsymbol{G}^{\mathrm{T}}\boldsymbol{\lambda}_2(t)\|_2^2\big).
\end{aligned} \tag{9.15}
$$

9.3.2 Theoretical Analyses

In this section, essential theorems are given and proved to demonstrate the validity of the proposed scheme (9.5) and the convergence of the proposed noise-tolerant dynamic neural network (9.14) in eliminating the effect of noises.

Theorem 9.1 *Under the noiseless condition, in the proposed scheme (9.5), $\boldsymbol{f}(t)$ globally converges to the minimum value while satisfying the constraints. That is, under the noiseless condition, the dexterous hand can use optimal force $\boldsymbol{f}(t)$ to achieve the grasp of the target.*

Proof In the absence of noises, (9.13) can be rewritten as $\dot{\boldsymbol{\delta}}(t) = -\gamma\boldsymbol{\delta}(t) - \kappa\int_0^t \boldsymbol{\delta}(t)\mathrm{d}t$, of which the component in each dimension is expressed as follows:

$$
\dot{\boldsymbol{\delta}}_i(t) = -\gamma\boldsymbol{\delta}_i(t) - \kappa\int_0^t \boldsymbol{\delta}_i(t)\mathrm{d}t. \tag{9.16}
$$

For the i-th dimensional component of $\dot{\boldsymbol{\delta}}(t)$, the Lyapunov function is defined as

$$
v_i(t) = \boldsymbol{\delta}_i^2(t) + \kappa(\int_0^t \boldsymbol{\delta}_i(\tau)\mathrm{d}\tau)^2. \tag{9.17}
$$

It can be seen that this Lyapunov function $v_i(t)$ is positive definite, i.e., $\forall\boldsymbol{\delta}_i(t),\ \int_0^t \boldsymbol{\delta}_i(\tau)\mathrm{d}\tau : v_i(t) \geq 0$. Then, the derivative of the Lyapunov function is calculated as follows:

$$
\begin{aligned}
\frac{\mathrm{d}v_i(t)}{\mathrm{d}t} &= 2\boldsymbol{\delta}_i(t)\dot{\boldsymbol{\delta}}_i(t) + 2\kappa\boldsymbol{\delta}_i(t)\int_0^t \boldsymbol{\delta}_i(\tau)\mathrm{d}\tau \\
&= -2\gamma\boldsymbol{\delta}_i^2(t) \leq 0.
\end{aligned} \tag{9.18}
$$

Combining (9.17), (9.18), and Lyapunov theorem, it can be obtained that $\boldsymbol{\delta}_i(t)$ converges globally to 0, which means that $\boldsymbol{\delta}(t)$ converges globally to 0 and $\boldsymbol{f}(t)$ takes its optimal value at time t. This completes ends the proof. ■

However, noises are inevitable in the actual control process, so it is necessary to analyze the convergence of the proposed solver under different noise conditions. The convergence of the proposed solver in the case of linear noises and random noises is proved below.

Theorem 9.2 *Under the linear noise condition ($\boldsymbol{\eta}(t) = \boldsymbol{k}t$), in the proposed scheme (9.5), $\boldsymbol{f}(t)$ globally converges to the minimum value while satisfying the constraints. That is, under the linear noise condition ($\boldsymbol{\eta}(t) = \boldsymbol{k}t$), the dexterous hand can use optimal force $\boldsymbol{f}(t)$ to achieve the grasp of the target.*

Proof Considering that $\boldsymbol{\eta}(t)$ is a linear noise, applying a one-sided Laplace transform to the i-th dimension of (9.13) yields

$$s\psi_i(s) - \psi_i(0) = -\gamma\psi_i(s) - \frac{\kappa}{s}\psi_i(s) + \frac{\iota_i(s)}{s^2}, \tag{9.19}$$

where $\psi_i(s)$ and $\iota_i(s)$ are the Laplace transform of $\boldsymbol{\delta}_i(t)$ and $\boldsymbol{\eta}_i(t)$, respectively. Using the final value theorem on (9.19) leads to

$$\begin{aligned} \lim_{t\to\infty} \psi_i(t) &= \lim_{s\to 0} s\psi_i(s) \\ &= \lim_{s\to 0} \frac{s^2(\psi_i)(0) + \boldsymbol{k}_i/s^2}{s^2 + s\gamma + \kappa} = \frac{\boldsymbol{k}_i}{\kappa}. \end{aligned} \tag{9.20}$$

From the above equation, it follows that $\lim_{t\to\infty} \|\boldsymbol{\delta}_i(t)\|_2 = \|\boldsymbol{k}\|_2/\kappa$. That is, as long as κ is sufficiently large, $\boldsymbol{\delta}_i(t)$ is sufficiently small, and $\boldsymbol{\delta}(t)$ is sufficiently small, which means that $\boldsymbol{\delta}(t)$ converges globally to 0 and $\boldsymbol{f}(t)$ takes its optimal value at time t. This completes the proof. ■

Theorem 9.3 *Under the bounded random noise condition, in the proposed scheme (9.5), $\boldsymbol{f}(t)$ globally converges to the minimum value while satisfying the constraints. That is, under the bounded random noise condition, the dexterous hand can use optimal force $\boldsymbol{f}(t)$ to achieve the grasp of the target.*

Proof Considering $\boldsymbol{\eta}(t)$ as a bounded random noise, applying the Laplace transform to the i-th dimension of (9.13) and transforming the equation yields

$$\psi_i(s) = \mathfrak{L}(s)\iota_i(s), \tag{9.21}$$

where $\mathfrak{L}(s) = s/(s^2 + \gamma s + \kappa)$. Let $s = jw$, then $|\mathfrak{L}(jw)|^2$ is denoted as

$$|\mathfrak{L}(jw)|^2 = \frac{1}{w^2 - 2\kappa + \frac{\kappa^2}{w^2} + \gamma^2}. \tag{9.22}$$

According to the definition of the mean value inequality, it follows that $|\mathfrak{L}(jw)|^2$ takes to its maximum value $1/\gamma^2$ when $w = \kappa$, i.e., $|\mathfrak{L}(jw)|$ takes to its maximum value $1/\gamma$. It can be seen that the value of γ is taken to be inversely proportional to $\mathfrak{L}(s)$. Combining the above derivation with (9.21), the relationship between $\boldsymbol{\delta}_i(t)$ and $\boldsymbol{\eta}_i(t)$ is expressed as follows:

$$\boldsymbol{\delta}_i(t) = \int \boldsymbol{\eta}_i(t-\tau)\mathfrak{L}(\tau)\mathrm{d}\tau. \tag{9.23}$$

The max value of $\boldsymbol{\delta}_i(t)$ is expressed as

$$\begin{aligned} \max|\boldsymbol{\delta}_i(t)| &= \max\left|\int \boldsymbol{\eta}_i(t-\tau)\mathfrak{L}(\tau)\mathrm{d}\tau\right| \\ &\le \boldsymbol{\eta}^+\left|\max\int|\mathfrak{L}(\tau)\mathrm{d}\tau|\right|, \end{aligned} \tag{9.24}$$

where $\boldsymbol{\eta}^+$ is the upper bound of $\boldsymbol{\eta}(t)$. From the above derivation, it is clear that $\mathfrak{L}(s)$ imposes a constraint on $\boldsymbol{\eta}(t)$. Therefore, when γ is large enough, $\boldsymbol{\delta}(t)$ converges globally to 0 and $\boldsymbol{f}(t)$ takes its optimal value. This completes the proof. ■

9.4 Experiments and Comparisons

In this section, the validity of the proposed NTGFO scheme is verified through simulations carried out on the MATLAB platform and the CoppeliaSim platform, and physical experiments carried out on the Franka Emika Panda robot equipped with a three-fingered dexterous hand. Comparative experiments with other schemes are also carried out to illustrate the superiority of the proposed NTGFO scheme.

9.4.1 *Simulations and Experiments*

Both simulations and experiments are based on a Franka Emika Panda robot and a DH-Robotics DH-3 three-fingered dexterous hand. The simulation task is designed to rotate a square target of mass m and side length r_o through 360° using the Franka Emika Panda robot fitted with the three-fingered dexterous hand. Task execution time is 5 s. The noise $\boldsymbol{\eta}(t)$ is designed as a linear noise, which takes the value $20 \times t \times \mathrm{ones}(18, 1)$. The parameter settings are shown in Table 9.2. By determining the coordinates of contact points of the multi-fingered dexterous hand with the target $\boldsymbol{r}_1$, $\boldsymbol{r}_2$, and $\boldsymbol{r}_3$, the grasp transformation matrix $\boldsymbol{G}$ is expressed as

$$\boldsymbol{G} = \begin{bmatrix} 0 & 1 & 0 & 0 & 0 & -1 & 1 & 0 & 0 \\ 0 & 0 & 1 & 1 & 0 & 0 & 0 & 0 & -1 \\ 1 & 0 & 0 & 0 & -1 & 0 & 0 & 1 & 0 \\ -1 & 0 & 0 & 0 & 0 & 0 & 0 & 1 & 0 \\ 0 & 0 & 0 & 0 & 1 & 0 & 0 & 0 & 0 \\ 0 & 1 & 0 & 1 & 0 & 0 & -1 & 0 & 0 \end{bmatrix}. \tag{9.25}$$

The simulation results are shown in Figs. 9.1, 9.2, and 9.3. Figure 9.1a illustrates the grasping forces obtained by the NTGFO scheme. Besides, Fig. 9.1b represents the error between the Z-axis grasping force obtained by the NTGFO

Table 9.2 Parameter settings

n	3
μ_i	0.6
r_o	0.1 m
m	0.07 kg
$\boldsymbol{r}_1$	(-0.05, 0.00, 0.00) m
$\boldsymbol{r}_2$	(0.00, 0.05, 0.00) m
$\boldsymbol{r}_3$	(0.05, 0.00, 0.00) m
γ	10^4
κ	10^4
$\eta(t)$	$20 \times t \times \text{ones}(18, 1)$
$\boldsymbol{K}$	$I_{3n \times 3n}$
ω	$\mathbf{0}_{3n}$

scheme and the desired Z-axis grasping force. Figure 9.1c shows the energy change $E(\boldsymbol{f}(t), \boldsymbol{\lambda}_1(t), \boldsymbol{\lambda}_2(t))$ of the NTGFO scheme during the task.

To verify the feasibility of the proposed scheme, a simulation on the CoppeliaSim platform and a physical experiment are also executed with the same parameter settings as in the simulation. The model of the three-fingered dexterous hand employed in both the simulation and experiment are depicted in Fig. 9.4. Each finger of this three-fingered dexterous hand consists of three joints and a force sensor. The results of the experiment are shown in Fig. 9.5. Both the results of the simulation and physical experiment show that the proposed NTGFO scheme can grasp the target with minimum force while keeping the target and the three-fingered dexterous hand from slipping relative to each other. To verify the generalisability of the proposed NTGFO scheme, a 3-second long experiment in which a multi-fingered dexterous hand grasps a target in circular motion is also carried out, the results of the experiment are shown in Fig. 9.6.

9.4.2 Comparisons

To illustrate the superiority of the proposed scheme, its comparisons with other schemes are shown in Table 9.3. On the one hand, it can be observed that traditional control models such as PID control and sliding mode control do not perform well in the face of nonlinear systems and dynamic changes due to their nature of non-mathematical modeling. Reinforcement learning models, on the other hand, are commonly used for controlling the gripping posture, with less research on controlling the gripping force. Moreover, they require training before applications and perform poorly in the face of untrained targets. General optimization schemes, however, usually do not consider the interference of noises and perform poorly in the face of noise. Compared with the above control schemes, the proposed NTGFO scheme can han-

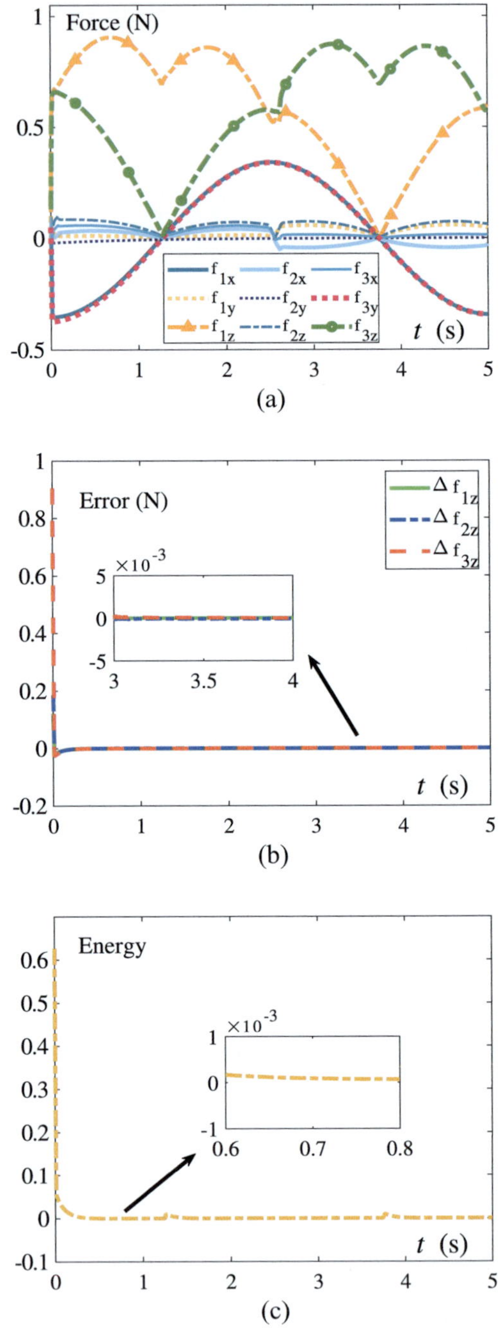

Fig. 9.1 Simulation results of rotating a square target grasped by a three-finger dexterous hand using the proposed NTGFO scheme. **a** Gripping forces. **b** The error between the Z-axis grasping force and the desired Z-axis grasping force. **c** Energy change during the task

Fig. 9.2 Simulation results of rotating a square target grasped by a three-finger dexterous hand using the scheme in Xia et al. (2004). **a** Gripping forces. **b** The error between the Z-axis grasping force and the desired Z-axis grasping force. **c** Energy change during the task

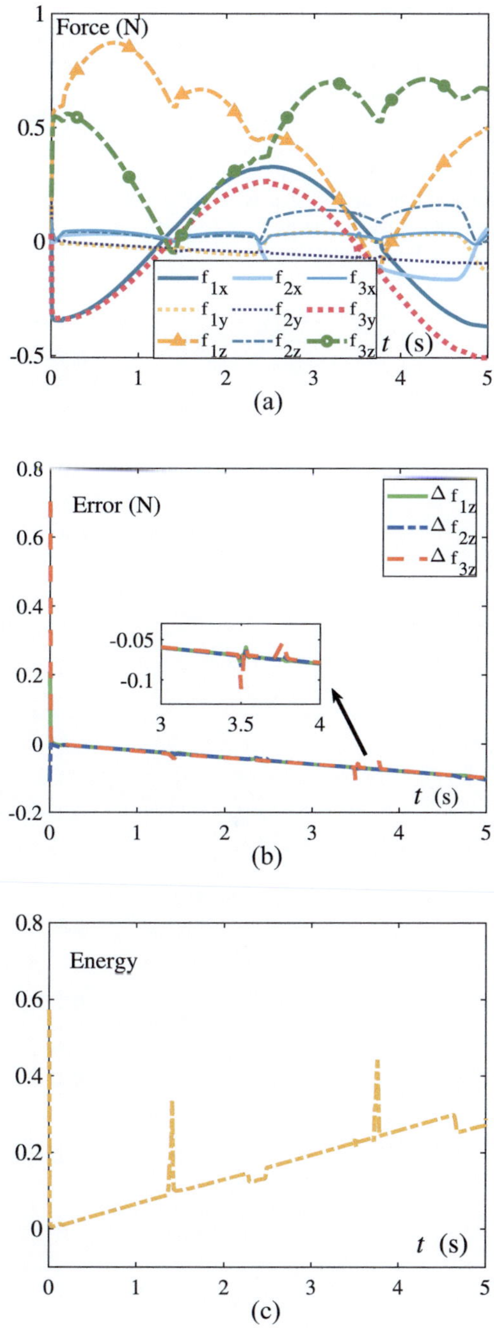

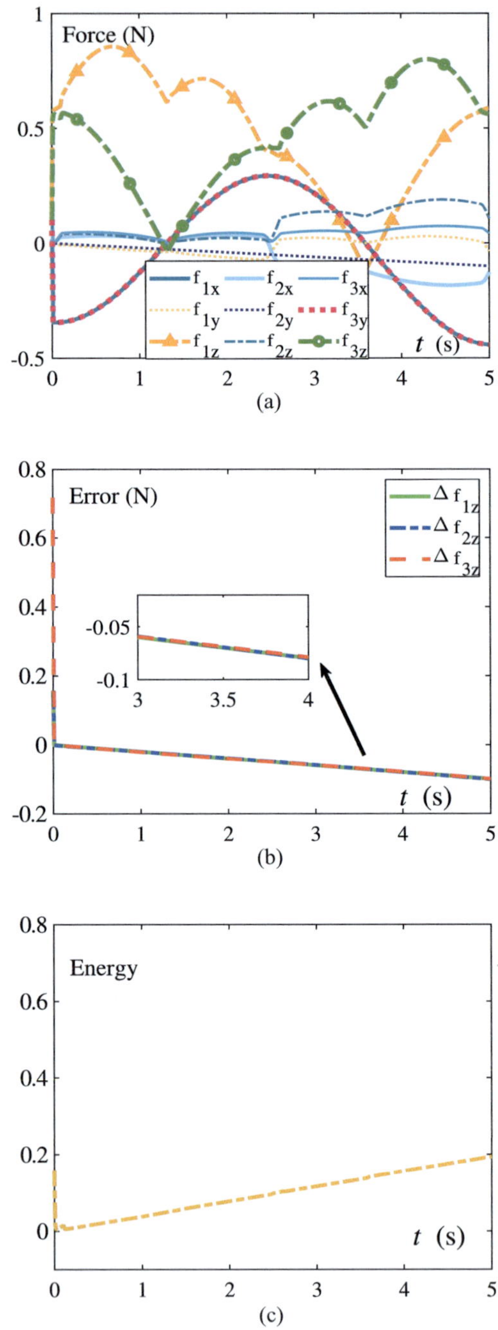

Fig. 9.3 Simulation results of rotating a square target grasped by a three-finger dexterous hand using the scheme in Nazemi (2019). **a** Gripping forces. **b** The error between the Z-axis grasping force and the desired Z-axis grasping force. **c** Energy change during the task

Fig. 9.4 Overall view, front view, and top view of a DH-Robotics DH-3 three-fingered dexterous hand

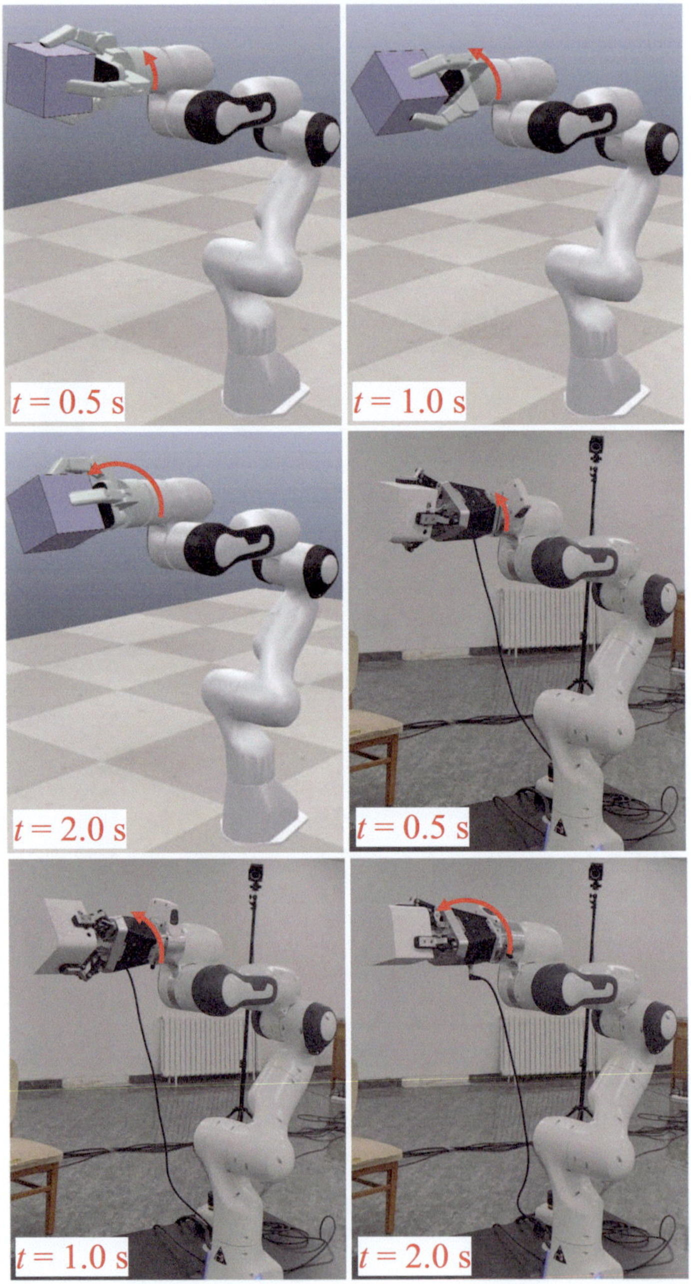

Fig. 9.5 Simulation and experiment snapshots of a rotation task of grasping a square target, performed by a Franka Panda robot fitted with a DH-Robotics DH-3 three-fingered dexterous hand

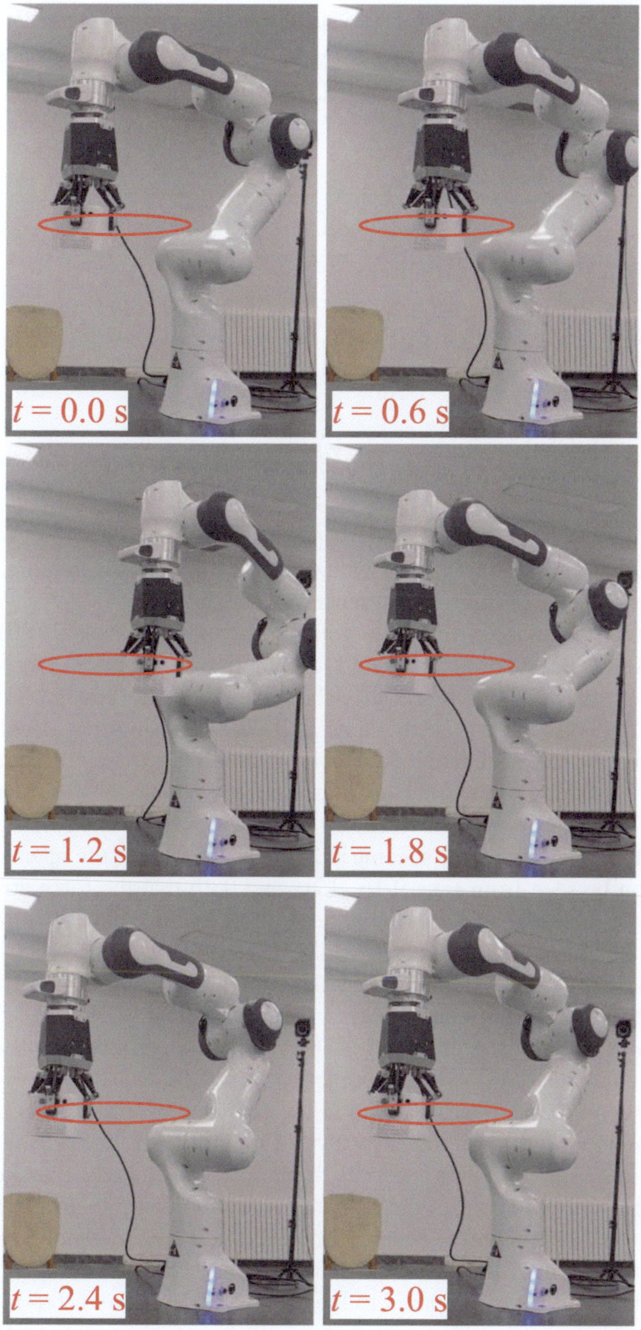

Fig. 9.6 Experiment snapshots of a circular motion task of grasping a square target, performed by a Franka Panda robot fitted with a DH-Robotics DH-3 three-fingered dexterous hand

Table 9.3 Comparisons among different optimal gripping force schemes

Scheme	Approach to control	Mathematical model based	Training required	Noise tolerance
Proposed	Optimization	Yes	No	Yes
Scheme in Liu et al. (2023a)	Reinforcement learning	Yes	Yes	No
Scheme in Nazemi (2019)	Optimization	Yes	No	No
Scheme in Xia et al. (2004)	Optimization	Yes	No	No
Scheme in Zhang and Xu (2017)	Sliding mode control	No	No	No
Scheme in Xu (2015)	Proportional integral differential	No	No	No

dle nonlinear systems and dynamics well, without the need for training, and shows robustness in the face of noise interference.

Besides, comparative experiments with the scheme in Nazemi (2019) and the scheme in Xia et al. (2004) are executed for quantitative analyses. The convergence parameters of the NTGFO scheme, the scheme in Xia et al. (2004), and the scheme in Nazemi (2019) are taken to be the same as 1000. The results are shown in Figs. 9.2 and 9.3, where Figs. 9.2a–c and 9.3a–c correspond to Fig. 9.1a–c, respectively, representing the grasping force, force error, and energy obtained by the scheme in Xia et al. (2004) and the scheme in Nazemi (2019). Comparing the results of the proposed NTGFO scheme with the scheme in Xia et al. (2004) and the scheme in Nazemi (2019), it can be seen from Figs. 9.1b, 9.2b, and 9.3b that the force error obtained by the scheme in Xia et al. (2004) and the scheme in Nazemi (2019) increase over time due to the interference of noises, while the force error of the proposed NTGFO scheme converges to 0. It demonstrates the superior performance of the proposed NTGFO scheme in the presence of noise. From Figs. 9.1c, 9.2c, and 9.3c, it can be seen that the energy of the proposed NTGFO scheme converges to 0 because the force error converges to 0, while the energy of schemes in Xia et al. (2004) and Nazemi (2019) increase over time due to the presence of force errors. In addition, it can be observed that when the derivative of the grasping force changes abruptly, the scheme in Xia et al. (2004) due to its limitation in solving the dynamic problem, it shows the inability to control the grasping force and energy during the execution of the task well. Although the scheme in Nazemi (2019) does not have an abrupt change in the grasping force error and energy when there is an abrupt change in the grasping force derivative compared to the scheme in Xia et al. (2004), its grasping energy cannot be minimized because it does not take into account the effect of noise.

In contrast, the proposed NTGFO scheme can cope with the dynamic problem well, i.e., it can still control the grasping force and energy well when the derivative of the grasping force changes abruptly. It also minimizes the grasping energy even under the influence of noise. This demonstrates the superiority of the proposed NTGFO scheme.

9.5 Conclusions

This chapter has proposed a noise-tolerant grasping force optimization (NTGFO) scheme to solve the grasping force optimization (GFO) problem in noisy environments. The scheme consists of an optimal grasping force model and a noise-tolerant dynamic neural network (NTDNN). The proposed NTGFO scheme is able to control the multi-fingered dexterous hand to grasp the target with optimal grasping force under noises and without training. In summary, this chapter has provided an effective grasping force control scheme for grasping tasks with increasingly diverse grasping targets and scenarios so that the multi-fingered dexterous hand is able to apply optimal grasping force to the target in complex scenarios. In future works, this scheme will be combined with multi-finger dexterous hand grasping attitude control models, thus realizing the optimal force and the optimal grasping attitude to grasp the target to increase the success rate of grasping.

References

Bi F, Luo X, Shen B, Dong H, Wang Z (2023) Proximal alternating-direction-method-of-multipliers-incorporated nonnegative latent factor analysis. IEEE/CAA J Autom Sin 10(6):1388–1406

Buss M, Hashimoto H, Moore JB (1996) Dextrous hand grasping force optimization. IEEE Trans Robot Autom 12(3):406–418

Chen Z, Wu Q, Hong C, Zhang X (2021) Multi-fingered grasping force optimization based on generalized penalty-function concepts. Robot Auton Syst 135:103672

Gazeau JP, Zeghloul S, Ramirez G (2005) Manipulation with a polyarticulated mechanical hand: a new efficient real-time method for computing fingertip forces for a global manipulation strategy. Robotica 23(4):479–490

Hasegawa T, Waita H, Kawakami T, Takemura Y, Ishikawa T, Kimura Y (2022) Powerful and dexterous multi-finger hand using dynamical pulley mechanism. In: IEEE international conference on robotics and automation (ICRA), pp 707–713

Heydarabad SM, Milella F, Davis S, Nefiti-Meziani S (2017) High-performing adaptive grasp for a robotic gripper using super twisting sliding mode control. In: IEEE international conference on robotics and automation (ICRA), pp 5852–5857

Hu Z, Zheng Y, Pan J (2023) Grasping living objects with adversarial behaviors using inverse reinforcement learning. IEEE Trans Robot 39(2):1151–1163

Jin L, Wei L, Li S (2023) Gradient-based differential neural-solution to time-dependent nonlinear optimization. IEEE Trans Autom Control 68(1):620–627

Jin L, Su Z, Fu D, Xiao X (2024) Coevolutionary neural solution for nonconvex optimization with noise tolerance. IEEE Trans Neural Netw Learn Syst 35(12):17571–17581

Kinderlehrer D, Stampacchia G (2000) An introduction to variational inequalities and their applications. In: Society for industrial and applied mathematics (SIAM)

Klein CA, Kittivatcharapong S (1990) Optimal force distribution for the legs of a walking machine with friction cone constraints. IEEE Trans Robot Autom 6(1):73–85

Lian Y, Xiao X, Zhang J, Jin L, Yu J, Sun Z (2024) Neural dynamics for cooperative motion control of omnidirectional mobile manipulators in the presence of noises: a distributed approach. IEEE/CAA J Autom Sin 11(7):1605–1620

Liu GP (2023) Tracking control of multi-agent systems using a networked predictive PID tracking scheme. IEEE/CAA J Autom Sin 10(1):216–225

Liu G, Xu J, Li Z (2004) On geometric algorithms for real-time grasping force optimization. IEEE Trans Control Syst Technol 12(6):843–859

Liu F, Sun F, Fang B, Li X, Sun S, Liu H (2023a) Hybrid robotic grasping with a soft multimodal gripper and a deep multistage learning scheme. IEEE Trans Robot 39(3):2379–2399

Liu J, Sun W, Liu C, Zhang X, Fu Q (2023b) Robotic continuous grasping system by shape transformer-guided multiobject category-level 6-D pose estimation. IEEE Trans Ind Inform 19(11):11171–11181

Lobo MS, Vandenberghe L, Boyd S, Lebret H (1998) Applications of second-order cone programming. Linear Algebr Appl 284(1–3):193–228

Maeno T, Hiromitsu S, Kawai T (2000) Control of grasping force by detecting stick/slip distribution at the curved surface of an elastic finger. In: IEEE international conference on robotics and automation (ICRA), pp 3895–3900

Marghitu D B (2009) Direct dynamics: Newton–Euler equations of motion. In: Mechanisms and robots analysis with MATLAB, pp 183–207

Matsuura Y, Hiraoka N, Kojima K, Yanokura I, Yoshioka H, Okada K (2022) Development of a multi-fingered hand with a multi-step locking mechanism for carrying heavy objects by a humanoid robot. In: IEEE-RAS international conference on humanoid robots (ICHR), pp 794–800

Nanayakkara T, Jiang A, Fernandez MDRA, Liu H, Althoefer K, Bimbo J (2016) Stable grip control on soft objects with time-varying stiffness. IEEE Trans Robot 32(3):626–637

Nazemi A (2019) A new collaborate neuro-dynamic framework for solving convex second order cone programming problems with an application in multi-fingered robotic hands. Appl Intell 49(10):3512–3523

Peng XB, Andrychowicz M, Zaremba W, Abbeel P (2018) Sim-to-real transfer of robotic control with dynamics randomization. In: IEEE international conference on robotics and automation (ICRA), pp 3803–3810

Wang D, Gao N, Liu D, Li J, Lewis FL (2024) Recent progress in reinforcement learning and adaptive dynamic programming for advanced control applications. IEEE/CAA J Autom Sin 11(1):18–36

Xia Y, Wang J, Fok LM (2004) Grasping-force optimization for multifingered robotic hands using a recurrent neural network. IEEE Trans Robot Autom 20(3):549–554

Xie Z, Jin L (2022) Hybrid control of orientation and position for redundant manipulators using neural network. IEEE Trans Syst Man Cybern Syst 53(5):2737–2747

Xie Z, Jin L, Luo X, Hu B, Li S (2021) An acceleration-level data-driven repetitive motion planning scheme for kinematic control of robots with unknown structure. IEEE Trans Syst Man Cybern Syst 52(9):5679–5691

Xu Q (2015) Robust impedance control of a compliant microgripper for high-speed position/force regulation. IEEE Trans Ind Electron 62(2):1201–1209

Yan J, Su Z, Ma X, Jin L (2025) An obstacle avoidance scheme for manipulators aided by noise-tolerant neural dynamics. IEEE Trans Ind Inform 21(1):703–712

Yang Y, Ni Z, Gao M, Zhang J, Tao D (2022) Collaborative pushing and grasping of tightly stacked objects via deep reinforcement learning. IEEE/CAA J Autom Sin 9(1):135–145

Zhang Y, Xu Q (2017) Adaptive sliding mode control with parameter estimation and Kalman filter for precision motion control of a piezo-driven microgripper. IEEE Trans Control Syst Technol 25(2):728–735
Zheng Y, Lin MC, Manocha D (2012) On computing reliable optimal grasping forces. IEEE Trans Robot 28(3):619–633

Chapter 10
Data-Driven RC^{2}M Control With DNN

10.1 Introduction

Robot manipulators are extensively utilized across various domains, particularly redundant manipulators with additional DOFs Zhang et al. (2023), Guo et al. (2022), Liu et al. (2024). The extra DOFs of redundant manipulators provide high flexibility for remote center of motion (RCM) tasks Kastritsi and Doulgeri (2021), Wang and Tan (2023), Lu et al. (2022). RCM can be regarded as a particular case of operations with space restricted Zhang et al. (2022), Yan et al. (2024), Saeidi et al. (2022). Derived from practical applications, the RCM characterizes a scenario where a tool passes through a fixed point distinct from the operating point Gao et al. (2021), Jin et al. (2025), Chen et al. (2021). Specifically, the manipulator is required to execute the RCM task through a tiny hole on the boundary between the exterior and interior spaces, with the hole's position serving as the RCM point. Diverse motion planning schemes provide the basis for redundant manipulators to execute RCM tasks Su et al. (2021), Wang et al. (2023), Chen et al. (2024). Typical RCM applications include manipulator-assisted surgery and robotic nasal sampling, characterized by restricted operating spaces.

Research into RCM-featured manipulator control and associated methods continues to grow rapidly. A surgical endoscope system incorporating visual servoing and RCM constraints is solved by a finite-time convergent neural network Li et al. (2020), utilizing commercial Universal Robots and a cable-driven endoscope. Moreover, a dual neural network (DNN) controller is utilized for endoscopic robots operating under the RCM condition Cui et al. (2022). To optimize the configurations of the surgical instrument's placement, a framework is implemented to meet RCM and additional constraints Cursi et al. (2022). During the COVID-19 epidemic, swab sampling serves as an essential prerequisite for nucleic acid testing while exposing healthcare workers to potential infection risks. Considering the characteristics of the sampling motion, an RCM-compliant remote robotic nasal sampling system is developed with position and orientation control Takeuchi et al. (2022).

M. Liu et al., *Dynamic Neural Networks for Motion Control of Redundant Manipulators*,
Intelligent Control and Learning Systems 21,
https://doi.org/10.1007/978-981-96-9144-9_10

Data-driven technology offers a viable solution due to its strong capability to handle problems with unknown or unclear structural information Yang et al. (2023), Yao et al. (2024), Fotiadis et al. (2024). The valid information extracted from data is utilized for the manipulator control, with the manipulator-environment interaction subsequently influencing new data generation. A model-unknown control method is developed in Khan et al. (2021), enabling simultaneous learning and control through data-driven operation. Data-driven sparse motion planning is also applied to redundant manipulators Jin et al. (2023). Additionally, a data-driven approach resolves a motion-force control problem in Fan et al. (2022), while a framework enables optimal endoscope view control through data-driven methods in Li et al. (2021). By leveraging optimization methods for quadratic programming problems, data-driven technologies can be further enhanced. Building upon prior works, this chapter applies data-driven techniques to estimate the modified manipulator's Jacobian matrix using end-effector position and velocity data, effectively addressing inaccuracies in the attached end-effector's structural information.

The rest of the chapter is organized as follows: Sect. 10.2 presents the forward kinematics of the redundant manipulator, the foundational principles of this chapter's data-driven methodology, and the RCM formulation. Section 10.3 formulates the problem and proposes a remote center of cyclic motion (RC^{2}M) scheme which is solved via a DNN model, along with theoretical analyses of its learning and control capabilities. Then, simulative and practical experiments are conducted to demonstrate the effectiveness of the constructed scheme, with comparative results provided in Sect. 10.4. Section 10.5 draws a conclusion. Main contributions of this chapter are summarized here.

- An RC^{2}M scheme which incorporates data-driven technology with a cyclic motion index to ensure posture recovery is constructed in this chapter, enabling system operation despite inaccuracies in the end-effector's structural information.
- A DNN model is developed to address the RC^{2}M scheme, with simulation and experiment results demonstrating the constructed scheme's superior performance over conventional approaches.

10.2 Preliminaries

This section presents fundamental explanations of redundant manipulator forward kinematics, data-driven technology, and RCM characteristics as essential preliminaries.

10.2.1 Forward Kinematics and Data-Driven Technology

Based on the forward kinematics of the redundant manipulator, the mapping relationship from the joint space $\boldsymbol{\theta} \in \mathbb{R}^n$ to the Cartesian space $\boldsymbol{r} \in \mathbb{R}^m$ $(m < n)$ is analogous to (8.2) and can be expressed as $\Psi(\boldsymbol{\theta}) = \boldsymbol{r}$, where Ψ represents a nonlinear transformation. By differentiating the preceding equation, an affine system is obtained to facilitate the redundancy resolution, expressed as $J(\boldsymbol{\theta})\dot{\boldsymbol{\theta}} = \dot{\boldsymbol{r}}$, where $J(\boldsymbol{\theta}) = \partial\Psi(\boldsymbol{\theta})/\partial\boldsymbol{\theta} \in \mathbb{R}^{n\times m}$ is the Jacobian matrix; $\dot{\boldsymbol{\theta}}$ is the velocity of the joint and $\dot{\boldsymbol{r}}$ is that of the end-effector.

To address errors induced by the attached end-effector, data-driven technology is employed to achieve control of the redundant manipulator despite its imprecise structural information. In $\widetilde{J}\dot{\boldsymbol{\theta}} = \bar{\dot{\boldsymbol{r}}}$, $\widetilde{J}$ denotes the estimated value of the Jacobian matrix, and $\bar{\dot{\boldsymbol{r}}}$ denotes the estimated velocity of the end-effector. $\bar{\dot{\boldsymbol{r}}}$ approaches $\dot{\boldsymbol{r}}$ once $\widetilde{J}$ is close to J. Then, $\chi = \|\bar{\dot{\boldsymbol{r}}} - \dot{\boldsymbol{r}}\|_2^2/2$ is used to minimize the estimation error of the Jacobian matrix, where $\|\cdot\|_2$ means the Euclidean norm of a vector. Differentiating χ with respect to $\widetilde{J}$ yields

$$\varrho\dot{\widetilde{J}} = -\frac{\partial\chi}{\partial\widetilde{J}} = (\dot{\boldsymbol{r}} - \widetilde{J}\dot{\boldsymbol{\theta}})\dot{\boldsymbol{\theta}}^{\mathrm{T}}, \tag{10.1}$$

where $\varrho > 0$ is a tiny value denoting the step length, $(\cdot)^{\mathrm{T}}$ is the transpose of a vector or a matrix, and $\dot{\widetilde{J}}$ is the derivative of $\widetilde{J}$. Due to the absence of J in expression (10.1), a redundant manipulator with inaccurate structural information can be controlled by obtaining the current values of $\dot{\boldsymbol{r}}$ and $\dot{\boldsymbol{\theta}}$.

10.2.2 Remote Center of Motion

In order to describe the principle of RCM, two vital points, A and B on the last link of a redundant manipulator are marked, as shown in the last subgraph of Fig. 10.1. Variable $\boldsymbol{r}_{\mathrm{A}} \in \mathbb{R}^m$ means the position of the end-effector abstracted as a point in the world coordinate, and $\boldsymbol{r}_{\mathrm{B}} \in \mathbb{R}^m$ is for the position of the last joint. In other words, the part between $\boldsymbol{r}_{\mathrm{A}}$ and $\boldsymbol{r}_{\mathrm{B}}$ represents the last link of the redundant manipulator, which shuttles through the RCM point. To describe the kinematic relation between A and B, velocities of them are expressed with the forward kinematics as $\dot{\boldsymbol{r}}_{\mathrm{A}} = J_{\mathrm{A}}\dot{\boldsymbol{\theta}}$ and $\dot{\boldsymbol{r}}_{\mathrm{B}} = J_{\mathrm{B}}\dot{\boldsymbol{\theta}}$, where $J_{\mathrm{A}} \in \mathbb{R}^{m\times n}$ and $J_{\mathrm{B}} \in \mathbb{R}^{m\times(n-1)}$ denote the Jacobian matrix of the end-effector and the $n-1$ link, respectively. According to the definition of RCM, the link AB is supposed to go through the RCM point (point C) consistently. There exists a relationship between the link AB and point C naturally, expressed as

$$\boldsymbol{r}_{\mathrm{C}} - \boldsymbol{r}_{\mathrm{B}} = k(\boldsymbol{r}_{\mathrm{A}} - \boldsymbol{r}_{\mathrm{B}}), \tag{10.2}$$

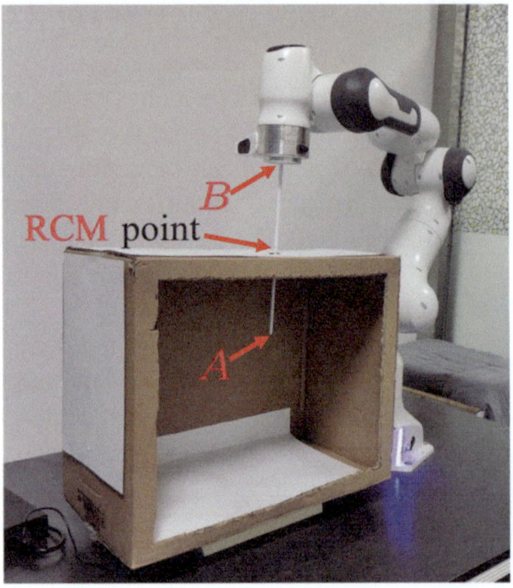

Fig. 10.1 Schematic diagram of RCM

where k denotes a parameter adjusting the position of point C on link AB, ranging from 0 to 1. Considering two edge cases: When k is equal to 0, the link AB has crossed to point C with joint B working as the RCM point; when k is equal to 1, the motion mode degrades to self motion of the end-effector.

10.3 Problem Clarification and Scheme Construction

In this section, an RC^{2}M optimization problem is formulated from the kinematic perspective. Subsequently, a DNN model is developed using a data-driven technology to further enhance processing accuracy. Additionally, necessary explanations regarding the validity of the constructed scheme are provided.

10.3.1 Problem Formulation

Based on the above preliminaries, the position and orientation information of a redundant manipulator is determined by the joint angle $\boldsymbol{\theta}$. The connection between the redundant manipulator and the RCM point is denoted by k, which indicates the RCM point on the specific link of the redundant manipulator's links as $\boldsymbol{r}_{\mathrm{C}} - \boldsymbol{r}_{\mathrm{B}} = k(\boldsymbol{r}_{\mathrm{A}} - \boldsymbol{r}_{\mathrm{B}})$. Differentiating the equation obtains:

$$\dot{\boldsymbol{r}}_{\mathrm{C}} - \dot{\boldsymbol{r}}_{\mathrm{B}} = \dot{k}(\boldsymbol{r}_{\mathrm{A}} - \boldsymbol{r}_{\mathrm{B}}) + k(\dot{\boldsymbol{r}}_{\mathrm{A}} - \dot{\boldsymbol{r}}_{\mathrm{B}}), \tag{10.3}$$

which can be generated:

$$\dot{\boldsymbol{r}}_{\mathrm{C}} - J_{\mathrm{B}}\dot{\boldsymbol{\theta}} = \dot{k}(\boldsymbol{r}_{\mathrm{A}} - \boldsymbol{r}_{\mathrm{B}}) + k(J_{\mathrm{A}} - J_{\mathrm{B}})\dot{\boldsymbol{\theta}}, \tag{10.4}$$

where $\dot{k}$ is the derivative of k. So far, the RCM constraint at velocity level is acquired as

$$\dot{k}(\boldsymbol{r}_{\mathrm{A}} - \boldsymbol{r}_{\mathrm{B}}) + (kJ_{\mathrm{A}} - (1-k)J_{\mathrm{B}})\dot{\boldsymbol{\theta}} - \dot{\boldsymbol{r}}_{\mathrm{C}} = 0. \tag{10.5}$$

The joint-drift problem cannot be ignored to achieve the repetitive motion of a redundant robot. Particularly, readjusting the redundant manipulator's position and orientation to the initial state becomes difficult under the RCM constraint. Therefore, the typical RCM control scheme is augmented through the addition of a cyclic motion index. The corresponding optimization problem for RC^{2}M is formulated as follows:

$$\begin{aligned} \arg\min_{\dot{\boldsymbol{\theta}},\dot{k}}\ & \frac{1}{2}\dot{\boldsymbol{\theta}}^{\mathrm{T}}\dot{\boldsymbol{\theta}} + \frac{1}{2}m_2\dot{k}^2 + \gamma(\boldsymbol{\theta} - \boldsymbol{\theta}_0)^{\mathrm{T}}\dot{\boldsymbol{\theta}} \\ \text{s.t. } & \dot{k}(\boldsymbol{r}_{\mathrm{A}} - \boldsymbol{r}_{\mathrm{B}}) + (kJ_{\mathrm{A}} + (1-k)J_{\mathrm{B}})\dot{\boldsymbol{\theta}} - \dot{\boldsymbol{r}}_{\mathrm{C}} = 0 \\ & J_{\mathrm{A}}\dot{\boldsymbol{\theta}} - \dot{\boldsymbol{r}}_{\mathrm{d}} + m_3(\boldsymbol{r}_{\mathrm{A}} - \boldsymbol{r}_{\mathrm{d}}) = 0, \end{aligned} \tag{10.6}$$

where γ denotes the parameter of the cyclic motion index, and $\gamma(\boldsymbol{\theta} - \boldsymbol{\theta}_0)^{\mathrm{T}}\dot{\boldsymbol{\theta}}$ is used to guide the joint angles of the redundant manipulator to recover to their initial states; m_2, m_3 represent two positive scaling coefficients which influence the convergence performance; $\boldsymbol{r}_{\mathrm{d}}$ and $\dot{\boldsymbol{r}}_{\mathrm{d}}$ denote the position and velocity of the desired trajectory. $J_{\mathrm{A}}\dot{\boldsymbol{\theta}} - \dot{\boldsymbol{r}}_{\mathrm{d}}$ keeps the velocity of the end-effector with the expectation, and $\boldsymbol{r}_{\mathrm{A}} - \boldsymbol{r}_{\mathrm{d}}$ works as a feedback to guarantee the tracking accuracy. The point C on the extension of segment AB is expected to remain stationary to avoid needless collision during the task. This is reflected in $m_2\dot{k}^2/2$ of the objective function, which aims to minimize the change of the RCM indicator k, and $\dot{\boldsymbol{r}}_{\mathrm{C}}$ in the equality constraint should be formulated as a convergent expression to minimize positional deviation of the RCM point:

$$\dot{\boldsymbol{r}}_{\mathrm{C}} = -m_1(\boldsymbol{r}_{\mathrm{C}} - \boldsymbol{r}_{\mathrm{C}}(0)), \tag{10.7}$$

where m_1 is a positive parameter to control the convergence of $\boldsymbol{r}_{\mathrm{C}}$.

10.3.2 DNN-Based Data-Driven Resolution

Before deriving the redundancy resolution for the stated control scheme, the scheme integrates data-driven technology (10.1) to address deviations arising from imprecise end-effector structural information. The data-driven RC^{2}M optimization problem is further expressed as

$$
\begin{aligned}
\min_{\dot{\boldsymbol{\theta}},\,\dot{k}}\ & \frac{1}{2}\dot{\boldsymbol{\theta}}^{\mathrm{T}}\dot{\boldsymbol{\theta}} + \frac{1}{2}m_2\dot{k}^2 + \gamma(\boldsymbol{\theta}-\boldsymbol{\theta}_0)^{\mathrm{T}}\dot{\boldsymbol{\theta}} \\
\text{s.t. } & \dot{k}(\boldsymbol{r}_{\mathrm{A}}-\boldsymbol{r}_{\mathrm{B}}) + (k\widetilde{J}_{\mathrm{A}} + (1-k)J_{\mathrm{B}})\dot{\boldsymbol{\theta}} - \dot{\boldsymbol{r}}_{\mathrm{C}} = 0 \\
& \widetilde{J}_{\mathrm{A}}\dot{\boldsymbol{\theta}} - \dot{\boldsymbol{r}}_{\mathrm{d}} + m_3(\boldsymbol{r}_{\mathrm{A}}-\boldsymbol{r}_{\mathrm{d}}) = 0 \\
& \varrho\dot{\widetilde{J}}_{\mathrm{A}} = (\dot{\boldsymbol{r}}_{\mathrm{A}} - \widetilde{J}_{\mathrm{A}}\dot{\boldsymbol{\theta}})\dot{\boldsymbol{\theta}}^{\mathrm{T}}.
\end{aligned}
\tag{10.8}
$$

It is worth pointing out that the data-driven technology is utilized to estimate the Jacobian matrix J_{A} of the end of the extra link, while the structural information of the original redundant manipulator J_{B} is already known.

As the RC^{2}M control scheme (10.8) constructed, a DNN-based data-driven resolution for the scheme is presented with a Lagrange function established firstly:

$$
\begin{aligned}
\mathfrak{L} =& \frac{1}{2}\dot{\boldsymbol{\theta}}^{\mathrm{T}}\dot{\boldsymbol{\theta}} + \frac{1}{2}m_2\dot{k}^2 + \gamma(\boldsymbol{\theta}-\boldsymbol{\theta}_0)^{\mathrm{T}}\dot{\boldsymbol{\theta}} + \boldsymbol{\lambda}_1^{\mathrm{T}}(\dot{k}(\boldsymbol{r}_{\mathrm{A}}-\boldsymbol{r}_{\mathrm{B}}) + (k\widetilde{J}_{\mathrm{A}} + (1-k)J_{\mathrm{B}})\dot{\boldsymbol{\theta}} \\
&+ m_1(\boldsymbol{r}_{\mathrm{C}} - \boldsymbol{r}_{\mathrm{C}}(0))) + \boldsymbol{\lambda}_1^{\mathrm{T}}(\widetilde{J}_{\mathrm{A}}\dot{\boldsymbol{\theta}} - \dot{\boldsymbol{r}}_{\mathrm{d}} + m_3(\boldsymbol{r}_{\mathrm{A}}-\boldsymbol{r}_{\mathrm{d}})),
\end{aligned}
\tag{10.9}
$$

where $\boldsymbol{\lambda}_1 \in \mathbb{R}^m$ and $\boldsymbol{\lambda}_2 \in \mathbb{R}^m$ represent the Lagrange multiplier vectors for two corresponding equality constraints. Then, taking partial derivatives to $\mathfrak{L}$ with respect to $\dot{\boldsymbol{\theta}}, \dot{k}, \boldsymbol{\lambda}_1, \boldsymbol{\lambda}_2$ respectively, we have:

$$
\begin{cases}
\dfrac{\partial\mathscr{L}}{\partial\dot{\boldsymbol{\theta}}} = -\dot{\boldsymbol{\theta}} - \gamma(\boldsymbol{\theta}-\boldsymbol{\theta}_0) - (k\widetilde{J}_{\mathrm{A}} + (1-k)J_{\mathrm{B}})^{\mathrm{T}}\boldsymbol{\lambda}_1 - \widetilde{J}_{\mathrm{A}}^{\mathrm{T}}\boldsymbol{\lambda}_2 \\
\dfrac{\partial\mathscr{L}}{\partial\dot{k}} = -m_2\dot{k} - \boldsymbol{\lambda}_1^{\mathrm{T}}(\boldsymbol{r}_{\mathrm{A}}-\boldsymbol{r}_{\mathrm{B}}) \\
\dfrac{\partial\mathscr{L}}{\partial\boldsymbol{\lambda}_1} = \dot{k}(\boldsymbol{r}_{\mathrm{A}}-\boldsymbol{r}_{\mathrm{B}}) + (k\widetilde{J}_{\mathrm{A}} + (1-k)J_{\mathrm{B}})\dot{\boldsymbol{\theta}} + m_1(\boldsymbol{r}_{\mathrm{C}} - \boldsymbol{r}_{\mathrm{C}}(0)) \\
\dfrac{\partial\mathscr{L}}{\partial\boldsymbol{\lambda}_2} = \widetilde{J}_{\mathrm{A}}\dot{\boldsymbol{\theta}} - \dot{\boldsymbol{r}}_{\mathrm{d}} + m_3(\boldsymbol{r}_{\mathrm{A}}-\boldsymbol{r}_{\mathrm{d}}).
\end{cases}
\tag{10.10}
$$

Finally, the DNN mdoel is developed as

$$
\begin{cases}
\zeta\ddot{\boldsymbol{\theta}} = -\dot{\bar{\boldsymbol{\theta}}} - \gamma(\boldsymbol{\theta}-\boldsymbol{\theta}_0) - (k\widetilde{J}_{\mathrm{A}} + (1-k)J_{\mathrm{B}})^{\mathrm{T}}\boldsymbol{\lambda}_1 - \widetilde{J}_{\mathrm{A}}^{\mathrm{T}}\boldsymbol{\lambda}_2 \\
\zeta\ddot{k} = -m_2\dot{k} - \boldsymbol{\lambda}_1^{\mathrm{T}}(\boldsymbol{r}_{\mathrm{A}}-\boldsymbol{r}_{\mathrm{B}}) \\
\zeta\dot{\boldsymbol{\lambda}}_1 = \dot{k}(\boldsymbol{r}_{\mathrm{A}}-\boldsymbol{r}_{\mathrm{B}}) + (k\widetilde{J}_{\mathrm{A}} + (1-k)J_{\mathrm{B}})\dot{\bar{\boldsymbol{\theta}}} + m_1(\boldsymbol{r}_{\mathrm{C}} - \boldsymbol{r}_{\mathrm{C}}(0)) \\
\zeta\dot{\boldsymbol{\lambda}}_2 = \widetilde{J}_{\mathrm{A}}\dot{\bar{\boldsymbol{\theta}}} - \dot{\boldsymbol{r}}_{\mathrm{d}} + m_3(\boldsymbol{r}_{\mathrm{A}}-\boldsymbol{r}_{\mathrm{d}}) \\
\varrho\dot{\widetilde{J}}_{\mathrm{A}} = (J_{\mathrm{A}}\dot{\boldsymbol{\theta}} - \widetilde{J}_{\mathrm{A}}\dot{\boldsymbol{\theta}})\dot{\boldsymbol{\theta}}^{\mathrm{T}} \\
\dot{\boldsymbol{\theta}} = \dot{\bar{\boldsymbol{\theta}}} + \boldsymbol{s}.
\end{cases}
\tag{10.11}
$$

where $0 < \zeta \ll 1$ is the convergence efficiency of the DNN model, which influences the speed of convergence. What is worth mentioning is that a zero-mean independent noise $\boldsymbol{s}$ is added to the joint velocities to guarantee a continuous learning process. The standard deviation of $\boldsymbol{s}$ is $\boldsymbol{\sigma}$ and $\boldsymbol{0} \le \|\boldsymbol{s}\|_2 \le \boldsymbol{s}_0$ with $\boldsymbol{s}_0$ being positive. The initial position of the RCM point is determined by both the initial value of k and the redundant manipulator's starting posture. When these variables are properly configured for the task, minimal positional offset of the RCM point is maintained.

10.4 Experiments and Comparisons

In this section, both simulations and physical experiments are performed to verify the RC^2M scheme. Trajectory tracking tasks are carried out in MATLAB and on a physical experimental demonstration system. Additionally, the constructed scheme is compared with latest RCM control methods.

10.4.1 Illustrative Simulation

This part offers some simulation results to validate the feasibility of the constructed method. According to experimental requirements, a slender rod whose length is 0.3 m is mounted as an end-effector of the Franka Emika Panda manipulator (hereafter "Franka manipulator"), simulating a practical surgical tool. The cylindrical axis of the rod is aligned with the rotational axis of the manipulator's final joint. The robotic system is controlled by scheme (10.11) to execute trajectory-tracking tasks in restricted workspaces. This configuration represents a realistic operational scenario. The redundant manipulator's initial joint configuration is specified as $\boldsymbol{\theta}_0 = [0; -0.785472; 0; -2.35425; 0; 1.57164; 0.785465]$ rad. Values of parameters are given: $k = 0.5$; $m_1 = 10$; $m_2 = 0.1$; $m_3 = 10$; $\zeta = 10^{-3}$; $\varrho = 10^{-5}$.

A trajectory-tracking simulation is designed to navigate through an aperture in a partition surface, with unknown end-effector structural parameters, as illustrated in Figs. 10.2 and 10.3. It should be emphasized that real-time end-effector velocity is obtained through integrated sensor measurements during task execution. Figure 10.2a demonstrates the motion trajectory generated during fish-shaped curve tracking by the end-effector. The actual trajectory closely matches the desired one in Fig. 10.2b. In this figure, the initial segment of the actual trajectory near $X = 0.3$ exhibits a slight deviation from the desired trajectory, attributable to the estimation of the end-effector's structural information based on its velocity information. The corresponding angle position and joint-velocity profiles are presented in Fig. 10.2c and d, respectively. Each joint can be seen returning to the initial state as expected. The position drifting of the RCM point, shown in Fig. 10.3b, demonstrates the capability of the presented scheme (10.11) to keep the RCM point's position stationary. It can be observed in Fig. 10.3c that estimation errors of the Jacobian matrix fluctuate when ϱ is 10^{-5}, while Fig. 10.3d shows desirable estimation errors during the task with $\varrho = 10^{-6}$.

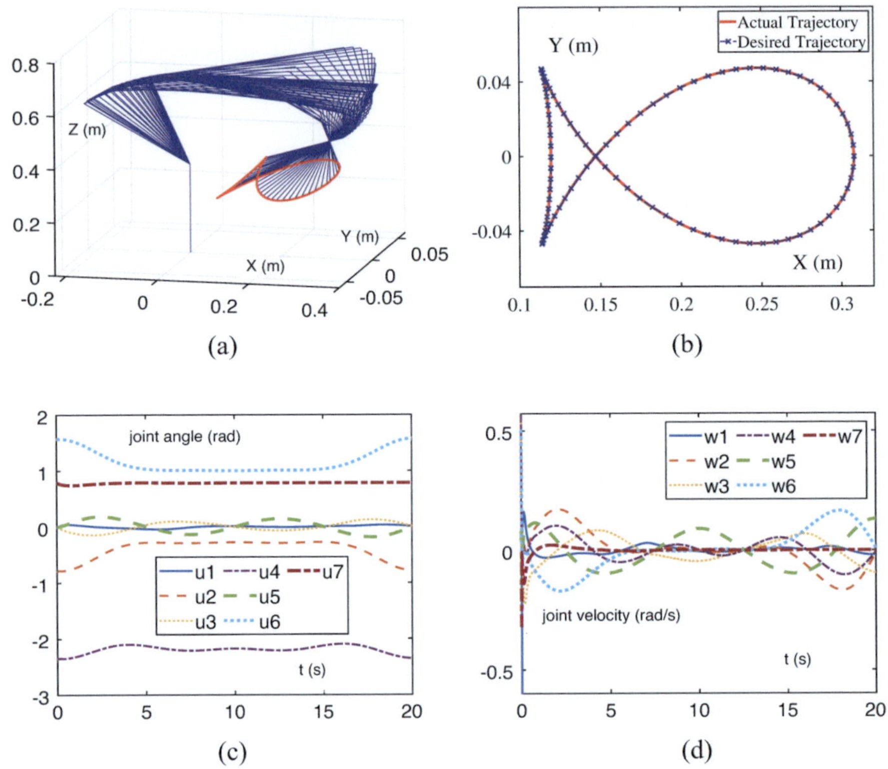

Fig. 10.2 Simulation results for the Franka manipulator equipped with a rod-shaped end-effector tracking a fish-shaped trajectory with $\varrho = 10^{-5}$. **a** Motion process. **b** Actual trajectory and desired trajectory. **c** Joint angle. **d** Joint velocity

Experimental verification of the fish-trajectory-tracking task with RCM constrains compliance is conducted (Fig. 10.4), utilizing a workspace including a box with a 25 mm square opening on its top surface. For enhanced visibility, one lateral panel of the box is eliminated. The end-effector dimensions in the physical manipulator system match exactly with those specified in the preceding simulation. As depicted in Fig. 10.4, a fish-shaped curve is manually drawn to illustrate the end-effector's trajectory. It can be observed that the end-effector does not collide with the obstacle around the RCM point throughout the entire experiment while the manipulator performs the task.

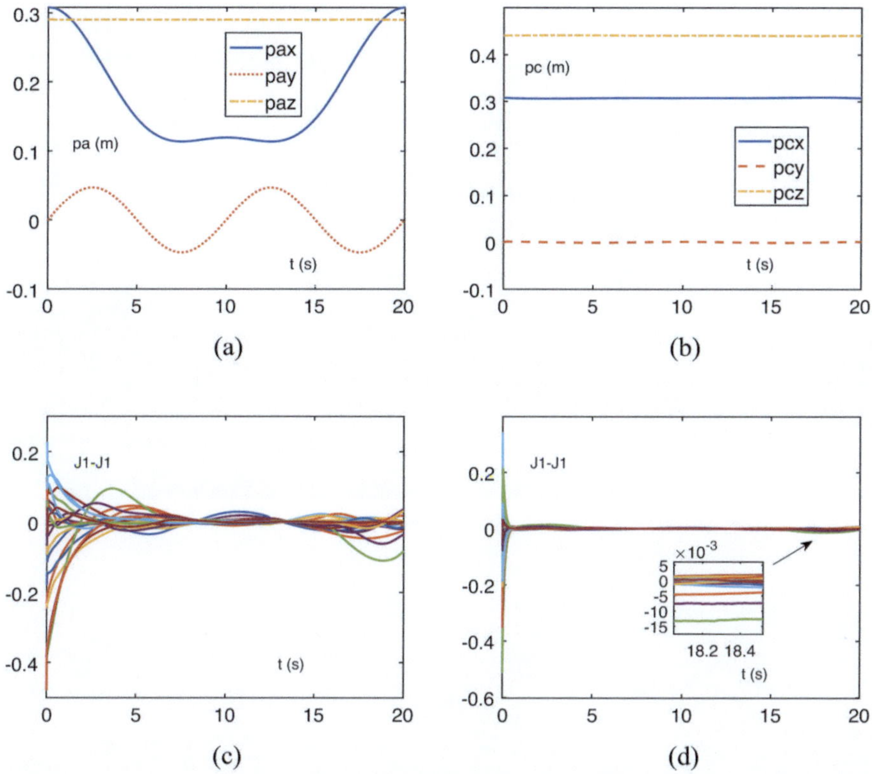

Fig. 10.3 Position and Jacobian matrix estimation errors of the Franka manipulator equipped with a rod-shaped end-effector tracking a fish-shaped trajectory with $\varrho = 10^{-5}$. **a** Position of the endmost operating point. **b** Position of the RCM point. **c** Jacobian matrix estimation errors with $\varrho = 10^{-5}$. **d** Jacobian matrix estimation errors with $\varrho = 10^{-6}$

10.4.2 Comparisons

Comparisons among the constructed scheme (10.11) and other representative RCM-related research are shown in Table 10.1 to highlight its novelty. Compared with other ones, scheme (10.11) achieves reliable operation under end-effector modeling uncertainties through its incorporated data-driven framework. Additionally, the cyclic motion index prevents joint-drift accumulation during repetitive kinematic cycles.

In addition to qualitative analysis, comparative performance tests are conducted for quantitative evaluation. The same path-tracking task is performed using the latest RCM kinematic control method Li and Li (2022). This method, featuring a simplified neural network structure, is designed to address the redundancy issue with RCM constraints. As demonstrated in Figs. 10.5 and 10.6, the simulation results confirm that the manipulator controlled via the method from Li and Li (2022) successfully accom-

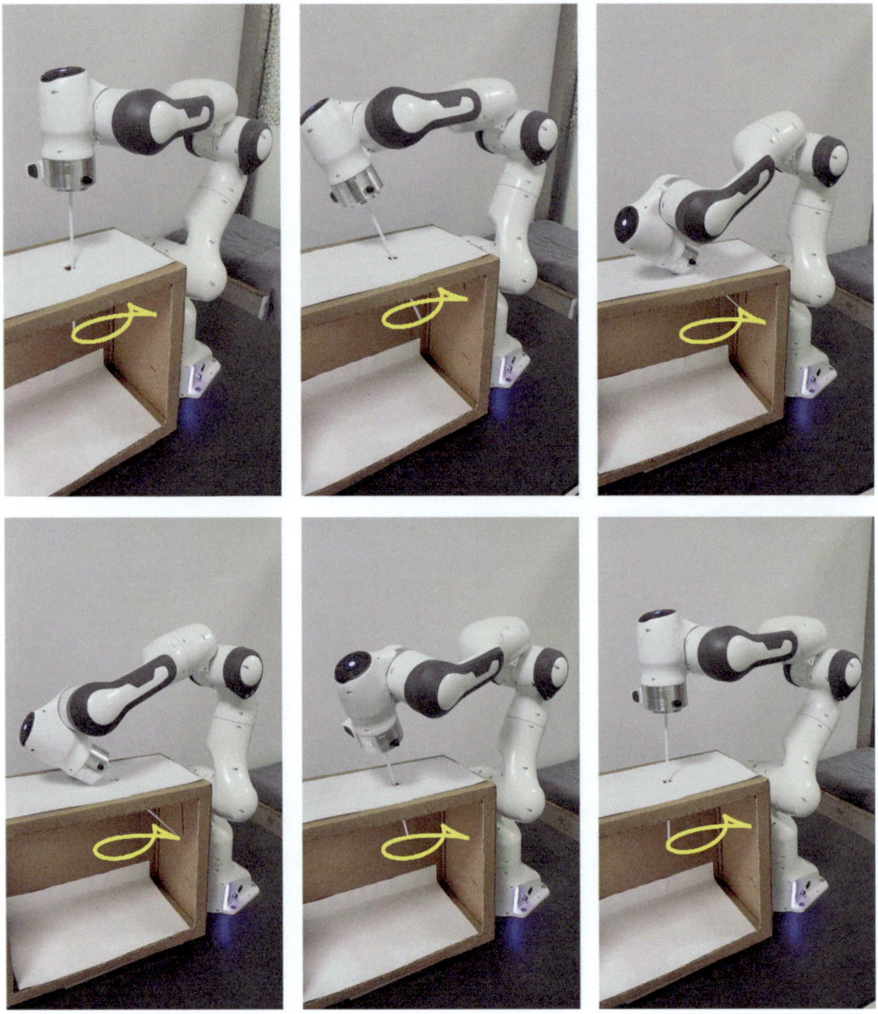

Fig. 10.4 Snapshots for the tracking task of the Franka manipulator driven by the presented scheme

plishes fish-shaped path tracking. Comparative end-effector position error profiles for both methods are documented in Figs. 10.5a and 10.6a. Figure 10.5a demonstrates rapid convergence of Y-axis positional errors induced by Jacobian matrix inaccuracies during the learning process, as achieved by the constructed scheme (10.11). However, the position error in Fig. 10.6a from the method in Li and Li (2022) is an order of magnitude higher than that of the constructed scheme (10.11).

Table 10.1 Comparisons on features among different RCM schemes

	Solver	Auxiliary control	Cyclic motion	Task type	Structural information
Scheme (10.11)	DNN	Data driven	Available	Trajectory tracking	Partially unknown
Scheme in Cui et al. (2022)	Dual neural network	Vision servoing	NA	Target tracking	Known
Scheme in Li and Li (2022)	Recurrent neural network	NA	NA	Trajectory tracking	Known
Scheme in Li et al. (2022)	Gradient neural network	Vision servoing	NA	Target tracking	Known
Scheme in Kastritsi and Doulgeri (2021)	Admittance model	Human guided	NA	Hands-on procedure	Known
Scheme in Wu et al. (2022)	Admittance model	NA	NA	Uterus manipulation	Known
Scheme in Wu et al. (2022)	Admittance model	NA	NA	Uterus manipulation	Known
Scheme in Liu et al. (2024)	Recurrent neural network	NA	NA	Trajectory tracking	Known

"NA" means that the relevant element is not available in the corresponding paper

In addition to the position error of the endmost operating point, a key index is the end-effector's position in the tiny hole plane during the task. Unlike RCM point position drift, operational inaccuracies near the workspace aperture pose critical failure risks, including potential damage to target objects. The comparative experimental results quantifying deviation magnitude for both methods are presented in Figs. 10.5 and 10.6. In the comparison, the constructed scheme (10.11) achieves significantly reduced deviation levels relative to Li and Li (2022), corresponding to lower collision risk during task execution. As evidenced by comparative results, the terminal operating point's position error undergoes rapid transient variations during initial operation of scheme (10.11), coinciding with the data-driven estimation phase for Jacobian matrix parameters. After that, the presented scheme (10.11) performs the task accurately.

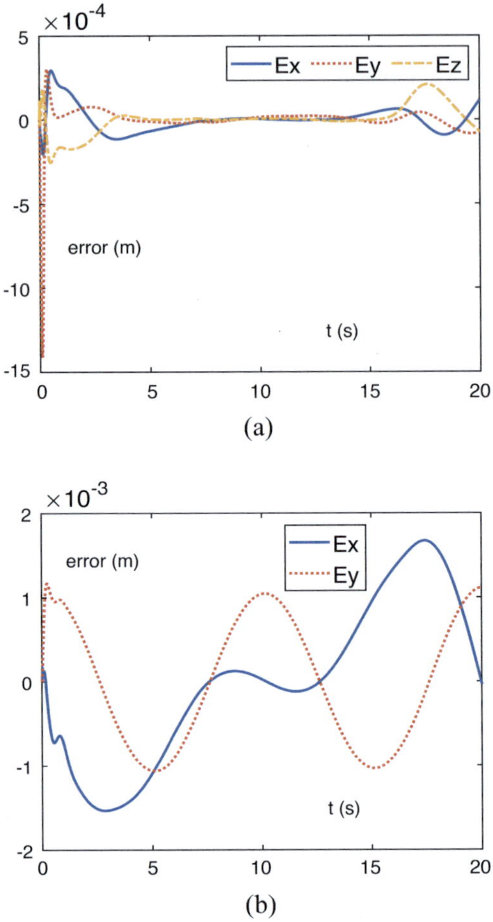

Fig. 10.5 Position error of the endmost operating point and the end-effector in the plane of the operating hole synthesized by (10.11). **a** Position error of the endmost operating point. **b** Position drifting of the end-effector in the plane of the operating hole

10.5 Conclusion

This chapter has presented a data-driven RC^{2}M control scheme for redundant manipulators, specifically designed to address end-effector structural parameter uncertainty. The objective function includes the minimum velocity norm, the minimum deviation of the RCM point norm, and the cyclic motion index. Meanwhile, presented scheme takes into account the RCM constraint, end-effector motion control, and Jacobian matrix estimation. Furthermore, an efficient DNN model has been

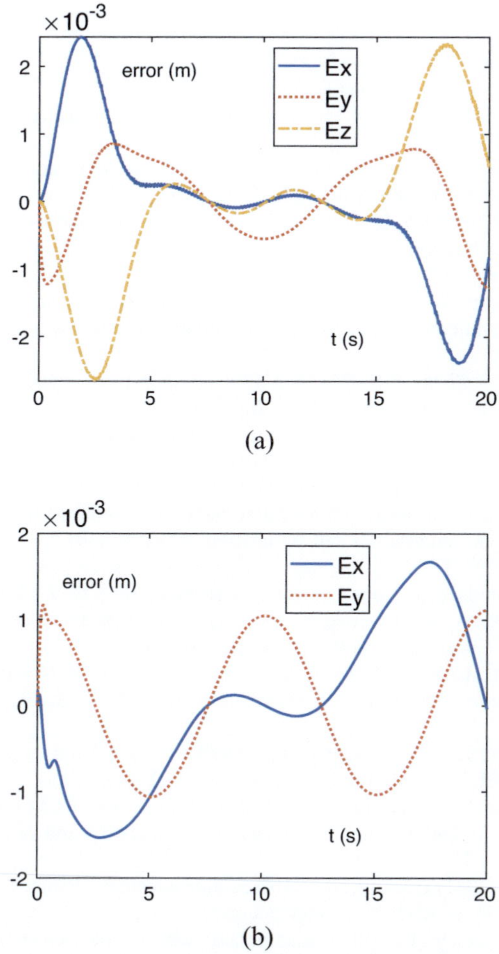

Fig. 10.6 Position error of the endmost operating point and the end-effector in the plane of the operating hole synthesized by Li and Li (2022). **a** Position error of the endmost operating point. **b** Position drifting of the end-effector in the plane of the operating hole

derived to enhance operation accuracy, posture recovery, and reduce RCM position deviation. In summary, this work offers an efficient solution for RC^{2}M control problems involving inaccurate end-effector structural information, thereby enhancing the practicality of RCM-type tasks. Future work will involve experimental validation of the scheme on manipulator-assisted minimally invasive surgical platforms as a promising research direction.

References

Chen W, Zhou J, Cheng SS, Lu Y, Zhong F, Gao Y, Wang Y, Xue L (2021) Tele-operated oropharyngeal swab (TOOS) robot enabled by TSS soft hand for safe and effective sampling. IEEE Trans Med Robot Bionics 3(4):1040–1053

Chen X, Jin L, Hu B (2024) A cerebellum-inspired control scheme for kinematic control of redundant manipulators. IEEE Trans Ind Electron 71(7):7542–7550

Cui Z, Li W, Zhang X, Chiu PWY, Li Z (2022) Accelerated dual neural network controller for visual servoing of flexible endoscopic robot with tracking error, joint motion, and RCM constraints. IEEE Trans Ind Electron 69(9):9246–9257

Cursi F, Bai W, Yeatman EM, Kormushev P (2022) Optimization of surgical robotic instrument mounting in a macro-micro manipulator setup for improving task execution. IEEE Trans Robot 38(5):2858–2874

Fan J, Jin L, Xie Z, Li S, Zheng Y (2022) Data-driven motion-force control scheme for redundant manipulators: a kinematic perspective. IEEE Trans Ind Inform 18(8):5338–5447

Fotiadis F, Vamvoudakis KG, Jiang ZP (2024) Data-driven actuator allocation for actuator redundant systems. IEEE Trans Autom Control 69(4):2249–2264

Gao H, Xiao X, Qiu L, Meng MQ-H, King NKK, Ren H (2021) Remote-center-of-motion recommendation toward brain needle intervention using deep reinforcement learning. In: IEEE international conference on robotics and automation, pp 8295–8301

Guo K, Su H, Yang C (2022) A small opening workspace control strategy for redundant manipulator based on RCM method. IEEE Trans Control Syst Technol 30(6):2717–2725

Jin L, Zhao J, Li S (2023) A data-driven sparse motion planning scheme for redundant manipulators. IEEE Trans Circuits Syst II Express Briefs 70(7):2600–2604

Jin L, Zhao J, Chen L, Li S (2025) Collective neural dynamics for sparse motion planning of redundant manipulators without Hessian matrix inversion. IEEE Trans Neural Netw Learn Syst. 36(3):4326–4335

Kastritsi T, Doulgeri Z (2021) A controller to impose a RCM for hands-on robotic-assisted minimally invasive surgery. IEEE Trans Med Robot Bionics 3(2):392–401

Kastritsi T, Doulgeri Z (2021) A control method for time-variant RCM constraint in hands-on RAMIS procedures. In: 29th mediterranean conference on control and automation (MED), Bari, pp 729–734

Khan AH, Li S, Cao X (2021) Tracking control of redundant manipulator under active remote center-of-motion constraints: An RNN-based metaheuristic approach. Sci China Inf Sci 64:1869–1919

Li Z, Li S (2022) Kinematic control of manipulator with remote center of motion constraints synthesised by a simplified recurrent neural network. Neural Process Lett 54(2):1035–1054

Li W, Chiu PWY, Li Z (2020) An accelerated finite-time convergent neural network for visual servoing of a flexible surgical endoscope with physical and RCM constraints. IEEE Trans Neural Netw Learn Syst 31(12):5272–5284

Li W, Han L, Xiao X, Liao B, Chen P (2022) A gradient-based neural network accelerated for vision-based control of an RCM-constrained surgical endoscope robot. Neural Comput Appl 34(2):1329–1343

Li B, Lu B, Lu Y, Dou Q, Liu Y-H (2021) Data-driven holistic framework for automated laparoscope optimal view control with learning-based depth perception. In: IEEE international conference on robotics and automation, pp 12366–12372

Liu M, Liu K, Zhu P, Zhang G, Ma X, Shang M (2024) Data-driven remote center of cyclic motion (RC^{2}M) control for redundant robots with rod-shaped end-effector. IEEE Trans Ind Inform 20(4):6772–6780

Lu B, Li B, Dou Q, Liu Y (2022) A unified monocular camera-based and pattern-free hand-to-eye calibration algorithm for surgical robots with RCM constraints. IEEE/ASME Trans Mechatron 27(6):5124–5135

Saeidi H, Opfermann JD, Kam M, Wei S, Leonard S, Hsieh MH, Kang JU, Krieger A (2022) Autonomous robotic laparoscopic surgery for intestinal anastomosis. Sci Robot 7(62):eabj2908

Su H, Mariani A, Ovur SE, Menciassi A, Ferrigno G, De Momi E (2021) Toward teaching by demonstration for robot-assisted minimally invasive surgery. IEEE Trans Autom Sci Eng 18(2):484–494

Takeuchi M, Hironaka Y, Aoyama T, Hasegawa Y (2022) Intuitive remote robotic nasal sampling by orientation control with variable RCM in limited space. IEEE Trans Med Robot Bionics 4(3):646–655

Wang X, Tan N (2023) Cerebellum-inspired model-free tracking control and visual servoing of a rigid-flexible hybrid robotic endoscope with RCM constraints. IEEE Trans Ind Electron 70(12):12626–12635

Wang Y, Han Y, Tang Z, Liu W, Wang P (2023) Reciprocal of exponential varying-parameter RNN solving repetitive tracking control problems with tolerance of random initial error compounded with noise perturbation. IEEE Trans Ind Inform 19(11):10998–11007

Wu J, Chen W, Guo D, Ma G, Wang Z, He Y, Liu YH (2022) Robot-enabled uterus manipulator for laparoscopic hysterectomy with soft RCM constraints: design, control, and evaluation. IEEE Trans Med Robot Bionics 4(3):656–666

Yan J, Jin L, Hu B (2024) Data-driven model predictive control for redundant manipulators with unknown model. IEEE Trans Cybern 54(10):5901–5911

Yang W, Li S, Li Z, Luo X (2023) Highly-accurate manipulator calibration via extended Kalman filter-incorporated residual neural network. IEEE Trans Ind Inform 19(11):10831–10841

Yao Z, Xu F, Jiang GP, Yao J (2024) Data-driven control of hydraulic manipulators by reinforcement learning. IEEE/ASME Trans Mechatron 29(4):2673–2684

Zhang J, Jin L, Yang C (2022) Distributed cooperative kinematic control of multiple robotic manipulators with an improved communication efficiency. IEEE/ASME Trans Mechatron 27(1):149–158

Zhang M, Tong W, Li P, Hou Y, Xu X, Zhu L, Wu EQ (2023) Robust neural dynamics method for redundant robot manipulator control with physical constraints. IEEE Trans Ind Inform 19(12):11721–11729

Conclusions

Redundant manipulators mark a significant advancement in the robotics field, offering unparalleled flexibility through their extra degrees of freedom. This flexibility enables them to perform complex, multi-objective tasks with exceptional precision, making them ideal for dynamic and high-precision environments such as medical surgery, delicate assembly processes, and fine object handling. Their ability to adjust dynamically to ever-changing conditions allows for optimized secondary objectives, such as energy efficiency, joint wear reduction, and system stability, while ensuring high levels of performance in primary tasks like trajectory tracking and force control. However, the inherent complexity of redundant manipulators cannot be overlooked. The challenges of motion planning, system modeling, and control arise from the non-unique nature of inverse kinematics solutions, which demand sophisticated optimization strategies. The selection of optimal joint configurations requires careful consideration of secondary performance indices, including energy consumption, repeatability, and stability. To address these challenges, optimization frameworks or techniques such as quadratic programming (QP) and model predictive control (MPC) have proven useful. These approaches enable real-time optimization, ensuring that both primary tasks and secondary constraints are effectively managed. Moreover, the integration of dynamic neural networks (DNN) holds great promise, offering solutions to the computational difficulties posed by the high-dimensional and time-varying nature of redundant manipulator systems. The versatility of redundant manipulators is further underscored by advanced control strategies, including index-oriented robot control, posture-considered robot control, and multi-fingered dexterous gripping. These techniques enable precise interaction with the environment, whether through fine force control, stable tool positioning (e.g., remote center of motion, RCM), or dynamic grasping strategies. The continued development of these strategies ensures that redundant manipulators can address a wide range of tasks with high levels of adaptability and precision.

M. Liu et al., *Dynamic Neural Networks for Motion Control of Redundant Manipulators*, Intelligent Control and Learning Systems 21,
https://doi.org/10.1007/978-981-96-9144-9

In conclusion, while the complexity of redundant manipulators poses significant challenges, ongoing advancements in optimization techniques and control strategies hold the key to unlocking their full potential. These innovations will ensure that redundant manipulators continue to lead the field of robotics, pushing the boundaries of what is possible in terms of efficiency, adaptability, and precision. As these systems become increasingly integrated into diverse industries, their ability to perform flexible, adaptive, and highly accurate tasks will remain a cornerstone of progress in robotics. The future of redundant manipulators is bright, with boundless opportunities for enhancing their capabilities and applications in an ever-evolving technological landscape.

MIX
Papier aus verantwortungsvollen Quellen
Paper from responsible sources
FSC® C105338

If you have any concerns about our products,
you can contact us on
ProductSafety@springernature.com

In case Publisher is established outside the EU,
the EU authorized representative is:
Springer Nature Customer Service Center GmbH
Europaplatz 3, 69115 Heidelberg, Germany

Printed by Libri Plureos GmbH
in Hamburg, Germany